Plane Trigonometry

Plane

Trigonometry

Allyn J. Washington
Carolyn E. Edmond
Dutchess Community College

Benjamin/Cummings Publishing Company

Menlo Park, California
Reading, Massachusetts
London • Amsterdam
Don Mills, Ontario • Sydney

Trigonometry is useful in numerous theoretical and practical types of applications. In the photograph on the cover and title page, we see a number of ways in which trigonometry may be applied. The angle of depression of the boat, the structure of the bridge, and the direction of the boat are among those which involve the use of trigonometry.

Design	Design Office/Bruce Kortebein
Photographs	Cover and Title Page: Will Connell
	Endsheets: detail from the *New Dictionary of Arts and Sciences*
	Chapter 1: ©IGN, Paris, 1945
	Chapter 2: Marshall Berman
	Chapter 3: Wide World Photo
	Chapter 4: Erich Hartmann, Magnum Photos
	Chapter 5: NASA
	Chapter 6: courtesy of Tektronix, Inc.
	Chapter 7: Marshall Berman
	Chapter 8: Julie Lundquist, Van Cleve Photography

Copyright © 1977 by Cummings Publishing Company, Inc.
Philippines copyright 1977.

All rights reserved. No part of this publication may be reproduced, stored in a retrieval system, or transmitted, in any form or by any means, electronic, mechanical, photocopying, recording, or otherwise, without the prior written permission of the publisher.
Printed in the United States of America.
Published simultaneously in Canada.
Library of Congress Catalog Card Number 76-7883

ISBN 0-8465-8622-3
FGHIJKLMN-HA-8987654321

Preface

This book is intended primarily for students who are enrolled in courses in which an understanding of the fundamental concepts of trigonometry is developed. Numerous applications related to everyday life are included primarily to indicate where and how trigonometric techniques are used. The material presented here is sufficient for a one-semester course.

The approach used is basically an intuitive one. It is not unduly mathematically rigorous, although all appropriate terms and concepts are introduced as needed and are given an intuitive or algebraic foundation. The aim is to help students develop a feeling for trigonometric methods, and not simply to have a collection of formulas when they have completed their work in the text.

Although some background in algebra and geometry is assumed, Chapter 1 reviews concepts required for the development of trigonometry. The instructor may omit this chapter and begin the course with Chapter 2.

Chapter 2 introduces the trigonometric functions using only first-quadrant angles. This allows the student to become familiar with the concepts of the trigonometric functions, the use of the trigonometric tables, and the applications of triangles having only acute angles. Chapter 3 expands the basic definitions to include angles other than first-quadrant angles and shows some applications using these angles. Chapter 4 discusses vectors and the solution of triangles which are not right triangles.

Chapter 5 deals with the theory of logarithms and their application as a calculating tool. The instructor may wish to omit the calculations with logarithms if calculators are to be used.

Graphing of the six trigonometric functions is covered in Chapter 6. The basic curves are discussed and graphed. Using the basic curves, the graphs of functions with various amplitudes, periods, and displacements are developed. The trigonometric identities and the complex numbers are developed in Chapters 7 and 8 respectively. The topics of radicals and significant digits are covered in the Appendices. This allows these topics to be introduced as the instructor wishes.

Another feature of this text is that it contains many more worked examples than are ordinarily included in such a book. These examples, totaling approximately 300, are used advantageously to introduce concepts, as well as to clarify and illustrate points made in the text. Also, there is extensive use of graphical methods.

At the end of each section there are exercises including stated problems. At the end of each chapter there are exercises which may be used either for additional problems or for review assignments. Over 2000 exercises appear in this book, and they are generally grouped such that there is an even-numbered exercise equivalent to each odd-numbered exercise.

The answers to the odd-numbered exercises are given at the back of the book. The answers to graphical problems and other types which are often not included in textbooks are included in thisitext.

The authors wish to thank the following reviewers for their valuable comments and suggestions: Frank Cerrato, City College of San Francisco; Richard Denning, Southern Technical Institute; Albert Grasser, Oakland Community College; Richard Marshall, Eastern Michigan University; Michael Monaco, City College of San Francisco; Haile Perry, Texas A&M University; and Gilbert Stork, Cuesta College.

Contents

Chapter 1
Functions and Graphs   3

    Introduction 5
1-1 Functions 5
    Exercises 1-1 10
1-2 Rectangular Coordinates 12
    Exercises 1-2 14
1-3 Pythagorean Theorem 16
    Exercises 1-3 19
1-4 The Graph of a Function 21
    Exercises 1-4 27
    Exercises for Chapter 1 28

Chapter 2
The Trigonometric Functions   33

2-1 Angles 35
    Exercises 2-1 37
2-2 Trigonometric Functions of Acute Angles 38
    Exercises 2-2 41
2-3 Values of the Trigonometric Functions 42
    Exercises 2-3 46

2-4	The Right Triangle	48
	Exercises 2-4	52
2-5	Applications of Right Triangles	54
	Exercises 2-5	56
	Exercises for Chapter 2	59

Chapter 3
Trigonometric Functions of Any Angle   65

3-1	Signs of the Trigonometric Functions	67
	Exercises 3-1	70
3-2	Trigonometric Functions of Any Angle	71
	Exercises 3-2	77
3-3	Radians	79
	Exercises 3-3	84
3-4	Applications of the Use of Radian Measure	85
	Exercises 3-4	88
3-5	Circular Functions	90
	Exercises 3-5	95
	Exercises for Chapter 3	96

Chapter 4
Vectors and Oblique Triangles   101

4-1	Vectors	103
	Exercises 4-1	110
4-2	Application of Vectors	112
	Exercises 4-2	116
4-3	Oblique Triangles, the Law of Sines	117
	Exercises 4-3	123
4-4	The Law of Cosines	125
	Exercises 4-4	130
4-5	Area of a Triangle	132
	Exercises 4-5	136
	Exercises for Chapter 4	137

Chapter 5
Logarithms   143

5-1 Definition of a Logarithm 145
    Exercises 5-1 147
5-2 Properties of Logarithms 148
    Exercises 5-2 154
5-3 Logarithms to the Base 10 155
    Exercises 5-3 160
5-4 Computations Using Logarithms 161
    Exercises 5-4 163
5-5 Logarithms of Trigonometric Functions 164
    Exercises 5-5 167
5-6 Logarithms to Bases Other Than 10 169
    Exercises 5-6 172
    Exercises for Chapter 5 173

Chapter 6
Graphs of the Trigonometric Functions   177

6-1 Graphs of $y = a \sin x$ and $y = a \cos x$ 179
    Exercises 6-1 185
6-2 Graphs of $y = a \sin bx$ and $y = a \cos bx$ 186
    Exercises 6-2 190
6-3 Graphs of $y = a \sin (bx + c)$ and
    $y = a \cos (bx + c)$ 190
    Exercises 6-3 195
6-4 Graphs of $y = \tan x$, $y = \cot x$, $y = \sec x$,
    $y = \csc x$ 196
    Exercises 6-4 202
6-5 Graphing by Addition of Ordinates 203
    Exercises 6-5 206
    Exercises for Chapter 6 208

Chapter 7
Basic Trigonometric Relationships — 211

- 7-1 Fundamental Trigonometric Identities 213
  - Exercises 7-1 220
- 7-2 Trigonometric Functions of the Sum and Difference of Two Angles 221
  - Exercises 7-2 226
- 7-3 Double-Angle Formulas 229
  - Exercises 7-3 232
- 7-4 Half-Angle Formulas 233
  - Exercises 7-4 236
- 7-5 Trigonometric Equations 238
  - Exercises 7-5 241
- 7-6 Introduction to the Inverse Trigonometric Functions 242
  - Exercises 7-6 244
- 7-7 The Inverse Trigonometric Functions 245
  - Exercises 7-7 250
  - Exercises for Chapter 7 252

Chapter 8
Complex Numbers — 257

- 8-1 Imaginary and Complex Numbers 259
  - Exercises 8-1 262
- 8-2 Basic Operations with Complex Numbers 264
  - Exercises 8-2 266
- 8-3 Graphical Representation of Complex Numbers . . . 267
  - Exercises 8-3 270
- 8-4 Polar Form of a Complex Number 270
  - Exercises 8-4 273
- 8-5 Exponential Form of a Complex Number 274
  - Exercises 8-5 276
- 8-6 Products, Quotients, Powers, and Roots of Complex Numbers 277
  - Exercises 8-6 283
  - Exercises for Chapter 8 284

Appendix A:	Exponents and Radicals	287
Appendix B:	Approximate Numbers and Significant Digits	303
Appendix C:	A Note Regarding Units of Measurement	311
Appendix D:	Tables	313
	Table 1: Squares and Square Roots	315
	Table 2: Four-Place Logarithms of Numbers	322
	Table 3: Four-Place Values of Functions and Radians	324
	Table 4: Four-Place Logarithms of Trigonometric Functions—Angle θ in Degrees	332
Answers to Odd-Numbered Exercises		341
Index		371

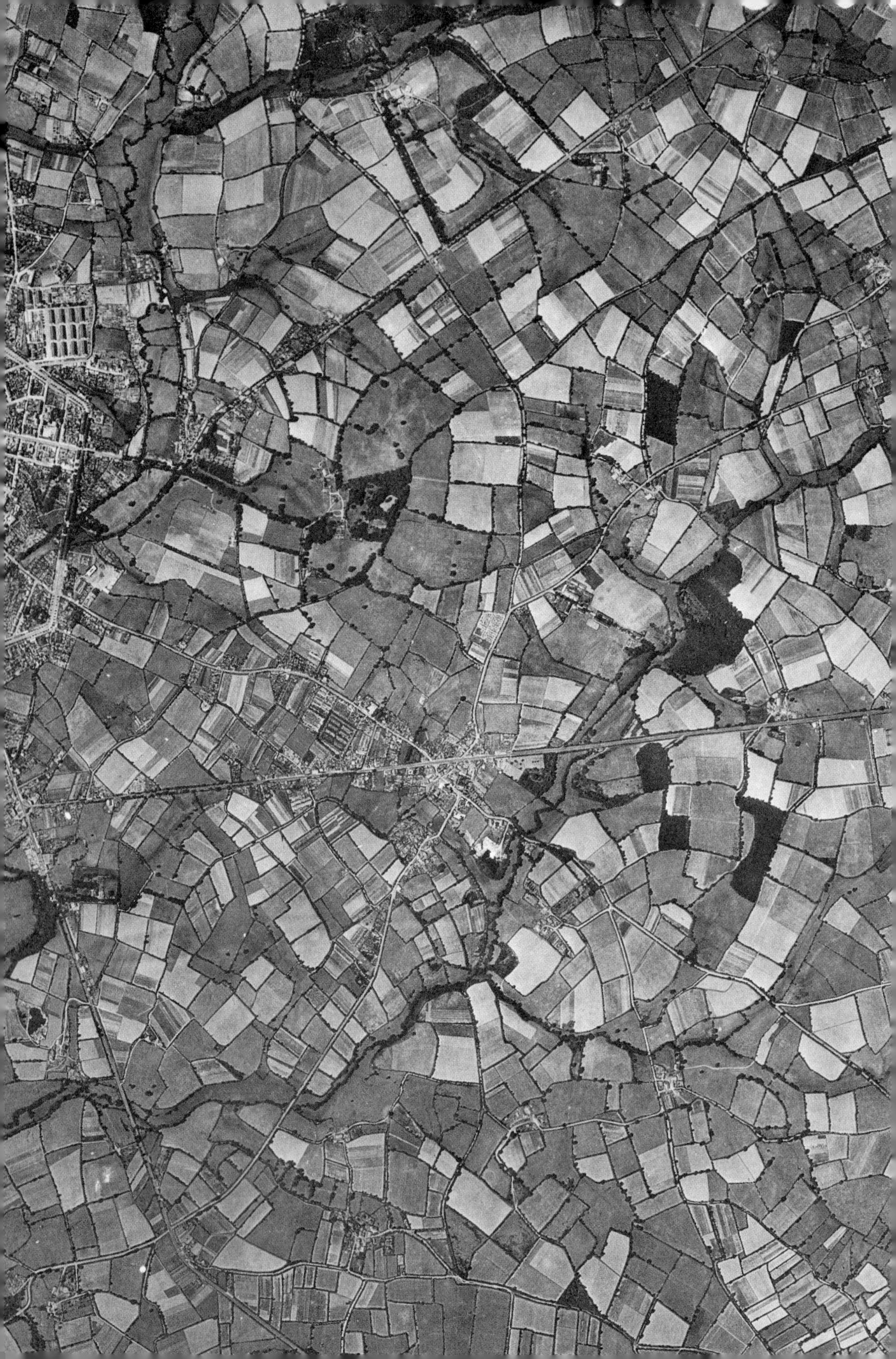

Functions and Graphs

We can see from this aerial photograph that locating an address in a community is similar to locating a point in a mathematical coordinate system.

Introduction

The solution of a great many applied problems involves the use of triangles, especially right triangles. Many of these applied situations are basically scientific in nature, but many others are found in everyday life. Problems that can be solved by the use of triangles include the determination of distances which cannot be measured directly, such as the widths of rivers and distances between various points in the universe. Also, problems involving building construction, forces on objects, and airplane headings lend themselves readily to solution by use of triangles.

The branch of mathematics which develops the methods for measuring triangles and solving related problems is trigonometry. The basic properties of triangles along with certain concepts of algebra are used in the development of trigonometry. Through the methods of trigonometry, applications in astronomy, surveying, navigation, architecture, and various fields of science may be solved. Trigonometry is one of the most practical and most applicable branches of mathematics.

1-1 Functions

Before we begin our study of trigonometry, we shall review certain basic mathematical topics that we shall find necessary in our development of trigonometry. In this section we discuss the concept of a function.

We know that quantities can be related by means of formulas. In most formulas one quantity is given in terms of one or more other quantities. Such formulas are often found in the study of natural phenomena where relationships can be found to exist among certain quantities.

If, for example, we study the distance d which an object travels at a constant rate r, we find that d and r are related to the time t of travel by the formula $d = rt$. Also, a study of the pressure p and volume V of a gas at constant temperature would show that $p = k/V$, where k is a constant.

Considerations such as these lead us to the concept of a *function*. **Whenever a relationship exists between two variables x and y such that for every value of x there is only one corresponding value of y, we say that y is a function of x.** Here we call the variable x the *independent variable* (since x may be chosen arbitrarily so long as the value chosen produces a real number for y). Also, y is the *dependent variable* (since, once the value of x is chosen, the value of y is determined—that is, y depends on x).

There are many ways to express functions. Every formula, such as those we have discussed, defines a function. Other ways to express functions are by means of tables, charts, and graphs.

Example A
In the equation $y = 2x$, we see that y is a function of x, since for each value of x there is only one value of y. If $x = 3$, for example, $y = 6$ and no other value. The dependent variable is y and the independent variable is x.

Example B
The volume of a cube of edge e is given by $V = e^3$. Here V is a function of e, since for each value of e there is one value of V. The dependent variable is V and the independent variable is e.

If the equation relating the volume and edge of a cube was written as $e = \sqrt[3]{V}$, that is, if the edge of a cube was expressed in terms of its volume, we would say that e is a function of V. In this case e would be the dependent variable and V the independent variable.

1-1 Functions

For convenience of notation, the phrase "function of x" is written as $f(x)$. This, in turn, is a simplification of the statement that "y is a function of x," so we may now write $y = f(x)$. Notice that $f(x)$ does *not* mean f times x. The symbols must be written in this form to indicate a function. [$f(x)$ is read as "f of x."]

Example C
If $y = 6x^3 - 5x$ we may say that y is a function of x, where this function $f(x)$ is $6x^3 - 5x$. We may also write $f(x) = 6x^3 - 5x$. It is common to write functions in this manner, rather than in the form $y = 6x^3 - 5x$. However, y and $f(x)$ represent the same expression.

One of the most important uses of this notation is to designate the value of a function for a particular value of the independent variable. That is, for the expression "the value of the function $f(x)$ when $x = a$" we may write $f(a)$.

Example D
If $f(x) = 3x^2 - 5$ and we wish to find $f(2)$, it is necessary to substitute 2 for x in the equation. Therefore, $f(2) = 3(2)^2 - 5 = 7$. Also, $f(-1) = 3(-1)^2 - 5 = -2$ and $f(c^2) = 3(c^2)^2 - 5 = 3c^4 - 5$.

In certain instances we need to define more than one function of x. Then we use different symbols to denote the functions. For example, $f(x)$ and $g(x)$ may represent different functions of x, such as $f(x) = 5x^2 - 3$ and $g(x) = 6x - 7$. Special functions are represented by special symbols. For example, in our study of trigonometry we shall come across the "sine of the angle ϕ," where the sine is a function of ϕ. This is designated by $\sin \phi$.

Example E
If $f(x) = \sqrt{4x} + x$ and $g(x) = ax^4 - 5x$, then $f(4) = 8$ and $g(4) = 256a - 20$.

Example F
If $g(t) = 4t^3 - 5t$, $h(t) = 6t + \sqrt{t}$, and $F(t) = 6t$, then $g(2) = 22$, $h(2) = 12 + \sqrt{2}$, and $F(2) = 12$.

A function may be looked upon as a set of instructions. These instructions tell us how to obtain the value of the dependent variable for a particular value of the independent variable, even if the set of instructions is expressed in literal symbols.

Example G
The function $f(x) = x^2 - 3x$ tells us to "square the value of the independent variable, multiply the value of the independent variable by 3, and subtract the second result from the first." An analogy would be a computer which was so programmed that when a number was fed into the computer it would square the number and then subtract 3 times the value of the number. This is represented in diagram form in Figure 1-1.

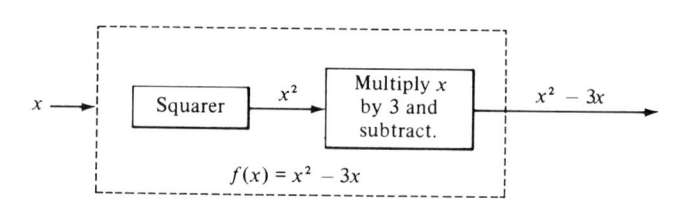

$x \longrightarrow$ [Squarer] $\xrightarrow{x^2}$ [Multiply x by 3 and subtract.] $\longrightarrow x^2 - 3x$

$f(x) = x^2 - 3x$

Figure 1-1

The functions $f(t) = t^2 - 3t$ and $f(n) = n^2 - 3n$ are the same as the function $f(x) = x^2 - 3x$, since the operations performed on the independent variable are the same. Although different literal symbols appear, this does not change the function.

In our definition of a function we noted that the independent variable may be chosen arbitrarily so long as the value chosen produces a real number

1-1 Functions

for the dependent variable. Since the square root of a negative number does not give a real number (it gives an imaginary number, which is discussed in detail in Chapter 8), we must exclude values of the independent variable which would produce the square root of a negative number. Also, since division by zero is an excluded operation, we must also exclude values of the independent variable which would produce division by zero.

Example H
For the function $f(x) = \sqrt{x - 2}$ we see that $f(2) = \sqrt{0} = 0$. However, if x is greater than 2 (written as $x > 2$) we have square roots of positive numbers, whereas if x is less than 2 (written as $x < 2$) we have square roots of negative numbers. Therefore, values of x are restricted to values equal to or greater than 2 (written as $x \geq 2$).

Example I
If $f(u) = \dfrac{u}{u - 4}$, this function is not defined if $u = 4$, since this would require division by zero. Therefore, the values of u are restricted to values other than 4.

We call all possible numbers which may be used for the independent variable the *domain* of the function; all possible numbers for the dependent variable are called the *range* of the function. Therefore, we see that we must first see what restrictions may exist for the independent variable. Then we must determine what values of the dependent variable result.

Example J
For the function $y = 2x$ the independent variable x may be any real number. Therefore, the domain of the function is all real numbers. Also, in using all real numbers for x, the resulting values of y will be all real numbers. Therefore, the range of the function is all real numbers.

Example K
For the function $y = x^2 + 1$ the independent variable x may be any real number. Therefore, the domain of the function is all real numbers. However, since the square of a negative number or positive number is positive, $x^2 + 1$ is greater than or equal to 1 for all values of x. Thus, the range of the function is all real numbers greater than or equal to 1.

Example L
For the function $y = \sqrt{x - 2}$ (see Example H) the domain is all real numbers greater than or equal to 2. The range is all real numbers which are positive or zero, since $\sqrt{x - 2}$ is defined to be positive for $x > 2$.

Exercises 1-1

In Exercises 1 through 12 determine the appropriate function.

1. Express the area A of a circle as a function of its radius r.
2. Express the area A of a circle as a function of its diameter d.
3. Express the circumference c of a circle as a function of its radius r.
4. Express the circumference c of a circle as a function of its diameter d.
5. Express the area A of a rectangle of width 5 as a function of its length l.
6. Express the volume V of a rectangular solid of width 5 and length 8 as a function of its height h.
7. Express the area A of a square as a function of its side s; express the side s of a square as a function of its area A.

1-1 Functions

8. Express the perimeter p of a square as a function of its side s; express the side s of a square as a function of its perimeter p.
9. A long-distance telephone call costs 77¢ for the first 3 min and 25¢ for each additional minute. Express the cost C of the call as a function of its length of time t. (Assume the call is at least 3 min in length.)
10. A taxi fare is 55¢ plus 10¢ for every $\frac{1}{5}$ mi traveled. Express the fare F as a function of the distance s traveled.
11. Express the simple interest I on $200 at 5 percent per year as a function of the number of years t.
12. A salesman earns $200 plus 3 percent commission on his monthly sales. Express his monthly income I as a function of his monthly dollar sales S.

In Exercises 13 through 24 evaluate the given functions.

13. Given $f(x) = 2x + 1$, find $f(1)$ and $f(-1)$.
14. Given $f(x) = 5x - 9$, find $f(2)$ and $f(-2)$.
15. Given $f(x) = 5 - 3x$, find $f(-2)$ and $f(4)$.
16. Given $f(x) = 7 - 2x$, find $f(5)$ and $f(-4)$.
17. Given $f(x) = x^2 - 9x$, find $f(3)$ and $f(-5)$.
18. Given $f(x) = 2x^3 - 7x$, find $f(1)$ and $f(\frac{1}{2})$.
19. Given $\phi(x) = 6x - x^2 + 2x^3$, find $\phi(1)$ and $\phi(-2)$.
20. Given $H(x) = x^4 - x + 3$, find $H(0)$ and $H(-3)$.
21. Given $g(t) = at^2 - a^2 t$, find $g(-\frac{1}{2})$ and $g(a)$.
22. Given $s(y) = 6\sqrt{y} - 3$, find $s(9)$ and $s(a^2)$.
23. Given $K(s) = 3s^2 - s + 6$, find $K(-s)$ and $K(2s)$.
24. Given $T(t) = 5t + 7$, find $T(-2t)$ and $T(t + 1)$.

In Exercises 25 through 28 state the instructions of the function in words as in Example G.

25. $f(x) = x^2 + 2$ 26. $f(x) = 2x - 6$
27. $g(y) = 6y - y^3$ 28. $\phi(s) = s^5 - 5s + 8$

In Exercises 29 through 32 determine the domain and the range of the function.

29. $y = 3x - 1$ 30. $y = 2x^2 - 3$
31. $y = \sqrt{x + 1}$ 32. $y = \sqrt{\dfrac{1}{x}}$

In Exercises 33 through 36 solve the given problems.

33. The volume of a cylinder of height 6 in. as a function of the radius of the base is given by $V = 6\pi r^2$. What is the volume of the cylinder if the base has a radius of 3 in.?

34. The distance s that a freely falling body travels as a function of time t is given by $s = 16t^2$, where s is measured in feet and t is measured in seconds. How far does an object fall in 2 seconds?

35. The Celsius temperature C as a function of the equivalent Fahrenheit temperature F is given by

$$C = \tfrac{5}{9}(F - 32).$$

What is the Celsius temperature equivalent to the Fahrenheit temperature of 77°? Of 12°?

36. The net profit P made on selling twenty $15 items as a function of the price p is $P = 20(p-15)$. What is the profit if the price is $28? If the price is $12?

1-2 Rectangular Coordinates

One of the most valuable ways of representing functions is by graphical representation. By using

1-2 Rectangular Coordinates

graphs we are able to obtain a picture of the function, and by using this picture we can learn a great deal about the function.

To make a graphical representation, we can let numbers be represented by points on a line. We use one line to represent x-values and another line to represent y-values. We do this most conveniently by placing the lines perpendicular to each other.

We place one line horizontally and label it the *x-axis*. The numbers of the set for the independent variable are normally placed on this axis. The other line we place vertically and label the *y*-axis. Normally the *y*-axis is used for values of the dependent variable. The point of intersection is called the *origin*, normally labeled zero. The four parts into which the plane is divided by the axes are called *quadrants*, which are numbered as in Figure 1-2.

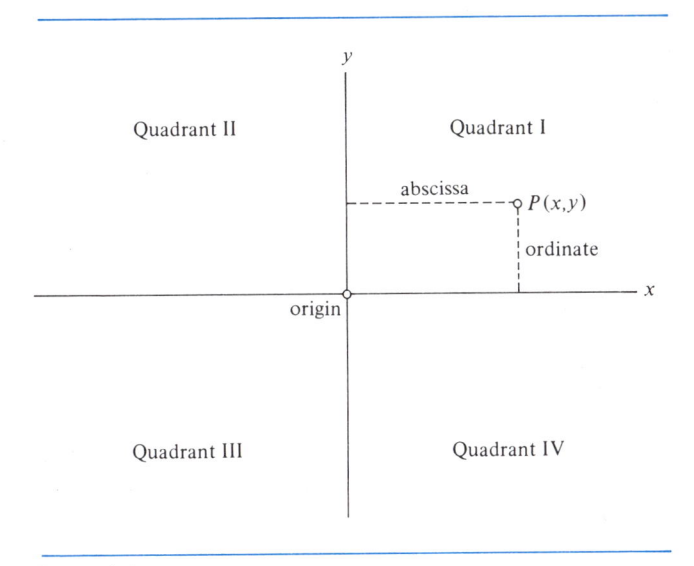

Figure 1-2

On the *x*-axis, positive values are to the right of the origin and negative values are to the left. On the *y*-axis, positive values are above the origin and negative values are below it. A point *P* in the plane is designated by the pair of numbers (x, y), where x

14 Functions and Graphs

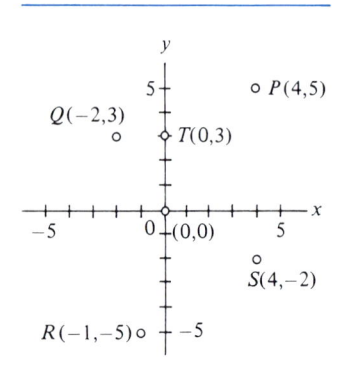

Figure 1-3

is the value of the independent variable and y is the corresponding value of the dependent variable. The x-value (called the *abscissa*) is the perpendicular distance of P from the y-axis. The y-value (called the *ordinate*) is the perpendicular distance of P from the x-axis. The values x and y together, written as (x, y), are the *coordinates* of the point P. This is the *rectangular coordinate system*.

Example A
The positions of points $P(4, 5)$, $Q(-2, 3)$, $R(-1, -5)$, $S(4, -2)$, and $T(0, 3)$ are shown in Figure 1-3. Note that this representation allows for *one point for any pair of values (x, y)*.

Example B
Three of the vertices of the rectangle in Figure 1-4 are $A(-3, -2)$, $B(4, -2)$, and $C(4, 1)$. What is the fourth vertex?

We use the fact that opposite sides of a rectangle are equal and parallel to find the solution. Since both vertices of the base AB of the rectangle have a y-coordinate of -2, the base is parallel to the x-axis. Therefore, the top of the rectangle must also be parallel to the x-axis. Thus, the vertices of the top must both have a y-coordinate of 1, since one of them has a y-coordinate of 1. In the same way the x-coordinates of the left side must both be -3. Therefore, the fourth vertex is $D(-3, 1)$.

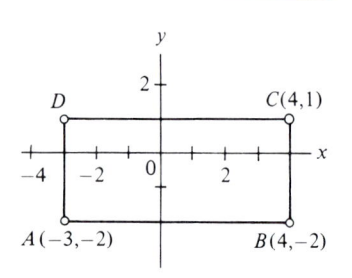

Figure 1-4

Example C
Where are all the points whose ordinates are 2?
All such points are 2 units above the x-axis; thus, the answer can be stated as "on a line 2 units above the x-axis."

Exercises 1-2

In Exercises 1 and 2 determine (at least approximately) the coordinates of the points specified in Figure 1-5.

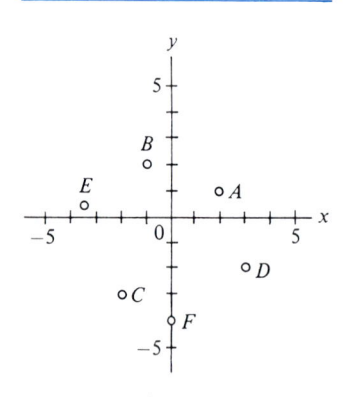

1. A, B, C
2. D, E, F

Figure 1-5

1-2 Rectangular Coordinates

In Exercises 3 and 4 plot (at least approximately) the given points.

3. $A(2, 7), B(-1, -2), C(-4, 2)$
4. $A(3, \frac{1}{2}), B(-6, 0), C(-\frac{5}{2}, -5)$

In Exercises 5 and 6 plot the given points and then join them in the order given by straight-line segments. Name the geometric figure formed.

5. $A(-1, 4), B(3, 4), C(1, -2)$
6. $A(-5, -2), B(4, -2), C(6, 3), D(-3, 3)$

In Exercises 7 and 8 find the indicated coordinates.

7. Three vertices of a rectangle are $(5, 2), (-1, 2)$, and $(-1, 4)$. What are the coordinates of the fourth vertex?
8. Two vertices of an equilateral triangle are $(7, 1)$ and $(2, 1)$. What is the abscissa of the third vertex?

In Exercises 9 through 20 answer the given questions.

9. Where are all the points whose abscissas are 1?
10. Where are all the points whose ordinates are -3?
11. Where are all the points whose abscissas equal their ordinates?
12. Where are all the points whose abscissas equal the negative of their ordinates?
13. What is the abscissa of all points on the y-axis?
14. What is the ordinate of all points on the x-axis?
15. Where are all the points for which $x > 0$?
16. Where are all the points for which $y < 0$?
17. Where are all points (x, y) for which $x > 0$ and $y < 0$?
18. Where are all points (x, y) for which $x < 0$ and $y > 1$?
19. In which quadrants is the ratio y/x positive?
20. In which quadrants is the ratio y/x negative?

1-3 Pythagorean Theorem

There are some occasions in our development of trigonometry when we wish to find the lengths of the sides of a right triangle. To do this we often make use of a very important concept called the *Pythagorean theorem*. In this section we develop this theorem and demonstrate some of its applications.

The Pythagorean theorem states: **In a right triangle, the square of the length of the hypotenuse equals the sum of the squares of the lengths of the other two sides.** We shall now show how this theorem may be found by use of areas of geometric figures. In Figure 1-6, a square of side c is inscribed in a square of sides $a + b$ as shown. The area of the outer square $(a + b)^2$ minus the areas of the four triangles of sides a, b, and c (the area of each of these triangles is $\frac{1}{2} ab$) equals the area of the inner square. This leads to:

$$(a + b)^2 - 4(\tfrac{1}{2} ab) = c^2$$
$$a^2 + 2ab + b^2 - 2ab = c^2$$

(1-1) $$a^2 + b^2 = c^2$$

The result is the equation which states the Pythagorean relation for each of the four triangles of sides a, b, and c. Therefore, in any right triangle with sides a, b, and c, where c is the hypotenuse, Equation (1-1) is valid.

We now present some examples illustrating the use of the Pythagorean theorem.

Example A

In a right triangle ABC with the right angle at C, $AC = 5.00$ and $AB = 13.0$. Find BC. See Figure 1-7.

The sides of the triangle are AC, BC, and AB, with AB the hypotenuse. Thus, the Pythagorean

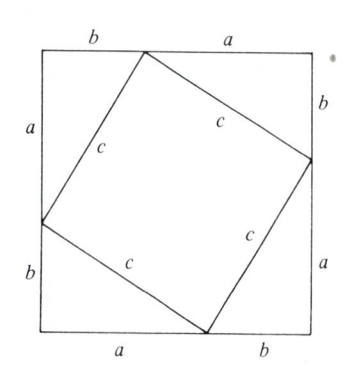

Figure 1-6

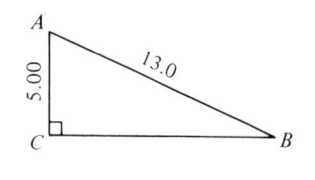

Figure 1-7

1-3 Pythagorean Theorem

theorem applied to this triangle is

$$(AC)^2 + (BC)^2 = (AB)^2.$$

Substituting the values for AC and AB, we have

$$(5.00)^2 + (BC)^2 = (13.0)^2$$
$$25.0 + (BC)^2 = 169$$
$$(BC)^2 = 144$$
$$BC = 12.0.$$

We use the principal square root of 144 since we know that the length of a side of a triangle is measured as a positive number.

Example B
Find the length of the hypotenuse of a right triangle whose other two sides are 65.0 in. and 92.0 in.

By using the Pythagorean theorem and letting c represent the hypotenuse as in Figure 1-8, we have

$$(65.0)^2 + (92.0)^2 = c^2$$
$$4225 + 8464 = c^2$$
$$12{,}689 = c^2$$
$$c = 113 \text{ in.}$$

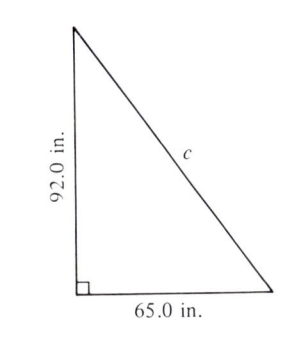

Figure 1-8

To find c in Example B, it was necessary to find the square root of 12,689. In Appendix D, there is a table of squares and square roots along with an explanation and illustrations of its use. Other means of determining square roots are logarithms (see Chapter 5), the slide rule, and certain electronic calculators.

The result in Example B was *rounded off* to three *significant digits*. If these terms are not familiar, a discussion of them may be found in Appendix B. In this section we shall round off numbers, when necessary, to three significant digits.

Example C
Find the length of the third side of a triangle which has a hypotenuse 0.0930 cm long and one leg 0.0410 cm long.

We let the third side be represented by a, as shown in Figure 1-9, and use the Pythagorean theorem.

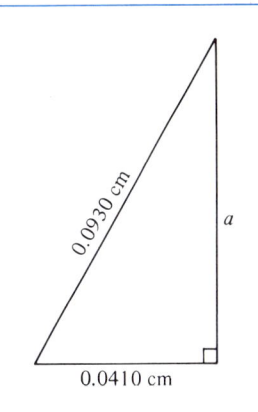

Figure 1-9

$$a^2 + (0.0410)^2 = (0.0930)^2$$
$$a^2 + 0.001681 = 0.008649$$
$$a^2 = 0.006968$$
$$a = 0.0835 \text{ cm}$$

Again we use the table of square roots, or some other appropriate method, to find $\sqrt{0.006968}$ in determining the value of a.

Example D
What is the maximum diameter of a circular table top which will fit through a door 25.0 in. by 78.0 in.? See Figure 1-10.

To fit through the door, the diameter of the table top can be no longer than the diagonal AC of the door. Using the Pythagorean theorem, we have $(AB)^2 + (BC)^2 = (AC)^2$. Substituting the given values of AB and BC, we have

$$(25.0)^2 + (78.0)^2 = (AC)^2$$
$$6709 = (AC)^2$$
$$AC = 81.9 \text{ in.}$$

Therefore, the diameter of the table top can be no greater than 81.9 in.

Example E
In the rectangular coordinate system, determine the distance between the points $A(4, 3)$ and $B(-2, 1)$.

By making a right triangle with AB as the hypotenuse and line segments parallel to the axes as legs, we see that the legs meet at $C(4, 1)$ as shown in Figure 1-11. We see that the length of the leg AC is the difference of the ordinates of points A and C.

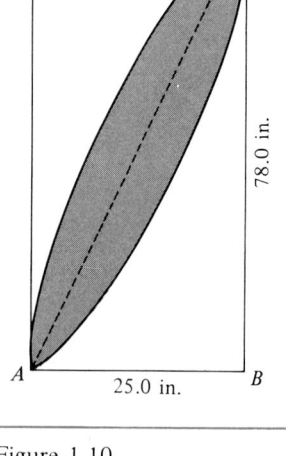

Figure 1-10

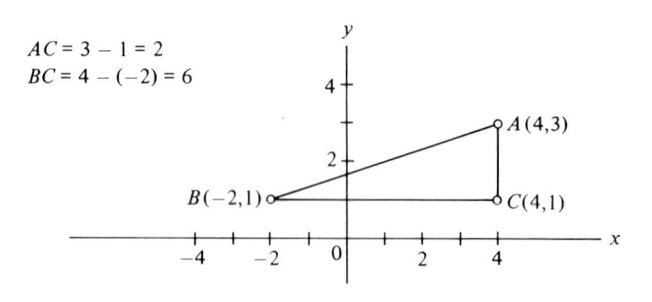

$AC = 3 - 1 = 2$
$BC = 4 - (-2) = 6$

Figure 1-11

1-3 Pythagorean Theorem

Also, the leg BC is 6 units long, which is the difference of the abscissas of points B and C. Therefore, the length of AB is found by use of the Pythagorean theorem:

$$\begin{aligned}(AB)^2 &= (AC)^2 + (BC)^2 \\ &= (3-1)^2 + [4-(-2)]^2 \\ &= 2^2 + 6^2 \\ &= 4 + 36 \\ &= 40.\end{aligned}$$

Thus, $AB = \sqrt{40} = 6.32$.

Exercises 1-3

In Exercises 1 through 8 find the square roots of the given numbers to three significant digits.

1. 267 ; 9730
2. 398 ; 5080
3. 1.76 ; 83.4
4. 9.42 ; 21.5
5. 0.579 ; 0.0143
6. 0.836 ; 0.0371
7. 0.00571 ; 0.000807
8. 0.00473 ; 0.000210

In Exercises 9 through 20 use the Pythagorean theorem to solve for the unknown side of the right triangle with legs a and b and hypotenuse c. If necessary, round off the result to three significant digits.

	a	b	c		a	b	c
9.	3.00	4.00		10.	8.00	15.0	
11.	3.10	5.90		12.	8.30	2.40	
13.	60.0		100	14.	30.0		90.0
15.	0.530		0.710	16.	0.210		0.430
17.		1.10	1.30	18.		1.50	1.80
19.		0.105	0.119	20.		0.124	0.155

Functions and Graphs

In Exercises 21 through 24 find the distance between the given points in the rectangular coordinate system.

21. $A(1, 2), B(4, 6)$ 22. $A(16, 12), B(1, 4)$
23. $A(3, -5), B(2, 4)$ 24. $A(-2, 3), B(5, -2)$

In Exercises 25 through 32 set up the given problems and solve by use of the Pythagorean theorem. Except where noted, determine values to the nearest tenth.

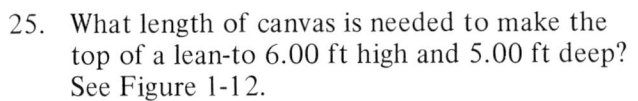

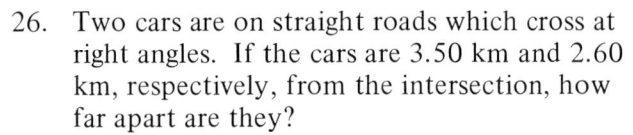

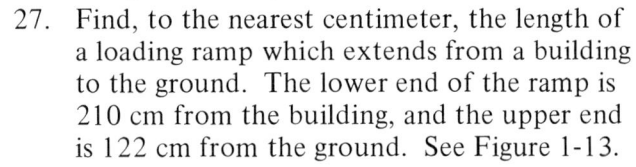

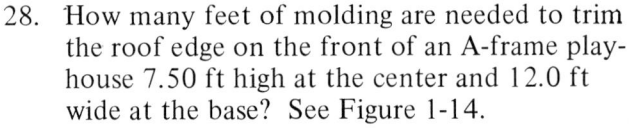

25. What length of canvas is needed to make the top of a lean-to 6.00 ft high and 5.00 ft deep? See Figure 1-12.

26. Two cars are on straight roads which cross at right angles. If the cars are 3.50 km and 2.60 km, respectively, from the intersection, how far apart are they?

27. Find, to the nearest centimeter, the length of a loading ramp which extends from a building to the ground. The lower end of the ramp is 210 cm from the building, and the upper end is 122 cm from the ground. See Figure 1-13.

28. How many feet of molding are needed to trim the roof edge on the front of an A-frame playhouse 7.50 ft high at the center and 12.0 ft wide at the base? See Figure 1-14.

29. What is the minimum inside diameter needed for a cylindrical casing to enclose a planter which is 2.40 ft square? See Figure 1-15.

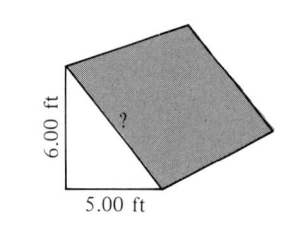

Figure 1-12

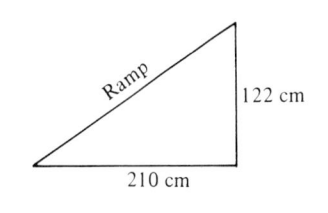

Figure 1-13

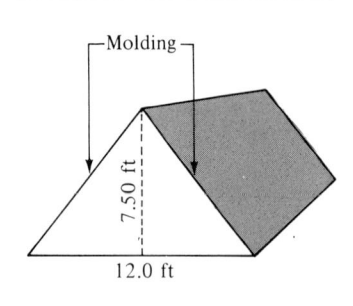

Figure 1-14

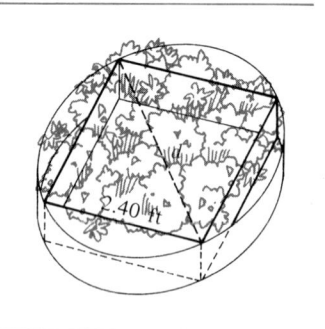

Figure 1-15

1-4 The Graph of a Function

30. What is the diameter of the smallest circular top which could cover the rectangular top on a table which is 1.50 m by 2.40 m?
31. What is the longest piece of pipe which can be stored in a cabinet 6.00 ft high, 3.00 ft wide, and 4.00 ft deep?
32. Find the length of the longest pole which can be packed in a trunk which is 2.10 ft by 3.40 ft by 4.80 ft.

1-4 The Graph of a Function

Now that we have introduced the concepts of a function and the rectangular coordinate system, we are in a position to determine the graph of a function. In this way we shall obtain a visual representation of a function, and be able to demonstrate some of the important properties of a function. Also, such graphs will be useful in representing functions we shall develop in trigonometry.

The graph of a function is the set of all points whose coordinates (x, y) satisfy the functional relationship $y = f(x)$. Since $y = f(x)$, we can write the coordinates of the points on the graph as $[x, f(x)]$. Writing the coordinates in this manner tells us exactly how to find them. We assume a certain value for x and then find the corresponding value of the function of x. These two numbers are the coordinates of the point.

Since there is no limit to the possible number of points which can be chosen, we normally select a few values of x, obtain the corresponding values of the function, and plot these points. These points are then connected by a *smooth* curve (not short straight lines from one point to the next), and are normally connected from left to right.

Example A

Graph the function $(3x - 5)$.

For purposes of graphing, we let y [or $f(x)$] $= 3x - 5$. We then let x take on various values and determine the corresponding values of y. If $x = 0$ we find that $f(0) = -5$, or $y = -5$. This means that the point $(0, -5)$ is on the graph of the function $3x - 5$. Choosing another value of x, for example 1, we find that $f(1) = -2$, or $y = -2$. This means that the point $(1, -2)$ is on the graph of the function $3x - 5$. Continuing to choose a few other values of x, we tabulate the results as shown in Figure 1-16. It is best to arrange the table so that the values of x increase; then there is no doubt how they are to be connected, for they are then connected in the order shown. Finally, we connect the points as shown in Figure 1-16, and see that the graph of the function $3x - 5$ is a straight line. We also note that there are no restrictions on values of x or the corresponding values of the function $y = f(x)$. Therefore, the domain is all real numbers and the range is also all real numbers.

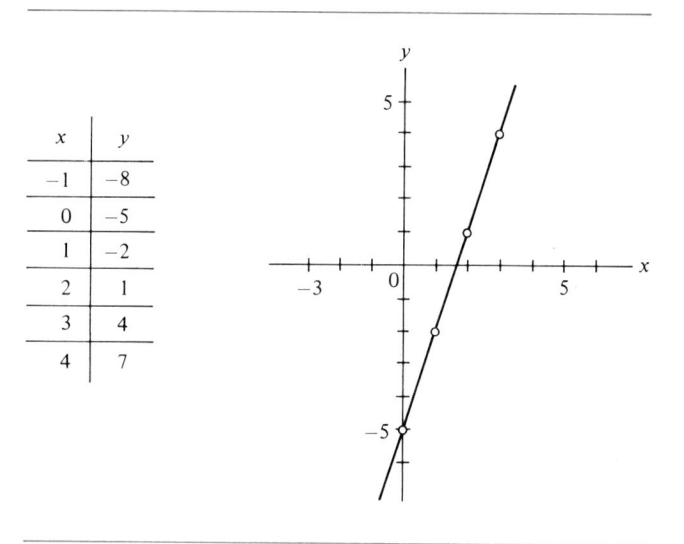

x	y
−1	−8
0	−5
1	−2
2	1
3	4
4	7

Figure 1-16

Example B
Graph the function $(2x^2 - 4)$.

First we represent the function by $f(x) = 2x^2 - 4$. Then we evaluate $f(x)$ for specified values of x and tabulate the values as shown in Figure 1-17. Then we plot and connect the points in a smooth curve as shown. We see that there are no restrictions on the values of x which may be chosen, but we also see that all resulting values of y are -4 or greater. Therefore, the domain of the function is all real numbers, and the range is all real numbers greater than or equal to -4.

Some special points should be noted. Since most common functions are smooth, any irregularities or sudden changes in the graph should be carefully checked. It usually helps to take values of x between those values where the question arises. Also, if the function is not defined for some value of x

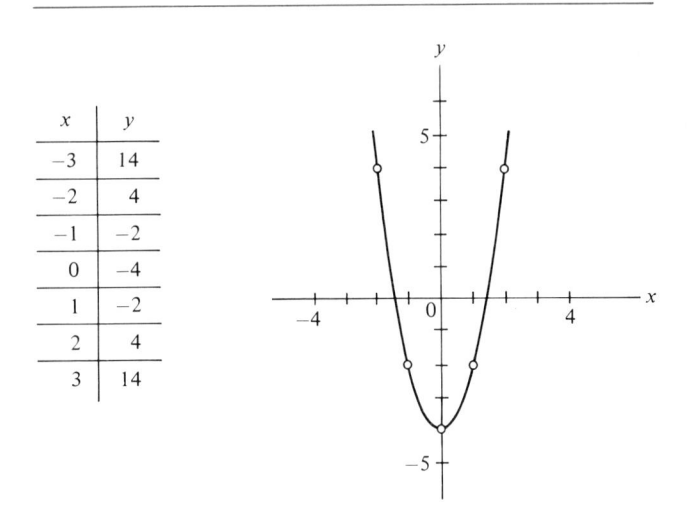

x	y
-3	14
-2	4
-1	-2
0	-4
1	-2
2	4
3	14

Figure 1-17

(remember, division by zero is not defined, and only real values of the variables are permissible), the function does not exist for that value of x. Finally, in applications we must be careful to plot values that are meaningful; often negative values for quantities such as time are not physically meaningful. The following examples illustrate these points.

Example C
Graph the function $y = x - x^2$.

First we determine the values in the table as shown in Figure 1-18. In plotting these points we note that $y = 0$ for both $x = 0$ and $x = 1$. Then a question arises: What happens between these values? Trying $x = \frac{1}{2}$, we find that $y = \frac{1}{4}$. Using this point completes the necessary information. Note that in plotting these graphs we do not stop the graph with the last point determined but indicate that the curve continues. Also, we note that the domain is all real numbers and the range is all real numbers less than or equal to $\frac{1}{4}$.

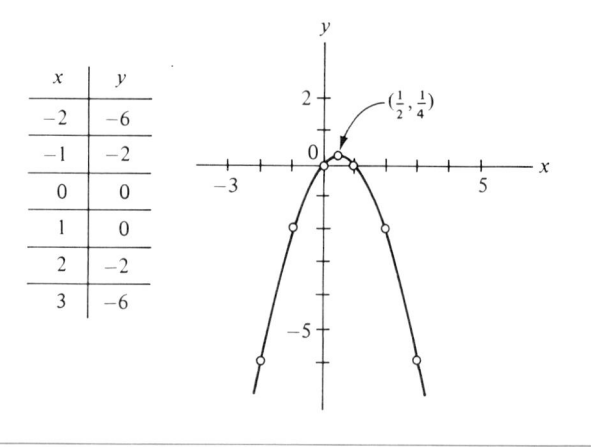

x	y
-2	-6
-1	-2
0	0
1	0
2	-2
3	-6

Figure 1-18

Example D

Graph the function $y = 1 + \dfrac{1}{x}$.

In finding the points on this graph as shown in Figure 1-19, we note that y is not defined if we try $x = 0$. Thus, we must be careful not to have any part of the curve cross the y-axis ($x = 0$). We choose points between $x = -1$ and $x = 1$ to choose values near $x = 0$, since we cannot let $x = 0$. Also, since x cannot equal zero, we see that the domain of the function is all real values except zero. From the graph, we see that the range is all real values except 1 (since y must always vary from 1 by the value of $1/x$, which cannot be zero).

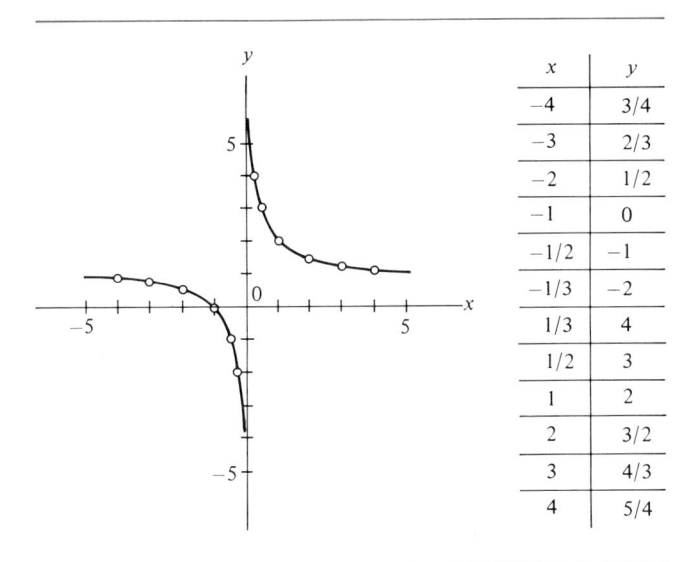

x	y
−4	3/4
−3	2/3
−2	1/2
−1	0
−1/2	−1
−1/3	−2
1/3	4
1/2	3
1	2
2	3/2
3	4/3
4	5/4

Figure 1-19

Example E
Graph the function $y = \sqrt{x + 1}$.

When finding the points for the graph, we may not let x equal any negative value less than -1, for all such values of x lead to imaginary values for y. Also, since we have the positive square root indicated, all values of y are positive. See Figure 1-20.

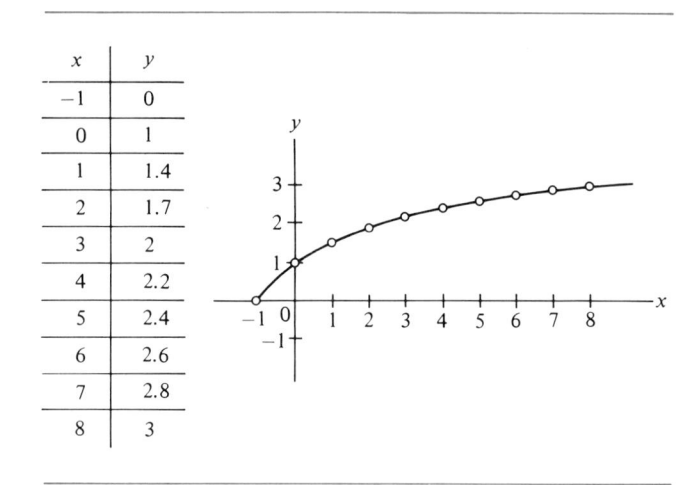

x	y
-1	0
0	1
1	1.4
2	1.7
3	2
4	2.2
5	2.4
6	2.6
7	2.8
8	3

Figure 1-20

Thus, we see that the domain of the function is all real numbers greater than or equal to -1, and the range of the function is all real positive numbers and zero.

Example F
The height s above the ground of an object t seconds after it was thrown into the air with a velocity of 64 ft/sec is given by the function

$$s = 64t - 16t^2.$$

Plot the height as a function of the time.

Since negative values for time and distance have no physical significance (the object was thrown from the ground and then returns to the ground), we need to plot t and s for positive (or zero) values only. The table of values thereby obtained is shown in Figure 1-21. We see that the domain is real values between and including 0 and 4, and the range is real numbers between and including 0 and 64.

1-4 The Graph of a Function

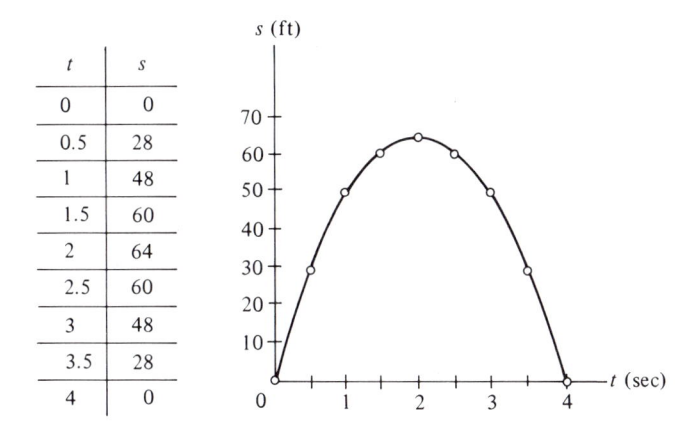

t	s
0	0
0.5	28
1	48
1.5	60
2	64
2.5	60
3	48
3.5	28
4	0

Figure 1-21

Exercises 1-4

In Exercises 1 through 24 graph the given functions.

1. $y = 2x - 4$
2. $y = 3x + 5$
3. $y = 5 - 3x$
4. $y = 7 - 2x$
5. $y = 4x + 1$
6. $y = 5x - 2$
7. $y = \frac{1}{2}x - 2$
8. $y = \frac{1}{3}x + 1$
9. $y = x^2 + 1$
10. $y = x^2 - 3$
11. $y = 3 - x^2$
12. $y = 8 - 2x^2$
13. $y = x^2 + 4x$
14. $y = 2x - x^2$
15. $y = x^2 - 3x + 1$
16. $y = x^2 + 3x + 2$
17. $y = \frac{1}{x}$
18. $y = \frac{2}{x + 1}$
19. $y = 1 + \frac{1}{x^2}$
20. $y = \frac{1}{x^2}$
21. $y = \sqrt{x}$
22. $y = \sqrt{x + 4}$
23. $y = \sqrt{1 - x}$
24. $y = \sqrt{4 - x}$

In Exercises 25 through 32 determine the domain and range of the given functions.

25. The function of Exercise 1.
26. The function of Exercise 4.

27. The function of Exercise 9.
28. The function of Exercise 12.
29. The function of Exercise 13.
30. The function of Exercise 16.
31. The function of Exercise 17.
32. The function of Exercise 24.

In Exercises 33 through 36 graph the given functions and determine the appropriate domain and range.

33. The velocity v (in feet per second) of an object under the influence of gravity, as a function of time t (in seconds), is given by $v = 32t$. Graph v as a function of t.
34. If $1000 is placed in an account earning 6 percent simple interest, the amount A in the account after t years is given by the function $A = 1000(1 + 0.06t)$. If the money is withdrawn from the account after 6 years, plot A as a function of t.
35. The surface area of a cube is given by $A = 6e^2$, where e is an edge of the cube. Plot A as a function of e.
36. If a rectangular tract of land has a perimeter of 600 m, its area as a function of its width w is $A = 300w - w^2$. Plot A as a function of w.

Exercises for Chapter 1

In Exercises 1 through 4 determine the appropriate function.

1. The sales tax in a certain state is 6 percent. Express the tax T on a certain item as a function of the cost C of the item.
2. Express the distance s in miles traveled in 5 hr as a function of the velocity v in miles per hour.

Exercises for Chapter 1

3. Fencing 1000 m long is to be used to enclose three sides of a rectangular field (fencing is not required on the fourth side). Express the area A of the field as a function of its length l (parallel to the fourth side).
4. Find the area A of a right triangle with a hypotenuse of 30 cm as a function of its base b.

In Exercises 5 through 12 evaluate the given functions.

5. Given $f(x) = 7x - 5$, find $f(3)$ and $f(-6)$.
6. Given $h(x) = 8 - 3x$, find $h(4)$ and $h(-2)$.
7. Given $g(y) = 5 - 2y^2$, find $g(2)$ and $g(-1)$.
8. Given $f(x) = \dfrac{4x^2 - x}{2}$, find $f(-1)$ and $f(-3)$.
9. Given $F(u) = u - 3u^2$, find $F(2u) - F(u)$.
10. Given $\phi(v) = \dfrac{3v + 2}{v}$, find $\phi(v+1) - \phi(v)$.
11. Given $f(x) = 3x^2 - 2x + 4$, find $f(x+h) - f(x)$.
12. Given $G(v) = v^3 + 2v^2 - 3v$, find $G(3+h) - G(3)$.

In Exercises 13 through 16 use the Pythagorean theorem to solve for the unknown side of the right triangle with legs a and b and hypotenuse c. Round off results to three significant digits.

13. $a = 12.0$, $b = 17.0$; find c.
14. $a = 350$, $b = 210$; find c.
15. $a = 4.85$, $c = 7.80$; find b.
16. $b = 0.750$, $c = 0.942$; find a.

In Exercises 17 through 20 find the distance between the given points in the rectangular coordinate system.

17. $A(2, 7)$, $B(6, 10)$
18. $A(-2, 1)$, $B(-10, 16)$
19. $A(-8, -7)$, $B(-2, -1)$
20. $A(-1, 5)$, $B(4, -2)$

In Exercises 21 through 32 graph the given functions.

21. $y = 5x - 10$
22. $y = 4 - x$
23. $y = 8 - 4x$
24. $y = \frac{1}{2}x - 3$
25. $y = 4x - x^2$
26. $y = 2x^2 - 6x$
27. $y = x^2 + x - 3$
28. $y = 2 - x - x^2$
29. $y = \dfrac{x}{x + 1}$
30. $y = \dfrac{4}{x - 2}$
31. $y = \sqrt{x + 2}$
32. $y = \sqrt{8 - 2x}$

In Exercises 33 through 36 determine the domain and range of the given functions.

33. The function of Exercise 25.
34. The function of Exercise 28.
35. The function of Exercise 29.
36. The function of Exercise 32.

In Exercises 37 through 44 answer the given questions.

37. Where are all the points (x, y) whose abscissas are 1 and for which $y > 0$?
38. Where are all the points (x, y) whose ordinates are 1 and for which $x < 0$?
39. The vertices of a rectangle are $(6, 5)$, $(-4, 5)$, and $(-4, 2)$. What are the coordinates of the fourth vertex?
40. Two vertices of a square are $(2, 1)$ and $(7, 1)$. If the ordinate of one of the other vertices is 6, what are the coordinates of the other vertices?
41. How wide is the base of the front of a tent for which there are 15.0 ft of canvas to drape over a center pole which is 5.50 ft above the ground? See Figure 1-22.
42. Two jets leave the same airport at the same time, one traveling due south at 600 mi/hr and the other due east at 1400 mi/hr. How far apart (to the nearest mile) are they after 30 min?

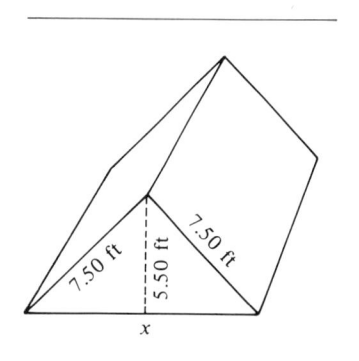

Figure 1-22

Exercises for Chapter 1

43. The perimeter p of a rectangle of length 20 cm and width w is $p = 40 + 2w$. Plot p as a function of w.

44. The distance h (in feet) above the surface of the earth of an object as a function of the time (in seconds) is given by $h = 60t - 16t^2$ if the object is given an initial upward velocity of 60 ft/sec. Plot the graph of h as a function of t.

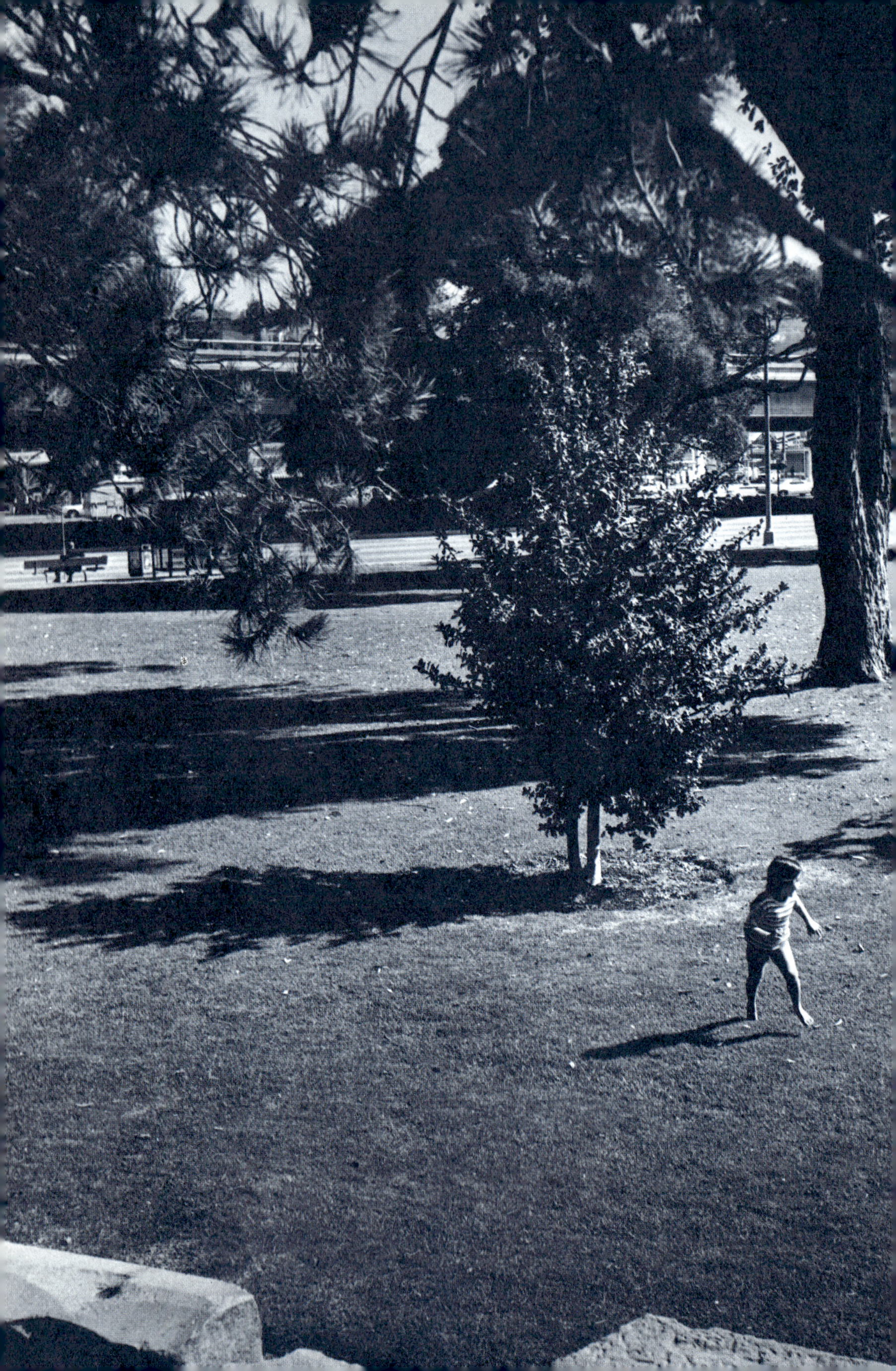

2

The Trigonometric Functions

Trigonometric functions are based on properties of similar triangles. As shown in this photograph, the lengths of the shadows are proportional to the heights of the objects. By use of the trigonometric functions, lengths and distances—including those which are not directly measurable—can be determined.

2-1 Angles

We are now ready to begin our study of trigonometry, the literal meaning of which is "triangle measurement." In Chapter 1 we noted some of the areas in which trigonometry may be used. In this chapter we use the fundamental properties of triangles to set up some very basic and useful functions involving their sides and angles. After introducing these trigonometric functions, we demonstrate some of their basic applications. At this point we begin by considering the concept of an angle.

An *angle* is defined as being generated by rotating a *half* line about its endpoint from an initial position to a terminal position. A half line is that portion of a line to one side of a fixed point on the line. We call the initial position of the half line the *initial side* and the terminal position of the half line the *terminal side*. The fixed point is the *vertex*. The angle itself is the amount of rotation from the initial side to the terminal side.

If the rotation of the terminal side from the initial side is counterclockwise, the angle is said to be *positive*. If the rotation is clockwise, the angle is *negative*. In Figure 2-1, angle 1 is positive and angle 2 is negative.

There are many symbols used to designate angles. Probably the most widely used are Greek letters such as θ (theta), ϕ (phi), α (alpha), and β (beta).

Two measurements of angles are widely used. These are the *degree* and the *radian*. A degree is defined as $\frac{1}{360}$ of a complete rotation. It is divided into 60 equal parts called *minutes*, and each minute is divided into 60 equal parts called *seconds*. The symbols °, ', and " are used to designate degrees, minutes, and seconds, respectively. Decimal parts of angles are also commonly used. The radian is discussed in Chapter 3.

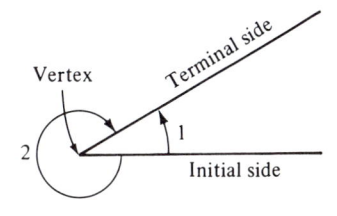

Figure 2-1

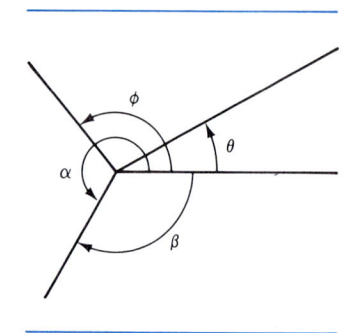

Figure 2-2

Example A
The angles $\theta = 30°$, $\phi = 140°$, $\alpha = 240°$, and $\beta = -120°$ are shown in Figure 2-2. **Note that angles α and β have the same initial and terminal sides. Such angles are called coterminal.**

Example B
To change the angle $43°24'$ to decimal form we use the fact that $1' = (\frac{1}{60})°$. Therefore, $24' = (\frac{24}{60}) = 0.4°$. Therefore,

$$43°24' = 43.4°.$$

To change $154.36°$ to an angle measured to the nearest minute, we use the fact that $1° = 60'$. Therefore, $0.36° = 0.36(60') = 21.6'$, which is $22'$ to the nearest minute. Therefore,

$$154.36° = 154°22'.$$

Example C
Determine the values of two angles which are coterminal with an angle of $145°32'$.

Since there are $360°$ in a complete rotation, we can find one coterminal angle by considering the angle which is $360°$ larger than the given angle. This gives us an angle of $505°32'$. Another method of finding a coterminal angle is to subtract $145°32'$ from $360°$ and then consider the resulting angle to be negative. This means that the original angle and the derived angle would make up one complete rotation when put together. This method leads us to the angle of $-214°28'$ (see Figure 2-3). These methods could be employed repeatedly to find other coterminal angles.

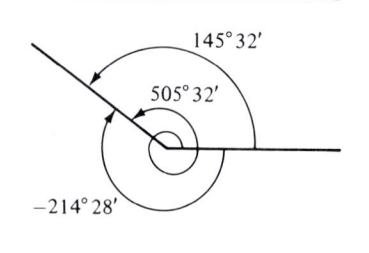

Figure 2-3

If the initial side of the angle is the positive *x*-axis and the vertex is at the origin, the angle is said to be in *standard position*. The angle is then determined by the position of the terminal side. **If the terminal side is in the first quadrant, the angle is called a *first-quadrant angle*.** Similar terms are used when the terminal side is in the other quadrants. **If the terminal side coincides with one of the axes, the angle is a *quadrantal angle*.** When an angle is in

2-1 Angles

standard position, the terminal side can be determined if we know any point, other than the origin, on the terminal side.

Example D

To draw a third-quadrant angle of 205°, we simply measure 205° from the positive x-axis and draw the terminal side. See angle α in Figure 2-4. Angle β is a quadrantal angle since its terminal side is the positive y-axis. Angle θ is in standard position and the terminal side is uniquely determined by knowing that it passes through the point (2, 1). The same terminal side passes through (4, 2) and $(\frac{11}{2}, \frac{11}{4})$, among other points. Knowing that the terminal side passes through any one of these points makes it possible to determine the terminal side.

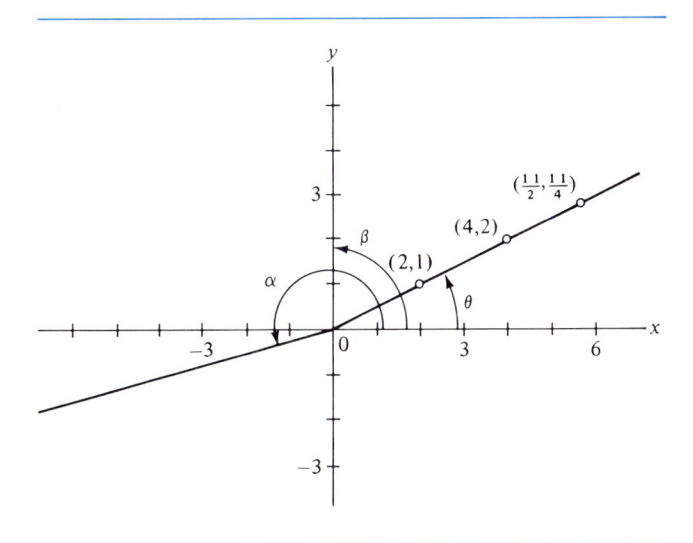

Figure 2-4

Exercises 2-1

In Exercises 1 through 4 draw the given angles.

1. 60°, 120°, −90°
2. 330°, −150°, 450°
3. 50°, −360°, −30°
4. 45°, 225°, −250°

In Exercises 5 through 12 determine one positive and one negative coterminal angle for each angle given.

5. 45°
6. 73°
7. 150°
8. 162°
9. 70°30′
10. 153°47′
11. 278.1°
12. 197.6°

In Exercises 13 through 16 change the given angles to equal angles expressed in decimal form.

13. 15°12′
14. 246°48′
15. 86°3′
16. 157°39′

In Exercises 17 through 20 change the given angles to equal angles expressed to the nearest minute.

17. 47.5°
18. 315.8°
19. 5.62°
20. 238.21°

2-2 Trigonometric Functions of Acute Angles

Very basic to the development of trigonometry are the trigonometric functions. To define these functions we place an angle θ in standard position and drop perpendicular lines from points on the terminal side to the x-axis as shown in Figure 2-5. In doing this we set up similar triangles, each with one vertex at the origin. Using the basic fact from geometry that corresponding sides of similar triangles are proportional, we may set up equal ratios of corresponding sides.

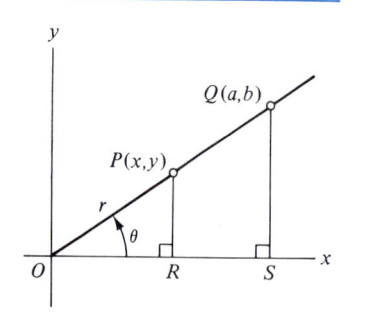

Figure 2-5

Example A

In Figure 2-5 triangles *OPR* and *OQS* are similar. Therefore, the ratio of the abscissa (x-value) to the ordinate (y-value) of points *P* and *Q* is the same in each triangle. We can state this as

$$\frac{x}{y} = \frac{a}{b}$$

2-2 Trigonometric Functions of Acute Angles

For any other point on the terminal side of θ that we might choose, the ratio of its abscissa to ordinate would still be the same as those already given.

The distance from the origin to a point on the terminal side is called the *radius vector* and is denoted r. It is the other important distance involved in the definitions of the trigonometric functions. For any angle θ in standard position, there are six ratios which may be set up. It can be seen that the value of these ratios depends on the position of the terminal side of the angle. Therefore, the ratios depend on the angle and are the functions of it. These functions are called the *trigonometric functions* and are defined as follows:

(2-1)

$$\sin\theta = \frac{\text{ordinate of } P}{\text{radius vector of } P} = \frac{y}{r}$$

$$\cosine\theta = \frac{\text{abscissa of } P}{\text{radius vector of } P} = \frac{x}{r}$$

$$\tangent\theta = \frac{\text{ordinate of } P}{\text{abscissa of } P} = \frac{y}{x}$$

$$\cotangent\theta = \frac{\text{abscissa of } P}{\text{ordinate of } P} = \frac{x}{y}$$

$$\secant\theta = \frac{\text{radius vector of } P}{\text{abscissa of } P} = \frac{r}{x}$$

$$\cosecant\theta = \frac{\text{radius vector of } P}{\text{ordinate of } P} = \frac{r}{y}$$

Although we are restricting our attention at this point to acute angles, the definitions of the trigonometric functions given above are general and may be used for angles of any magnitude. We consider the trigonometric functions of any angle in the next chapter.

For convenience, the names of the various functions are usually abbreviated to $\sin\theta$, $\cos\theta$, $\tan\theta$, cot θ, sec θ, and csc θ. The following example illustrates finding the trigonometric functions when the terminal side of the angle passes through a known point.

Example B
Determine the trigonometric functions of the angle with a terminal side passing through the point (3, 4).

By placing the angle in standard position, as shown in Figure 2-6, and drawing the terminal side through (3, 4), we find that $r = 5$ (by use of the Pythagorean theorem). Using the values $x = 3$, $y = 4$, and $r = 5$, we find that

$$\sin \theta = \tfrac{4}{5} \qquad \cos \theta = \tfrac{3}{5}$$
$$\tan \theta = \tfrac{4}{3} \qquad \cot \theta = \tfrac{3}{4}$$
$$\sec \theta = \tfrac{5}{3} \qquad \csc \theta = \tfrac{5}{4}$$

Note that a given function is not defined if the denominator is zero. The denominator is zero for tan θ and sec θ when $x = 0$ and for cot θ and csc θ when $y = 0$. This restricts the domain of these functions, for any value of θ which would give us a zero in the denominator would have to be excluded from the domain. When $\theta = 0°$, for example, the denominator of the cotangent function and the cosecant function is zero, and therefore $\theta = 0°$ is not in the domain of those functions. In all cases we assume that $r > 0$. If $r = 0$ there would be no terminal side and therefore no angle.

If one of the trigonometric functions is known, we can determine the other functions of the angle by using the Pythagorean theorem and the definitions of the functions. The following example illustrates the method.

Example C
If we know that $\sin \theta = \tfrac{2}{5}$ and that θ is a first-quadrant angle, we know that the ratio of the ordinate to the radius vector (of y to r) is 2 to 5. Therefore, the point on the terminal side for which y is 2 can be determined by use of the Pythagorean theorem. The x-value for this point is

$$x = \sqrt{5^2 - 2^2} = \sqrt{25 - 4} = \sqrt{21}.$$

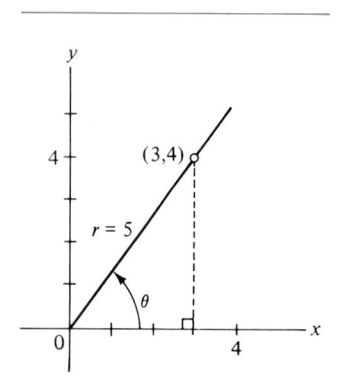

Figure 2-6

2-2 Trigonometric Functions of Acute Angles

Therefore, the point $(\sqrt{21}, 2)$ is on the terminal side of θ as shown in Figure 2-7. Using the values $x = \sqrt{21}$, $y = 2$, and $r = 5$, we now have the trigonometric functions of θ:

$$\cos \theta = \frac{\sqrt{21}}{5} \qquad \tan \theta = \frac{2}{\sqrt{21}} \qquad \cot \theta = \frac{\sqrt{21}}{2}$$

$$\sec \theta = \frac{5}{\sqrt{21}} \qquad \csc \theta = \frac{5}{2}.$$

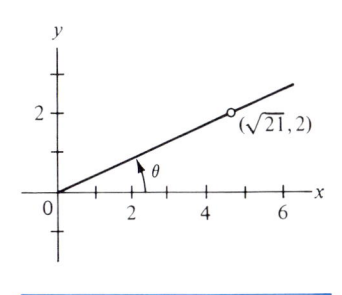

Figure 2-7

Exercises 2-2

In Exercises 1 through 8 determine the trigonometric functions of the angles whose terminal sides pass through the given points.

1. (4, 3)
2. (5, 12)
3. (15, 8)
4. (7, 24)
5. $(1, \sqrt{15})$
6. (1, 1)
7. (2, 5)
8. $(1, \frac{1}{2})$

In Exercises 9 through 16 use the given trigonometric functions to find the indicated trigonometric functions.

9. $\tan \theta = 1$; find $\sin \theta$ and $\sec \theta$.
10. $\sin \theta = \frac{1}{2}$; find $\cos \theta$ and $\csc \theta$.
11. $\cos \theta = \frac{2}{3}$; find $\tan \theta$ and $\cot \theta$.
12. $\sec \theta = 3$; find $\tan \theta$ and $\sin \theta$.
13. $\sin \theta = 0.7$; find $\cot \theta$ and $\csc \theta$.
14. $\cos \theta = \frac{5}{12}$; find $\sec \theta$ and $\tan \theta$.
15. $\cot \theta = 0.25$; find $\cos \theta$ and $\csc \theta$.
16. $\csc \theta = 1.2$; find $\cos \theta$ and $\cot \theta$.

In Exercises 17 and 18 each of the listed points is on the terminal side of an angle. Show that each of the indicated functions is the same for each of the points.

17. (3, 4), (6, 8), (4.5, 6); $\sin \theta$ and $\tan \theta$
18. (5, 12), (15, 36), (7.5, 18); $\cos \theta$ and $\cot \theta$

In Exercises 19 and 20 answer the given questions.

19. From the definitions of the trigonometric functions, it can be seen that csc θ is the reciprocal of sin θ. What function is the reciprocal of cos θ? Of cot θ?
20. Divide the expression for sin θ by that for cos θ. Is the result the expression for any of the other functions?

2-3 Values of the Trigonometric Functions

We can now calculate the trigonometric functions if we know one point on the terminal side of the angle. However, in practice it is more common to know the angle in degrees, for example, and to be required to find the functions of this angle. Therefore, we must be able to determine the trigonometric functions of angles in degrees.

One way to determine the functions of a given angle is to make a scale drawing. That is, we draw the angle in standard position using a protractor, and then measure the lengths of the values of x, y, and r for some point on the terminal side. By using the proper ratios we may determine the functions of this angle.

We may also use certain geometric facts to determine the functions of certain angles. The following two examples illustrate this procedure.

Example A

From geometry we know that the side opposite a 30° angle in a right triangle is one-half the hypotenuse. By using this fact and letting $y = 1$ and $r = 2$ (see Figure 2-8), we determine that sin 30° = $\frac{1}{2}$. Also, by use of the Pythagorean theorem we determine that $x = \sqrt{3}$. Thus, cos 30° = $\sqrt{3}/2$ and tan 30° = $1/\sqrt{3}$. In a similar way we may determine the values of the functions of 60° to be as follows: sin 60° = $\sqrt{3}/2$, cos 60° = $\frac{1}{2}$, and tan 60° = $\sqrt{3}$.

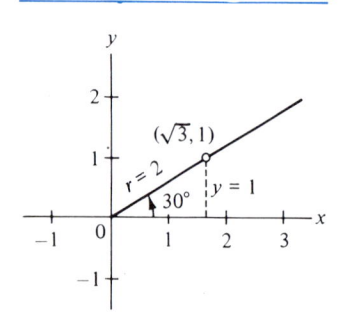

Figure 2-8

2-3 Values of the Trigonometric Functions

Example B
Determine sin 45°, cos 45°, and tan 45°.

From geometry we know that in an isosceles right triangle the angles are 45°, 45°, and 90°. We know that the sides are in proportion 1, 1, $\sqrt{2}$, respectively. Putting the 45° angle in standard position, we find $x = 1$, $y = 1$, and $r = \sqrt{2}$ (see Figure 2-9). From this we determine

$$\sin 45° = \frac{1}{\sqrt{2}} \qquad \cos 45° = \frac{1}{\sqrt{2}} \qquad \tan 45° = 1.$$

Summarizing the results for 30°, 45°, and 60°, we have the following table.

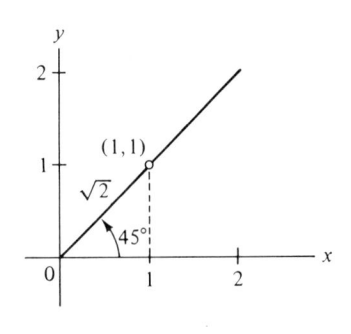

Figure 2-9

				Decimal Approximations		
θ	30°	45°	60°	30°	45°	60°
sin θ	1/2	$1/\sqrt{2}$	$\sqrt{3}/2$	0.500	0.707	0.866
cos θ	$\sqrt{3}/2$	$1/\sqrt{2}$	1/2	0.866	0.707	0.500
tan θ	$1/\sqrt{3}$	1	$\sqrt{3}$	0.577	1.000	1.732

The scale-drawing method is only approximate, and the geometric methods work only for certain angles. However, the values of the functions may be determined through more advanced methods. We now refer to Table 3 in Appendix D. To obtain the value of a function, we note that the angles from 0° to 45° are listed in the left-hand column and are read down. The angles from 45° to 90° are listed on the right-hand side and are read *up*. The functions for angles from 0° to 45° are listed along the top, and those for the angles from 45° to 90° are listed along the bottom.

Example C
Sin 34° is found under sin θ (at top) and to the right of 34° (at left). Sin 34° = 0.5592.

Cos 72° is found over cos θ (at bottom) and to the left of 72° (at right). Cos 72° = 0.3090.

Example D
Tan 42° 20' is found under tan θ (at top) and to the right of 20' (which appears under 42°).
Tan 42°20' = 0.9110.

Sin 64° 40' is found over sin θ (at bottom) and to the left of 40' (which appears *over* 64°).
Sin 64°40' = 0.9038.

Using Table 3, we can obtain values for the trigonometric functions of angles expressed to the nearest minute. Greater accuracy can be obtained by using tables which give five or more digits for the values of the functions or by using certain electronic calculators. Since the angles are given only to the nearest 10', it is necessary to use linear *interpolation* for angles given to the nearest minute which cannot be found directly. Interpolation assumes that if a particular angle lies between two of those listed in the table, then the functions of that angle are at the same proportional distance between the functions listed. This assumption is not strictly correct, although it is a very good approximation.

Example E
To find sin 23° 27' we must interpolate between sin 23° 20' and 23° 30'. Since 27' is $\frac{7}{10}$ of the way between 20' and 30', we assume that sin 23° 27' is $\frac{7}{10}$ of the way between sin 23° 20' and sin 23° 30'. These values are 0.3961 and 0.3987. The *tabular difference* between them is 26, and $\frac{7}{10}$ of this is 18 to the nearest integer. Adding 0.0018 to 0.3961, we obtain sin 23°27' = 0.3979. Another method of indicating the interpolation is shown in Figure 2-10.

From Figure 2-10 we see that

$$\frac{7}{10} = \frac{x}{26}$$

$$10x = 182$$

$$x = 18.2.$$

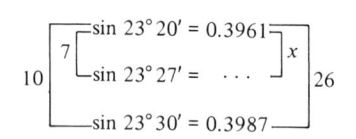

$$10 \begin{bmatrix} 7 \begin{bmatrix} \text{sin } 23° 20' = 0.3961 \\ \text{sin } 23° 27' = \cdots \end{bmatrix} x \\ \text{sin } 23° 30' = 0.3987 \end{bmatrix} 26$$

Figure 2-10

We want to determine the value of the function to the nearest four decimal places. To do this, we must

2-3 Values of the Trigonometric Functions

round off the value of x to the nearest integer. (For an explanation of rounding off, see Appendix B.) Therefore, $x = 18$ and

$\sin 23°27' = 0.3961 + 0.0018 = 0.3979$.

If an angle is given in degrees and decimal parts of a degree, we may first express the angle to the nearest minute and then find the value of the function from the table.

Example F
To find $\cos 76.3°$, we first determine that $0.3° = 18'$. Thus, we are to find $\cos 76°18'$. This means that we want the value $\frac{8}{10}$ of the way from $\cos 76°10'$ to $\cos 76°20'$. These values are 0.2391 and 0.2363, and their tabular difference is 28. Eight-tenths of this is 22 (to the nearest integer). Thus, subtracting this from 0.2391, we get $\cos 76°18' = 0.2369$. (The value of $\cos 76°10'$ is greater than $\cos 76°20'$ — the values of the cosine get *smaller* as θ gets larger.) This means that $\cos 76.3° = 0.2369$. Again we can indicate the interpolation as in Figure 2-11.
From Figure 2-11 we see that

$$\frac{8}{10} = \frac{x}{28}$$

$$x = 22.$$

If an electronic calculator is used, it may not be necessary to use interpolation to obtain additional accuracy. However, interpolation is a method which can be used with a great many tables, including those for which values may not be available on a calculator.

Not only are we able to use interpolation to find values of the functions if we know the angle, but we can also find the angle if we know the function. This requires that we reverse the procedures just mentioned.

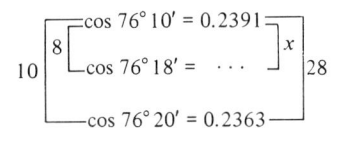

$$10 \begin{bmatrix} 8 \begin{bmatrix} \cos 76°10' = 0.2391 \\ \cos 76°18' = \cdots \end{bmatrix} x \\ \cos 76°20' = 0.2363 \end{bmatrix} 28$$

Figure 2-11

Example G
Given that $\sin \theta = 0.2616$, find θ.

We look for 0.2616, or the nearest number to it, in the columns for $\sin \theta$. Since this number appears exactly, we conclude that $\theta = 15°10'$.

Example H
Given that $\cos \theta = 0.8811$, find θ.

We find that this number lies between $\cos 28°10'$ and $\cos 28°20'$. These values are 0.8816 and 0.8802. The tabular difference between $\cos \theta$ and $\cos 28°10'$ is 5, and the tabular difference between the two values given in the table is 14. Thus $\cos \theta$ is assumed to be $\frac{5}{14}$ of the way from the first to the second. To the nearest tenth, $\frac{5}{14}$ is $\frac{4}{10}$. Hence $\theta = 28°14'$ (to the nearest minute). The solution of this type of problem can also be indicated by means of a figure such as Figure 2-12, from which we see that

$$\frac{5}{14} = \frac{x}{10}$$

$$x = 4.$$

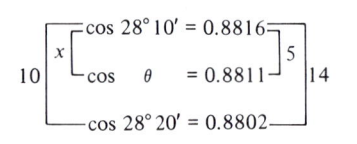

Figure 2-12

Exercises 2-3

In Exercises 1 through 4 use a protractor to draw the given angle. Measure off 10 units (centimeters are convenient) along the radius vector. Then measure the corresponding values of x and y. From these values determine the trigonometric functions of the angle.

1. 40°   2. 75°   3. 15°   4. 53°

In Exercises 5 through 12 find the value of each of the trigonometric functions without using Table 3 in Appendix D.

5. cot 45°   6. csc 45°   7. csc 30°   8. sec 30°
9. cot 60°   10. csc 60°   11. sec 60°   12. cot 30°

2-3 Values of the Trigonometric Functions

In Exercises 13 through 32 find the value of each of the trigonometric functions from Table 3 in Appendix D.

13. sin 19° 0'
14. cos 43° 0'
15. tan 67° 0'
16. cot 76° 0'
17. cos 22° 20'
18. tan 34° 50'
19. sec 56° 30'
20. csc 52° 10'
21. tan 28° 56'
22. cos 48° 44'
23. sin 63° 15'
24. sec 71° 47'
25. cot 7° 8'
26. csc 12° 14'
27. cot 65° 49'
28. sin 57° 52'
29. cos 26.5°
30. cot 83.5°
31. sin 75.6°
32. tan 43.1°

In Exercises 33 through 40 find θ to the nearest minute for each of the given trigonometric functions.

33. $\sin \theta = 0.0958$
34. $\cos \theta = 0.1593$
35. $\tan \theta = 3.108$
36. $\sin \theta = 0.9528$
37. $\cot \theta = 0.8070$
38. $\tan \theta = 0.6539$
39. $\sec \theta = 1.289$
40. $\csc \theta = 1.168$

In Exercises 41 through 44 find θ to the nearest tenth of a degree for each of the given trigonometric functions.

41. $\cos \theta = 0.8605$
42. $\sin \theta = 0.7044$
43. $\tan \theta = 5.284$
44. $\cot \theta = 0.0708$

In Exercises 45 and 46 solve the given problems.

45. When a projectile is fired into the air, its horizontal velocity v_x is related to the velocity v with which it is fired by the relation $v_x = v(\cos \theta)$, where θ is the angle between the horizontal and the direction in which it is fired. What is the horizontal velocity of a projectile fired with a velocity of 200 ft/sec at an angle of 36° 0' with respect to the horizontal? See Figure 2-13.

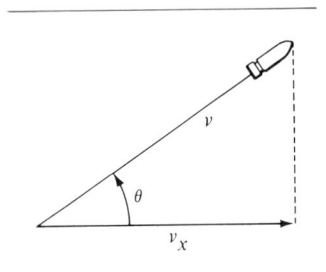

Figure 2-13

46. When a light ray enters glass from the air it bends somewhat toward a line perpendicular to the surface. The *index of refraction* of the glass is defined as

$$n = \frac{\sin i}{\sin r}$$

where i is the angle between the perpendicular and the ray in the air and r is the angle between the perpendicular and the ray in the glass. A typical case for glass is $i = 59°0'$ and $r = 34°0'$. Find the index of refraction for this case. See Figure 2-14.

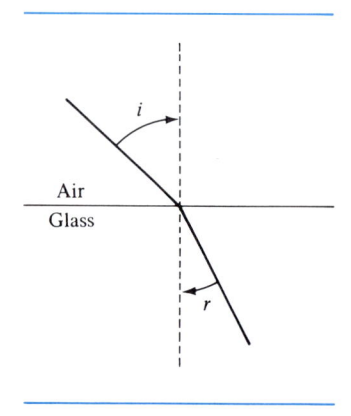

Figure 2-14

2-4 The Right Triangle

From geometry we know that a triangle, by definition, consists of three sides and has three angles. If one side and any other two of these six parts of the triangle are known, it is possible to determine the other three parts. One of the three known parts must be a side, for if we know only the three angles, we can conclude only that an entire set of similar triangles has those particular angles.

Example A
Assume that one side and two angles are known. Then we may determine the third angle by the fact that the sum of the angles of a triangle is always 180°. Of all the possible similar triangles having these three angles, we have the one with the particular side which is known. Only one triangle with these parts is possible.

To *solve* a triangle means that, when we are given three parts of a triangle (at least one a side), we are to find the other three parts. In this section we are going to demonstrate the method of solving a right triangle. Since one angle of the triangle

2-4 The Right Triangle

will be 90°, it is necessary to know one side and one other part. Also, we know that the sum of the three angles of a triangle is 180°, and this in turn tells us that the sum of the other two angles, both acute, is 90°. **Any two acute angles whose sum is 90° are said to be *complementary*.**

For consistency, when we are labeling the parts of the right triangle we shall use the letters A and B to denote the acute angles and C to denote the right angle. The letters a, b, and c will denote the sides opposite these angles, respectively. Thus, side c is the hypotenuse of the right triangle.

We shall find it convenient in solving right triangles to define the trigonometric functions of the acute angles in terms of the sides (see Figure 2-15). Placing the vertex of angle A at the origin and the vertex of angle C on the positive x-axis, we obtain the following definitions:

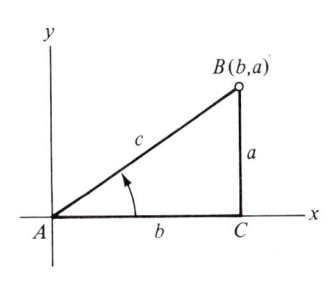

Figure 2-15

(2-2)

$$\sin A = \frac{y}{r} = \frac{a}{c} \qquad \cos A = \frac{x}{r} = \frac{b}{c} \qquad \tan A = \frac{y}{x} = \frac{a}{b}$$

$$\cot A = \frac{x}{y} = \frac{b}{a} \qquad \sec A = \frac{r}{x} = \frac{c}{b} \qquad \csc A = \frac{r}{y} = \frac{c}{a}$$

If we should place the vertex of angle B at the origin, instead of the vertex of angle A, we would obtain the following definitions for the functions of angle B:

(2-3)

$$\sin B = \frac{b}{c} \qquad \cos B = \frac{a}{c} \qquad \tan B = \frac{b}{a}$$

$$\cot B = \frac{a}{b} \qquad \sec B = \frac{c}{a} \qquad \csc B = \frac{c}{b}$$

Inspecting these results, we may generalize our definitions of the trigonometric functions of acute angles of a right triangle to be as follows:

(2-4)

$$\sin \alpha = \frac{\text{opposite side}}{\text{hypotenuse}} \qquad \cos \alpha = \frac{\text{adjacent side}}{\text{hypotenuse}} \qquad \tan \alpha = \frac{\text{opposite side}}{\text{adjacent side}}$$

$$\cot \alpha = \frac{\text{adjacent side}}{\text{opposite side}} \qquad \sec \alpha = \frac{\text{hypotenuse}}{\text{adjacent side}} \qquad \csc \alpha = \frac{\text{hypotenuse}}{\text{opposite side}}$$

Using the definitions in this form, we can solve right triangles without placing the angle in standard position. The angle need only be part of any right triangle.

We note from the preceding discussion that $\sin A = \cos B$. From this we conclude that **cofunctions of acute complementary angles are equal.** The sine function and cosine function are cofunctions, the tangent function and cotangent function are cofunctions, and the secant function and cosecant function are cofunctions. From this we can see how the tables of trigonometric functions are constructed. Since $\sin A = \cos (90° - A)$, the number representing either of these need appear only once in the tables.

Example B

Given $a = 4.00$, $b = 7.00$, and $c = \sqrt{65.0}$ ($C = 90°0'$), find $\sin A$, $\cos A$, and $\tan A$. (See Figure 2-16.)

$$\sin A = \frac{\text{side opposite angle } A}{\text{hypotenuse}} = \frac{4.00}{\sqrt{65.0}} = 0.496$$

$$\cos A = \frac{\text{side adjacent angle } A}{\text{hypotenuse}} = \frac{7.00}{\sqrt{65.0}} = 0.868$$

$$\tan A = \frac{\text{side opposite angle } A}{\text{side adjacent angle } A} = \frac{4.00}{7.00} = 0.571$$

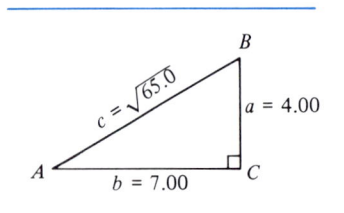

Figure 2-16

2-4 The Right Triangle

Example C
In Figure 2-16, the functions of B are

$$\sin B = \frac{\text{side opposite angle } B}{\text{hypotenuse}} = \frac{7.00}{\sqrt{65.0}} = 0.868$$

$$\cos B = \frac{\text{side adjacent angle } B}{\text{hypotenuse}} = \frac{4.00}{\sqrt{65.0}} = 0.496$$

$$\tan B = \frac{\text{side opposite angle } B}{\text{side adjacent angle } B} = \frac{7.00}{4.00} = 1.75.$$

We also note that $\sin A = \cos B$ and $\cos A = \sin B$.

We are now ready to solve right triangles. We do this by expressing the unknown parts in terms of the known parts, as the following examples illustrate. For consistency, all results are rounded off to three significant digits or to the nearest 10 minutes, or tenth of a degree, for angles.

Example D
Given $A = 50°0'$ and $b = 6.70$, solve the right triangle of which these are parts (see Figure 2-17).
Since $a/b = \tan A$, we have $a = b \tan A$. Thus,

$$a = 6.70(\tan 50°0')$$
$$= 6.70(1.192) = 7.99.$$

Since $A = 50°0'$, we have $B = 40°0'$. Since $b/c = \cos A$, we have

$$c = \frac{b}{\cos A} = \frac{6.70}{0.6428} = 10.4.$$

We have found that $a = 7.99$, $c = 10.4$, and $B = 40°0'$.

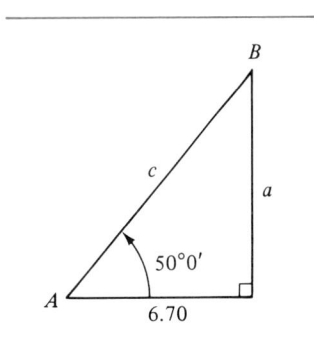

Figure 2-17

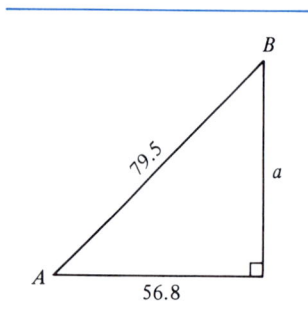

Figure 2-18

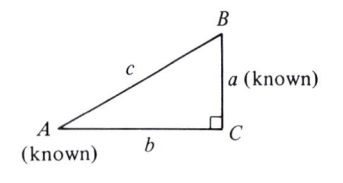

Figure 2-19

Example E
Given $b = 56.8$ and $c = 79.5$, solve the right triangle of which these are parts (see Figure 2-18).
Since $\cos A = b/c$, we have

$$\cos A = \frac{56.8}{79.5} = 0.7145.$$

This means that

$$A = 44°\,20'$$

and

$$B = 90°\,00' - 44°\,20' = 45°40'.$$

We solve for a by use of the Pythagorean theorem since in this way we can express a in terms of the given parts:

$$a^2 + b^2 = c^2 \quad \text{or} \quad a = \sqrt{c^2 - b^2}.$$

Thus,

$$a = \sqrt{79.5^2 - 56.8^2} = \sqrt{6320 - 3226}$$
$$= \sqrt{3094} = 55.6.$$

This means we have determined that $a = 55.6$, $A = 44°\,20'$, and $B = 45°\,40'$.

Example F
Assuming that A and a are known, express the unknown parts of the right triangle in terms of a and A (see Figure 2-19).
Since $a/b = \tan A$, we have

$$b = \frac{a}{\tan A}.$$

Since A is known,

$$B = 90°0' - A.$$

Since $a/c = \sin A$, we have

$$c = \frac{a}{\sin A}.$$

Exercises 2-4

In Exercises 1 through 4 draw appropriate figures and verify through observation the given conclusions.

2-4 The Right Triangle

1. All triangles with a 30° angle included between sides of 3 in. and 4 in. are congruent (corresponding angles and sides are equal).
2. All triangles with a side of 4 cm included between angles of 40° and 70° are congruent.
3. All right triangles with a hypotenuse of 5 cm and one leg of 3 cm are congruent.
4. Not all right triangles with acute angles of 40° and 50° are congruent.

In Exercises 5 through 12 solve the right triangles which have the given parts. Leave answers in radical form where appropriate. Refer to Figure 2-20.

5. $a = 8, B = 60°$
6. $a = 3, B = 45°$
7. $a = 7, b = 7$
8. $a = 3, b = 3\sqrt{3}$
9. $b = 5\sqrt{3}, c = 10$
10. $c = 20, B = 30°$
11. $a = 8, A = 60°$
12. $c = 4, A = 45°$

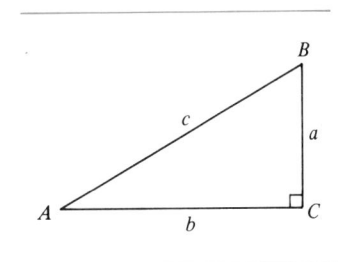

Figure 2-20

In Exercises 13 through 20 solve the right triangles which have the given parts. Express angles to the nearest 10'. Refer to Figure 2-20.

13. $A = 32°0', c = 56.8$
14. $B = 12°0', c = 18.0$
15. $a = 56.7, b = 44.0$
16. $a = 9.98, c = 12.6$
17. $B = 37°0', a = 0.886$
18. $B = 9°30', b = 0.0762$
19. $b = 86.7, c = 167$
20. $a = 6.85, b = 2.12$

In Exercises 21 through 24 solve the right triangles which have the given parts. Express angles to the nearest tenth of a degree. Refer to Figure 2-20.

21. $A = 77.8°, a = 6700$
22. $A = 18.4°, c = 8.97$
23. $a = 150, c = 345$
24. $a = 93.2, c = 124$

In Exercises 25 through 28 the parts listed refer to those in Figure 2-20 and are assumed to be known. Express the other parts in terms of these known parts.

25. b, c
26. a, b
27. A, c
28. B, a

2-5 Applications of Right Triangles

Many applied problems can be solved by setting up the solutions in terms of right triangles. These applications are essentially the same as solving right triangles, although it is usually one specific part of the triangle that we wish to determine. The following examples illustrate some of the basic applications of right triangles.

Example A

A tree has a shadow 17.0 ft long. From the point on the ground at the end of the shadow, the angle of elevation [*the angle between the horizontal and the line of sight, when the object is above the horizontal* (see Figure 2-21)] of the top of the tree is measured to be 52°0′. How tall is the tree?

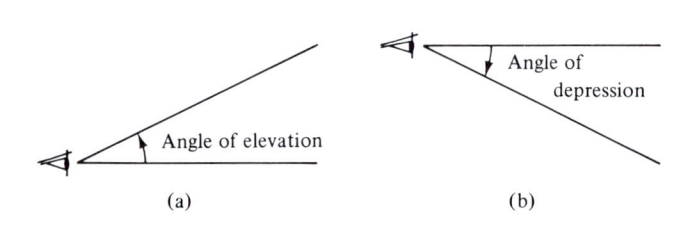

Figure 2-21

By drawing an appropriate figure as in Figure 2-22, we note the information given and that which is required. Here we have let h be the height of the tree. Thus, we see that

$$\frac{h}{17.0} = \tan 52°0′$$

or

$$\begin{aligned} h &= 17.0(\tan 52°0′) \\ &= 17.0(1.280) \\ &= 21.8 \text{ ft.} \end{aligned}$$

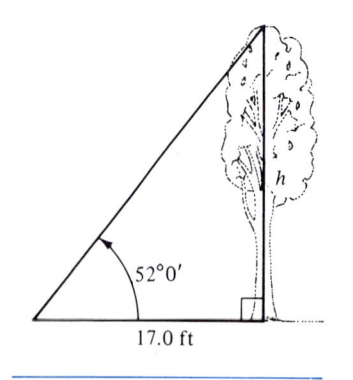

Figure 2-22

2-5 Applications of Right Triangles

Example B

From the roof of a building 46.0 ft high, the angle of depression [*the angle between the horizontal and the line of sight, when the object is below the horizontal* (see Figure 2-21)] of an object in the street is 74°0'. What is the distance of the observer from the object?

Again we draw an appropriate figure (Figure 2-23). Here we let d represent the required distance. From the figure we see that

$$\frac{46.0}{d} = \cos 16°0'$$

$$d = \frac{46.0}{\cos 16°0'} = \frac{46.0}{0.9613}$$

$$= 47.9 \text{ ft.}$$

Example C

A missile is launched at an angle of 26°30' with respect to the horizontal. If it travels in a straight line over level terrain for 2 min and its average speed is 6000 mi/hr, what is its altitude at this time?

In Figure 2-24, we have let h represent the altitude of the missile after 2 min (altitude is measured on a perpendicular). Also, we determine that the missile has flown 200 mi in a direct line from the launching site. This is found from the fact that it travels at 6000 mi/hr for $\frac{1}{30}$ hr (2 min) and from the fact that (6000 mi/hr) ($\frac{1}{30}$ hr) = 200 mi. Therefore,

$$\frac{h}{200} = \sin 26°30'$$

$$h = 200(\sin 26°30') = 200(0.4462)$$

$$= 89.2 \text{ mi.}$$

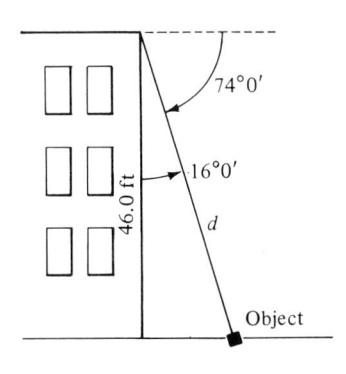

Figure 2-23

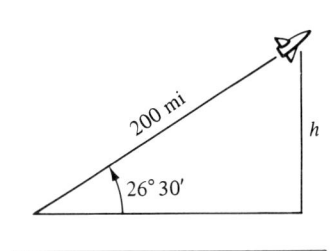

Figure 2-24

Example D
A television antenna is on the roof of a building. From a point on the ground 36.0 ft from the building the angles of elevation of the top and the bottom of the antenna are 51°0′ and 42°0′, respectively. How tall is the antenna?

In Figure 2-25 we let x represent the distance from the top of the building to the ground and y represent the distance from the top of the antenna to the ground. Therefore,

$$\frac{x}{36.0} = \tan 42°0'$$

$$x = 36.0(\tan 42°0') = 36.0(0.9004)$$
$$= 32.4 \text{ ft}$$

and

$$\frac{y}{36.0} = \tan 51°0'$$

$$y = 36.0(\tan 51°0') = 36.0(1.235)$$
$$= 44.5 \text{ ft.}$$

The length of the antenna is the difference of these distances, or 12.1 ft.

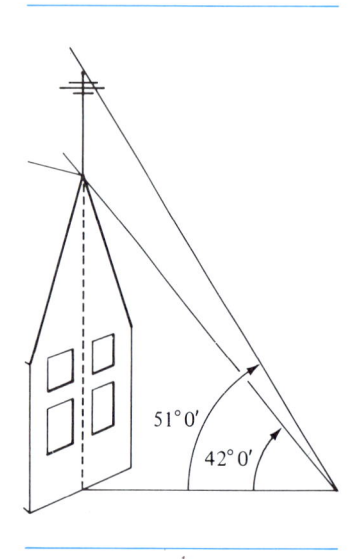

Figure 2-25

Exercises 2-5

In the following exercises express the answers to three significant digits, except where noted. Exercises 1 through 4 may be solved without the use of tables.

1. A rope is stretched from the top of a vertical pole to a point 17.6 ft from the foot of the pole. The rope makes an angle of 30°0′ with the pole. How tall is the pole?
2. A doghouse is 3.00 ft wide and has walls 2.00 ft high. If the pitch of the roof is 45°0′, how high is the doghouse in the center? See Figure 2-26.

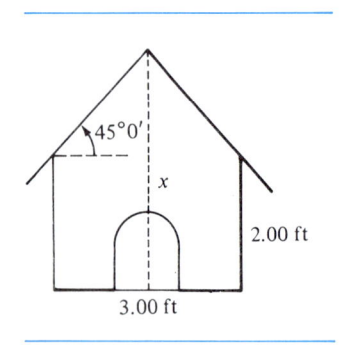

Figure 2-26

2-5 Applications of Right Triangles

3. What is the distance across the front of a corner shelf which extends 20.0 cm along each wall?

4. The top of a hexagonal (six-sided) table is to be covered by material in the form of six equilateral triangles. If the table top measures 60.0 cm from its center to the center of one of its sides, what is the length of the side of each of the equilateral triangles? See Figure 2-27.

5. A 20.0-ft ladder leans against the side of a house. The angle between the ground and ladder is 63° 0'. How far from the house is the foot of the ladder?

6. The length of a kite string (assumed straight) is 560 ft. The angle of elevation of the kite is 71° 30'. How high is the kite?

7. The angle of elevation of the sun is 48° 30' at the time a television tower casts a shadow 146 m long on level ground. Find the height of the tower.

8. From the top of a cliff 110 m high the angle of depression to a boat is 23° 0'. How far is the boat from the foot of the cliff?

9. A 30.0-ft steel girder is leaning against the side of a building. The foot of the girder is 22.0 ft from the building. Find the angle the girder makes with the ground to the nearest 10'.

10. A roadway rises 120 ft for every 2000 ft along the road. Find the angle of inclination of the roadway to the nearest 10'.

11. A man wishes to make a vertical trellis for roses. If the trellis is to be 2.25 m high, how long must each sidepiece be if it makes an angle of 80° 0' with the ground? See Figure 2-28.

12. A jet cruising at 870 km/hr climbs at an angle of 15° 30'. What is its gain in altitude in 2 min?

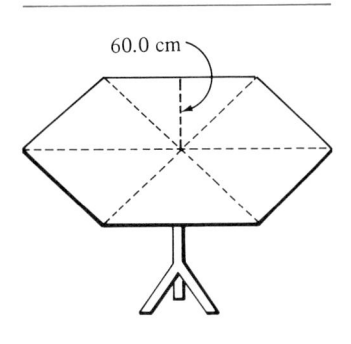

Figure 2-27

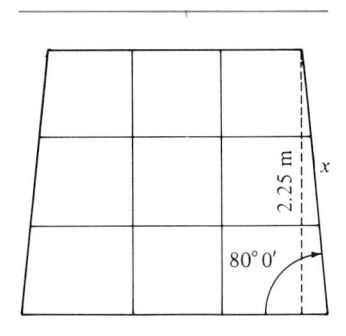

Figure 2-28

13. A floor slopes at 1°30' and one end of a 9.25-ft bar is raised from the floor such that the bar is level. How far above the floor is the raised end of the bar?

14. Ten rivets are equally spaced on the circumference of a circular plate. If the center-to-center distance between two rivets is 6.25 in., what is the radius of the circle?

15. A woman wishes to spotlight a painting on a wall. The center of the painting is 6.25 ft from the floor and the ceiling is 8.00 ft high. The spotlight is placed in the ceiling 2.75 ft from the wall. At what angle, to the nearest 0.1°, must the spotlight be placed?

16. A man is building a swimming pool 12.5 m long. The depth at one end is to be 1.00 m, and the depth at the other is to be 2.50 m. Find, to the nearest 0.1°, the slope of the bottom of the pool.

17. A large rock is on the edge of a cliff. From a point 310 ft from the foot of the cliff the angles of elevation of the top and bottom of the rock are 22.5° and 21.0°, respectively. Find the height of the rock.

18. A flagpole is atop a building. From a point on the ground 700 ft from the building, the angles of elevation of the top and bottom of the flagpole are 33.5° and 31.5°, respectively. How tall is the flagpole?

19. From points on opposite sides of a river the angles of elevation to the top of a 65.0-ft tree are 68°0' and 14°0'. The points and tree are on the same straight line, which is perpendicular to the river. How wide is the river? See Figure 2-29.

20. The angle of depression of a fire noticed directly north of a 74.5-ft fire tower is 5°30'. The angle of depression of a stream running east to west, and also north of the tower, is 13°0'. How far is the fire from the stream?

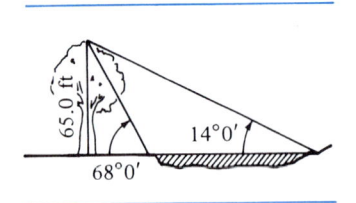

Figure 2-29

Exercises for Chapter 2 59

21. A man considers building a dormer on his house. What are the dimensions x and y of the dormer in his plans? See Figure 2-30.
22. A machine part is indicated in Figure 2-31. What is the angle θ?
23. A surveyor wishes to determine the width of a river. He sights a point B on the opposite side of the river from point C. He then measures off 400 ft from C to A such that C is a right angle. He then determines that angle $A = 56.1°$. How wide is the river?
24. An astronaut circling the moon at an altitude of 100 mi notes that the angle of depression of the horizon is 23.8°. What is the radius of the moon?

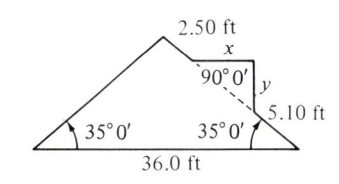

Figure 2-30

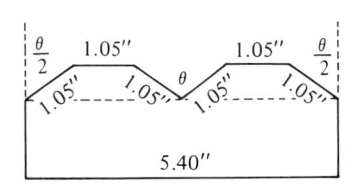

Figure 2-31

Exercises for Chapter 2

In Exercises 1 through 4 find the smallest positive angle and the smallest negative angle (numerically) coterminal with, but not equal to, the given angles.

1. 17°0'   2. 248°0'   3. −217.5°   4. −17°40'

In Exercises 5 through 8 change the given angles to equal angles expressed in decimal form.

5. 31°54'   6. 174°45'   7. 38°6'   8. 321°27'

In Exercises 9 through 12 change the given angles to equal angles expressed to the nearest minute.

9. 17.5°   10. 65.4°   11. 49.7°   12. 126.25°

In Exercises 13 through 16 determine the trigonometric functions of the angles whose terminal side passes through the given points.

13. (24, 7)   14. (5, 4)   15. (4, 4)   16. (1.2, 0.5)

In Exercises 17 through 20 use the given trigonometric functions to find the indicated trigonometric functions.

17. Given $\sin \theta = \frac{5}{13}$, find $\cos \theta$ and $\cot \theta$.
18. Given $\cos \theta = \frac{3}{8}$, find $\sin \theta$ and $\tan \theta$.
19. Given $\tan \theta = 2$, find $\cos \theta$ and $\csc \theta$.
20. Given $\cot \theta = 4$, find $\sin \theta$ and $\sec \theta$.

In Exercises 21 through 32 find the value of each of the given trigonometric functions.

21. $\sin 72° 0'$
22. $\cos 40° 10'$
23. $\tan 61° 20'$
24. $\csc 19° 30'$
25. $\tan 81° 15'$
26. $\cot 37° 17'$
27. $\cos 55° 53'$
28. $\sec 58° 54'$
29. $\cot 37.2°$
30. $\csc 49.9°$
31. $\sec 42.3°$
32. $\sin 12.7°$

In Exercises 33 through 40 find θ to the nearest minute for each of the given trigonometric functions.

33. $\sin \theta = 0.5324$
34. $\tan \theta = 1.265$
35. $\cos \theta = 0.4669$
36. $\sec \theta = 2.107$
37. $\cot \theta = 1.132$
38. $\cos \theta = 0.7365$
39. $\sin \theta = 0.8666$
40. $\csc \theta = 1.533$

In Exercises 41 through 44 find θ to the nearest tenth of a degree for each of the given trigonometric functions.

41. $\cos \theta = 0.9500$
42. $\sin \theta = 0.6305$
43. $\tan \theta = 1.574$
44. $\csc \theta = 2.755$

In Exercises 45 through 48 solve the right triangles which have the given parts. Leave answers in radical form where appropriate. Refer to Figure 2-32.

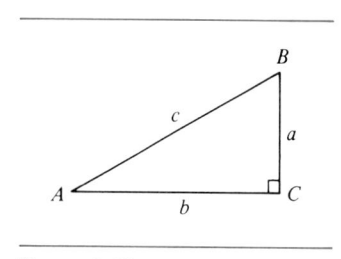

45. $A = 30°$, $c = 4$
46. $B = 45°$, $c = 8$
47. $B = 60°$, $b = 9$
48. $B = 30°$, $a = 7$

Figure 2-32

In Exercises 49 through 52 solve the right triangles which have the given parts. Express angles to the nearest 10'. Refer to Figure 2-32.

49. $A = 17° 0'$, $b = 6.00$
50. $B = 68° 10'$, $a = 1080$
51. $a = 81.0$, $b = 64.5$
52. $a = 1.06$, $c = 3.82$

Exercises for Chapter 2

In Exercises 53 through 56 solve the right triangles which have the given parts. Express angles to the nearest minute. Refer to Figure 2-32.

53. $A = 49°43'$, $c = 0.820$
54. $B = 4°26'$, $b = 5.60$
55. $a = 10.0$, $c = 15.0$
56. $a = 724$, $b = 852$

In Exercises 57 through 60 solve the right triangles which have the given parts. Express angles to the nearest 0.1°. Refer to Figure 2-32.

57. $A = 37.5°$, $a = 12.0$ 58. $B = 15.7°$, $c = 12.6$
59. $b = 6.50$, $c = 7.60$ 60. $a = 72.1$, $b = 14.3$

In Exercises 61 through 76 express answers to three significant digits, except where noted.

61. The angle of elevation of the sun is 35°0' at the time a post casts a shadow 18.0 ft long. Find the height of the post.
62. The angle of elevation of the top of a building from a point 793 ft from the foot of the building is 49°0'. Find the height of the building.
63. From the top of a lighthouse 52.0 m high the angle of depression of a boat at sea is 31°30'. Find the distance from the boat to the foot of the lighthouse.
64. An observer balloon 840 m above an airport finds that the angle of depression of a racetrack is 21°0'. How far is the racetrack from the airport?
65. A girl flying a kite lets out 250 ft of string, which makes an angle of 51°0' with the ground. Assuming that the string is straight, find the height of the kite above the ground.
66. A bridge is 32.5 ft above the surrounding area. If the angle of elevation of the approach is 4°40', how long is the approach? See Figure 2-33.

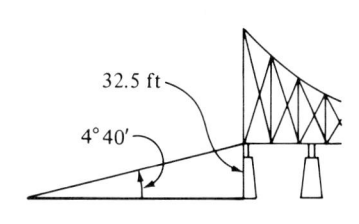

Figure 2-33

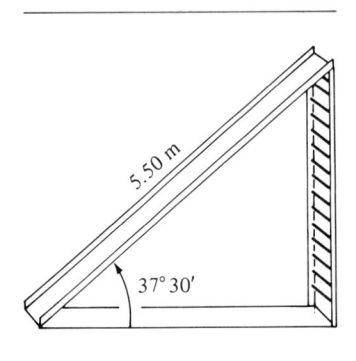

Figure 2-34

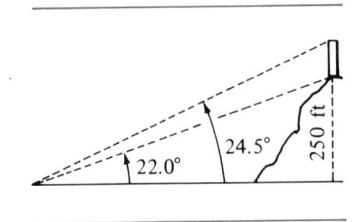

Figure 2-35

67. A child's slide is 5.50 m long and makes an angle of 37° 30' with the ground. How far is it from the bottom of the slide to the bottom of the vertical ladder? See Figure 2-34.

68. The windshield on an automobile is inclined 42.5° with respect to the horizontal. Assuming that the windshield is flat and rectangular, what is its area if it is 4.80 ft wide and the bottom is 1.50 ft in front of the top?

69. The span on a roof is 32.0 ft. Its rise is 7.50 ft at the center of the span. What angle, to the nearest 10', does the roof make with the horizontal?

70. The horizontal distance between the extreme positions of a certain pendulum is 6.50 in. If the length of the pendulum is 18.0 in., through what angle, to the nearest 10', does it swing?

71. From the ground the angles of elevation of the top and bottom of a tower on a hill are 24.5° and 22.0°, respectively. If the hill is 250 ft high, how tall is the tower? See Figure 2-35.

72. From a helicopter flying at 120 m, the angles of depression of a campsite and a search party are 71.0° and 50.0°, respectively. If the campsite and search party are on the same side of the helicopter and all are in a vertical plane, how far is the search party from the campsite?

73. An observer 3500 ft from the launchpad of a rocket measures the angle of elevation to the rocket soon after lift-off to be 54° 25'. How high, to the nearest foot, is the rocket, assuming that it has moved vertically?

74. The distance from ground level to the underside of a cloud is called the *ceiling*. One observer 1000 ft from a searchlight aimed vertically notes that the angle of elevation of the spot of light on a cloud is 76° 0'. What is the ceiling?

Exercises for Chapter 2

75. A ship's captain desiring to travel due south discovers that, because of a faulty instrument, he has gone 22.60 mi in a direction 4°17' east of south. How far from his course (to the east) is he?

76. A railroad embankment has a level top 22.50 ft wide, equal sloping sides of 14.50 ft, and a height of 7.25 ft. What is the width, to the nearest 0.01 ft, of the base of the embankment?

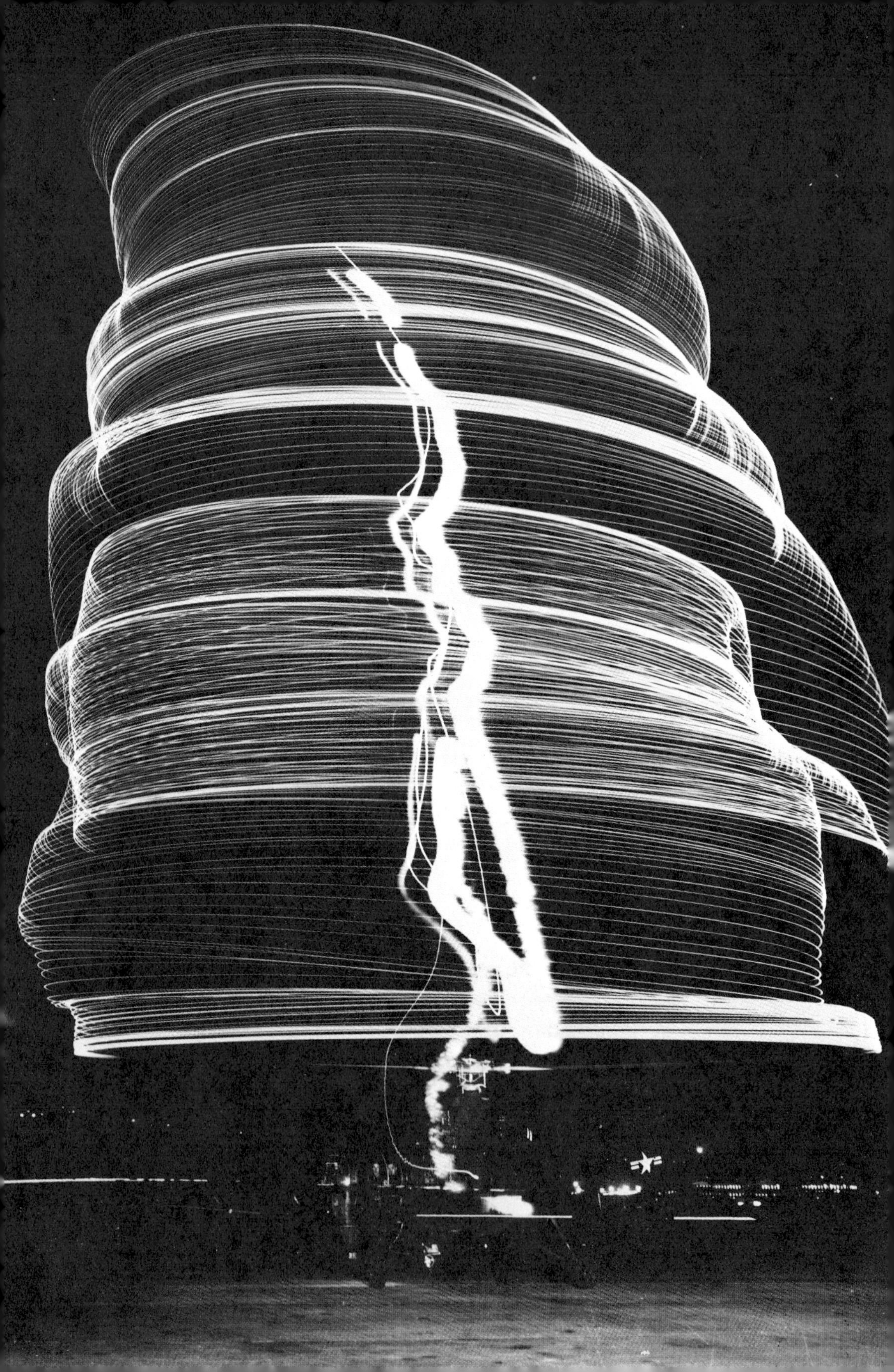

3

Trigonometric Functions of Any Angle

Trigonometry can also be used to determine the speed of an object moving in a circular path. The path followed by the tip of the helicopter blade can be easily observed in this photograph.

3-1 Signs of the Trigonometric Functions

When we were dealing with trigonometric functions in Chapter 2 we restricted ourselves primarily to the functions of acute angles measured in degrees. Since we did define the functions in general, we can use these same definitions for finding the functions of any possible angle. We shall not only find the trigonometric functions of angles measured in degrees, but we shall introduce radian measure as well.

In this section we determine the signs of the trigonometric functions in each of the four quadrants. We recall the definitions of the trigonometric functions given in Section 2-2:

(3-1)

$$\sin \theta = \frac{y}{r} \quad \cos \theta = \frac{x}{r} \quad \tan \theta = \frac{y}{x}$$

$$\cot \theta = \frac{x}{y} \quad \sec \theta = \frac{r}{x} \quad \csc \theta = \frac{r}{y}$$

We see that the functions are defined so long as we know the abscissa, ordinate, and radius vector of the terminal side of θ. Remembering that r is always considered positive, we can see that the various functions will vary in sign, depending on the signs of x and y.

If the terminal side of the angle is in the first or second quadrant, the value of $\sin \theta$ will be positive, but if the terminal side is in the third or fourth quadrant, $\sin \theta$ is negative. This is because y is positive if the point defining the terminal side is above the x-axis and negative if this point is below the x-axis.

Example A
The value of sin 20° is positive, since the terminal side of 20° is in the first quadrant. The value of sin 160° is positive, since the terminal side of 160° is in the second quadrant. The values of sin 200° and sin 340° are negative, since the terminal sides of these angles are in the third and fourth quadrants, respectively.

The sign of tan θ depends upon the ratio of y to x. In the first quadrant both x and y are positive, and therefore the ratio y/x is positive. In the third quadrant both x and y are negative, and therefore the ratio y/x is positive. In the second and fourth quadrants either x or y is positive and the other is negative, and so the ratio of y/x is negative.

Example B
The values of tan 20° and tan 200° are positive, since the terminal sides of these angles are in the first and third quadrants, respectively. The values of tan 160° and tan 340° are negative, since the terminal sides of these angles are in the second and fourth quadrants, respectively.

The sign of cos θ depends upon the sign of x. Since x is positive in the first and fourth quadrants, cos θ is positive in these quadrants. In the same way, cos θ is negative in the second and third quadrants.

Example C
The values of cos 20° and cos 340° are positive, since these angles are first- and fourth-quadrant angles, respectively. The values of cos 160° and cos 200° are negative, since these angles are second- and third-quadrant angles, respectively.

Since csc θ is defined in terms of r and y, as is sin θ, the sign of csc θ is the same as that of sin θ. For the same reason, cot θ has the same sign as

3-1 Signs of the Trigonometric Functions

$\tan \theta$ and $\sec \theta$ has the same sign as $\cos \theta$. A method for remembering the signs of the functions in the four quadrants is as follows: **All functions of first-quadrant angles are positive. The $\sin \theta$ and $\csc \theta$ are positive for second-quadrant angles. The $\tan \theta$ and $\cot \theta$ are positive for third-quadrant angles. The $\cos \theta$ and $\sec \theta$ are positive for fourth-quadrant angles. All others are negative.** See Figure 3-1. This discussion does not include the quadrantal angles, those angles with terminal sides on one of the axes. They are discussed later.

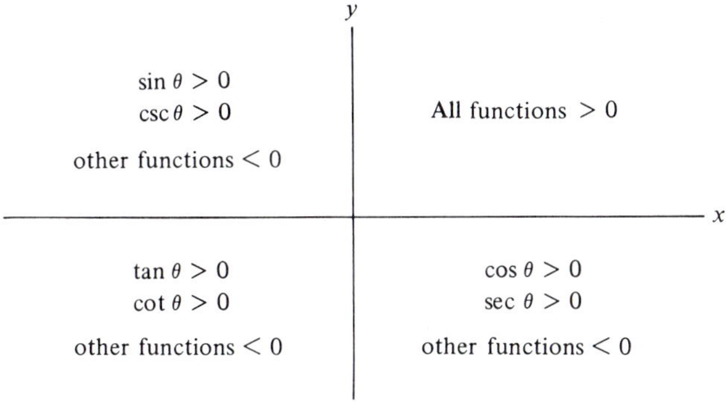

$\sin \theta > 0$ $\csc \theta > 0$ other functions < 0	All functions > 0
$\tan \theta > 0$ $\cot \theta > 0$ other functions < 0	$\cos \theta > 0$ $\sec \theta > 0$ other functions < 0

Figure 3-1

Example D
All the following are positive: $\sin 50°$, $\sin 150°$, $\sin(-200°)$, $\cos 300°$, $\cos(-40°)$, $\tan 220°$, $\tan(-100°)$, $\cot 260°$, $\cot(-310°)$, $\sec 280°$, $\sec(-37°)$, $\csc 140°$, and $\csc(-190°)$.

Example E
All the following are negative: $\sin 190°$, $\sin 325°$, $\cos 100°$, $\cos(-95°)$, $\tan 172°$, $\tan 295°$, $\cot 105°$, $\cot(-6°)$, $\sec 135°$, $\sec(-135°)$, $\csc 240°$, and $\csc 355°$.

Example F
Determine the trigonometric functions of θ if the terminal side of θ passes through $(-1, \sqrt{3})$.

We know that $x = -1$ and $y = +\sqrt{3}$, and from the Pythagorean theorem we find that $r = 2$. Therefore, the trigonometric functions of θ are

$$\sin \theta = +\frac{\sqrt{3}}{2} \qquad \cos \theta = -\frac{1}{2} \qquad \tan \theta = -\sqrt{3}$$

$$\cot \theta = -\frac{1}{\sqrt{3}} \qquad \sec \theta = -2 \qquad \csc \theta = +\frac{2}{\sqrt{3}}$$

We note that the point $(-1, \sqrt{3})$ is on the terminal side of a second-quadrant angle, and the signs of the functions of θ are those of a second-quadrant angle.

Exercises 3-1

In Exercises 1 through 8 determine the algebraic sign of the given trigonometric functions.

1. $\sin 60°$, $\cos 120°$, $\tan 320°$
2. $\tan 185°$, $\sec 115°$, $\sin(-36°)$
3. $\cos 300°$, $\csc 97°$, $\cot(-35°)$
4. $\sin 100°$, $\sec(-15°)$, $\cos 188°$
5. $\cot 186°$, $\sec 280°$, $\sin 470°$
6. $\tan(-91°)$, $\csc 87°$, $\cot 103°$
7. $\cos 700°$, $\tan(-560°)$, $\csc 530°$
8. $\sin 256°$, $\tan 321°$, $\cos(-370°)$

In Exercises 9 through 16 find the trigonometric functions of θ, where the terminal side of θ passes through the given point.

9. $(2, 1)$
10. $(-1, 1)$
11. $(-2, -3)$
12. $(4, -3)$
13. $(-5, 12)$
14. $(-3, -4)$
15. $(5, -2)$
16. $(3, 5)$

In Exercises 17 through 20 determine the quadrant in which the terminal side of θ lies, subject to the given conditions.

17. $\sin \theta$ is positive, $\cos \theta$ is negative.
18. $\tan \theta$ is positive, $\cos \theta$ is negative.
19. $\sec \theta$ is negative, $\cot \theta$ is negative.
20. $\cos \theta$ is positive, $\csc \theta$ is negative.

3-2 Trigonometric Functions of Any Angle

The trigonometric functions of acute angles were determined in Section 2-4, and in the last section we determined the signs of the trigonometric functions in each of the four quadrants. In this section we show how we can find the trigonometric functions of an angle of any magnitude. This information will be very important in Chapter 4 when we discuss oblique triangles and in Chapter 6 when we graph the trigonometric functions.

Any angle in standard position is coterminal with some positive angle less than 360°. Since the terminal sides of coterminal angles are the same, the trigonometric functions of coterminal angles are the same. Therefore, we need consider only the problem of finding the values of the trigonometric functions of positive angles less than 360°.

Example A
The following pairs of angles are coterminal.

390° and 30°   −60° and 300°
900° and 180° −150° and 210°

From this we conclude that the trigonometric functions of both angles in these pairs are equal. That is, for example, $\sin 390° = \sin 30°$ and $\tan(-150°) = \tan 210°$.

Considering the definitions of the functions, we see that the values of the functions depend only on

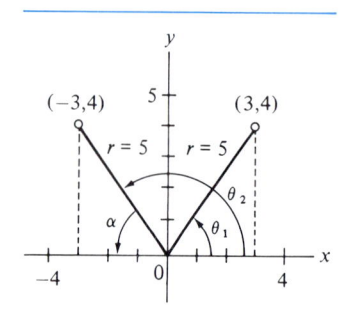

Figure 3-2

the values of x, y, and r. The values of the functions of second-quadrant angles are numerically equal to the functions of corresponding first-quadrant angles. Consider the angles shown in Figure 3-2. For angle θ_2 with terminal side passing through $(-3, 4)$, $\tan \theta_2 = -\frac{4}{3}$; for angle θ_1 with terminal side passing through $(3, 4)$ $\tan \theta_1 = \frac{4}{3}$. In Figure 3-2, we see that the triangles containing angles θ_1 and α are congruent, which means that θ_1 and α are equal. We know that the trigonometric functions of θ_1 and θ_2 are numerically equal. This means that:

(3-2) $\qquad |F(\theta_2)| = |F(\theta_1)| = |F(\alpha)|$

where F represents any of the trigonometric functions.

The angle labeled α is called the *reference angle*. The reference angle of a given angle is the acute angle formed by the terminal side of the angle and the x-axis.

Using Equation (3-2) and the fact that $\alpha = 180° - \theta_2$, we may conclude that the value of any trigonometric function of any second-quadrant angle is found from

(3-3) $\qquad F(\theta_2) = \pm F(180° - \theta_2) = \pm F(\alpha)$

The sign to be used depends on whether the *function* is positive or negative in the second quadrant.

Example B
In Figure 3-2, the trigonometric functions of θ_2 are as follows.

$$\sin \theta_2 = +\sin(180° - \theta_2)$$
$$= +\sin \alpha = +\sin \theta_1 = \tfrac{4}{5}$$
$$\cos \theta_2 = -\cos \theta_1 = -\tfrac{3}{5}$$
$$\tan \theta_2 = -\tfrac{4}{3} \qquad \cot \theta_2 = -\tfrac{3}{4}$$
$$\sec \theta_2 = -\tfrac{5}{3} \qquad \csc \theta_2 = +\tfrac{5}{4}$$

3-2 Trigonometric Functions of Any Angle

In the same way we may derive the formulas for finding the trigonometric functions of any third- or fourth-quadrant angle. Considering the angles shown in Figure 3-3, we see that the reference angle α is found by subtracting $180°$ from θ_3 and that functions of α and θ_1 are numerically equal. Therefore:

(3-4) $\qquad F(\theta_3) = \pm F(\theta_3 - 180°)$

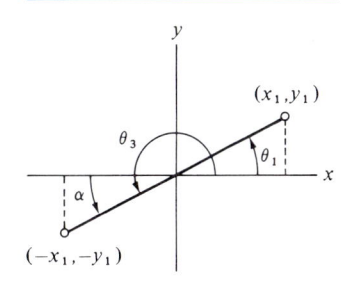

Figure 3-3

Considering the angles shown in Figure 3-4, we see that the reference angle α is found by subtracting θ_4 from $360°$. Therefore:

(3-5) $\qquad F(\theta_4) = \pm F(360° - \theta_4)$

Example C
In Figure 3-3, if $\theta_3 = 210°$ the trigonometric functions of θ_3 are found by using Equation (3-4) as follows.

$$\sin 210° = -\sin(210° - 180°) = -\sin 30°$$
$$= -0.5000$$
$$\cos 210° = -\cos 30° = -0.8660$$
$$\tan 210° = +0.5774 \qquad \cot 210° = +1.732$$
$$\sec 210° = -1.155 \qquad \csc 210° = -2.000$$

Example D
In Figure 3-4, if $\theta_4 = 315°$ the trigonometric functions of θ_4 are found by using Equation (3-5) as follows.

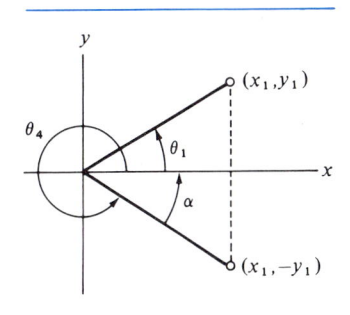

$$\sin 315° = -\sin(360° - 315°) = -\sin 45°$$
$$= -0.7071$$
$$\cos 315° = +\cos 45° = +0.7071$$
$$\tan 315° = -1.000 \qquad \cot 315° = -1.000$$
$$\sec 315° = +1.414 \qquad \csc 315° = -1.414$$

Figure 3-4

Example E
Other illustrations of the use of Equations (3-3), (3-4), and (3-5) are as follows.

$$\sin 160° = +\sin(180° - 160°) = \sin 20°$$
$$= 0.3420$$
$$\tan 110° = -\tan(180° - 110°) = -\tan 70°$$
$$= -2.747$$
$$\cos 225° = -\cos(225° - 180°) = -\cos 45°$$
$$= -0.7071$$
$$\cot 260° = +\cot(260° - 180°) = \cot 80°$$
$$= 0.1763$$
$$\sec 304° = +\sec(360° - 304°) = \sec 56°$$
$$= 1.788$$
$$\sin 357° = -\sin(360° - 357°) = -\sin 3°$$
$$= -0.0523$$

To express a trigonometric function of an angle in terms of a function of a positive acute angle less than 45°, we recall that cofunctions of complementary angles are equal. This is illustrated in the following example.

Example F

$$\cos 124° = -\cos 56° = -\sin 34°$$
$$\tan 232° = +\tan 52° = \cot 38°$$
$$\sec 295° = +\sec 65° = \csc 25°$$
$$\cot 276° = -\cot 84° = -\tan 6°$$
$$\sin 100° = +\sin 80° = \cos 10°$$
$$\csc 229° = -\csc 49° = -\sec 41°$$

If we know the value of one of the functions of an angle and the quadrant in which the terminal side of the angle lies, we may determine the values of the other functions of the angle. The following example illustrates the procedure.

Example G
Given that $\sin A = -\frac{3}{5}$ and $\cos A$ is negative, find the values of all the functions of A.

We know that A is a third-quadrant angle, for that is the only quadrant in which both the sine

3-2 Trigonometric Functions of Any Angle

and cosine are negative. From the definition of the sine of an angle we see that if we let $y = -3$ and $r = 5$ we have the ratio of y/r, which is equal to $-\frac{3}{5}$. Using the Pythagorean theorem and the fact that x must be negative in the third quadrant, we find that $x = -4$. Therefore, the remaining functions of A are as follows:

$$\cos A = -\frac{4}{5} \qquad \tan A = \frac{3}{4} \qquad \cot A = \frac{4}{3}$$
$$\sec A = -\frac{5}{4} \qquad \csc A = -\frac{5}{3}.$$

If a function of an angle is known, we may also determine the angle itself. The following examples illustrate how Equations (3-3) through (3-5) are used to determine θ when a function of θ is known.

Example H
Given that $\sin \theta = 0.2250$, find θ for $0° < \theta < 360°$.

Here we are to find any angles between $0°$ and $360°$ for which $\sin \theta = 0.2250$. Since $\sin \theta$ is positive for first- and second-quadrant angles, there will be two such angles. From the tables we find our reference angle to be $13°$. Therefore, the first-quadrant angle we want is $13°$. The second-quadrant angle is $180° - 13° = 167°$. Therefore, our two required angles are $13°$ and $167°$.

Example I
Given that $\tan \theta = 2.050$ and $\cos \theta < 0$, find θ for $0° < \theta < 360°$.

Since $\tan \theta$ is positive and $\cos \theta$ is negative, θ must be in the third quadrant. We note from the tables that $2.050 = \tan 64°$. Thus, $64°$ is our reference angle and $\theta = 180° + 64° = 244°$. Since the required angle is to be between $0°$ and $360°$, this is the only possible answer.

With the use of Equations (3-3) through (3-5) we may find the value of any function so long as the terminal side of the angle lies *in* one of the quadrants. This problem reduces to finding the function

Trigonometric Functions of Any Angle

of an acute angle. We are left with the case of the terminal side being along one of the axes, a *quadrantal angle*. Using the definitions of the functions and remembering that $r > 0$, we arrive at the values in the following table.

θ	$\sin\theta$	$\cos\theta$	$\tan\theta$	$\cot\theta$	$\sec\theta$	$\csc\theta$
0°	0.000	1.000	0.000	Undef.	1.000	Undef.
90°	1.000	0.000	Undef.	0.000	Undef.	1.000
180°	0.000	−1.000	0.000	Undef.	−1.000	Undef.
270°	−1.000	0.000	Undef.	0.000	Undef.	−1.000
360°	Same as the functions of 0° (same terminal side)					

The values in the table may be verified by referring to Figure 3-5.

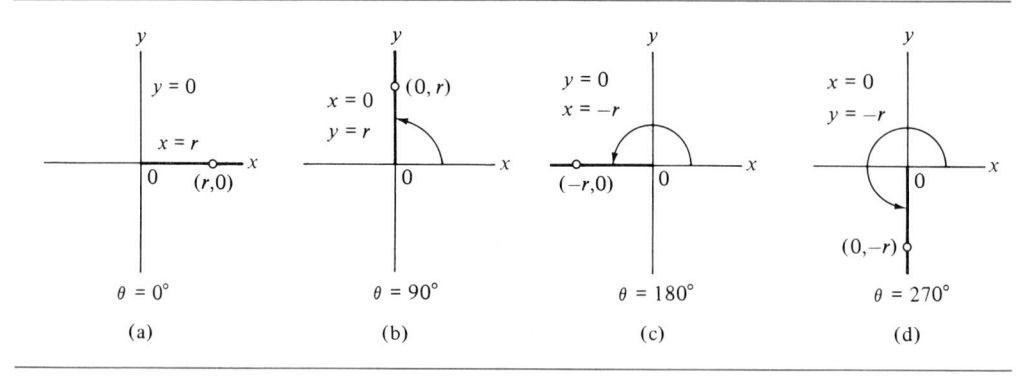

Figure 3-5

Example J

Since $\sin\theta = y/r$, from Figure 3-5(a) we see that $\sin 0° = 0/r = 0$.

Since $\tan\theta = y/x$, from Figure 3-5(b) we see that $\tan 90° = r/0$, which is undefined due to the division by zero.

Since $\cos\theta = x/r$, from Figure 3-5(c) we see that $\cos 180° = -r/r = -1$.

Since $\cot\theta = x/y$, from Figure 3-5(d) we see that $\cot 270° = 0/-r = 0$.

3-2 Trigonometric Functions of Any Angle

We now consider the trigonometric functions of negative angles. From Figure 3-6 we see that $\sin \theta = y/r$ and $\sin(-\theta) = -y/r$. From this we conclude that $\sin(-\theta) = -\sin \theta$. We may determine the other trigonometric functions of negative angles in a similar manner. Therefore, we have

(3-6)

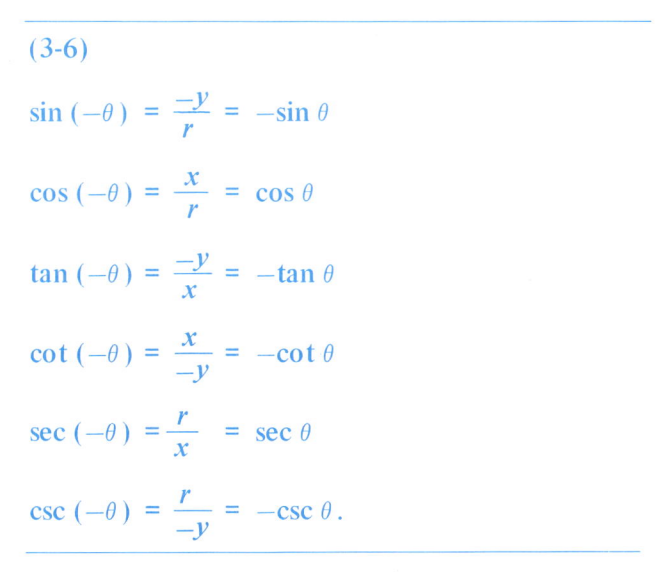

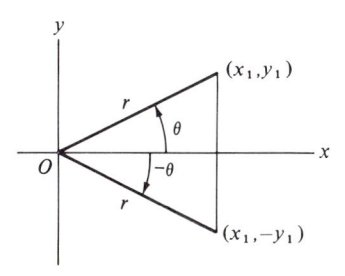

$$\sin(-\theta) = \frac{-y}{r} = -\sin \theta$$

$$\cos(-\theta) = \frac{x}{r} = \cos \theta$$

$$\tan(-\theta) = \frac{-y}{x} = -\tan \theta$$

$$\cot(-\theta) = \frac{x}{-y} = -\cot \theta$$

$$\sec(-\theta) = \frac{r}{x} = \sec \theta$$

$$\csc(-\theta) = \frac{r}{-y} = -\csc \theta.$$

Figure 3-6

Example K
By using Equations (3-6) we may express the following as functions of positive angles.

$$\sin(-60°) = -\sin 60°$$
$$\cos(-76°) = \cos 76°$$
$$\tan(-100°) = -\tan 100°$$

Exercises 3-2

In Exercises 1 through 8 express the given trigonometric functions in terms of the same function of a positive acute angle.

1. sin 160°, cos 220°
2. tan 91°, sec 345°
3. tan 105°, csc 302°

4. cos 190°, cot 290°
5. sin (−123°), cot 174°
6. sin 98°, sec (−315°)
7. cos 400°, tan (−400°)
8. tan 920°, csc (−550°)

In Exercises 9 through 16 express the given trigonometric functions in terms of a function of a positive acute angle less than 45°.

9. sin 110°, tan 265°
10. cos 283°, csc 160°
11. csc 238°, cos 104°
12. sin 240°, cot 301°
13. cot 115°, sec (−65°)
14. tan (−98°), sec 260°
15. cos (−120°), csc (−233°)
16. sec (−87°), sin (−109°)

In Exercises 17 through 28 determine the values of the given trigonometric functions by use of tables.

17. sin 195°
18. tan 311°
19. cos 106°
20. sin 254°
21. cot 136°
22. cos 297°
23. sec (−115°)
24. csc (−193°)
25. sin 322° 52′
26. cot 254° 17′
27. tan 118.6°
28. cos (−67.1°)

In Exercises 29 through 36 find θ when $0° < \theta < 360°$. In Exercises 33 and 34 express the answers to the nearest 1′, and in Exercises 35 and 36 express the answers to the nearest 0.1°.

29. tan θ = 0.5317
30. cos θ = 0.6428
31. sin θ = −0.8450
32. cot θ = −0.2432
33. sin θ = −0.9527
34. sec θ = 2.286
35. cot θ = −0.7164
36. tan θ = −2.664

In Exercises 37 through 40 determine the function which satisfies the given conditions.

37. Find tan θ when sin θ = −0.5736 and cos θ > 0.
38. Find sin θ when cos θ = 0.4226 and tan θ < 0.

39. Find $\cos \theta$ when $\tan \theta = -0.8098$ and $\csc \theta > 0$.
40. Find $\cot \theta$ when $\sec \theta = 1.122$ and $\sin \theta < 0$.

In Exercises 41 through 44 find all the functions of θ for the given conditions.

41. $\cos \theta = \frac{5}{13}$ and $\csc \theta > 0$
42. $\tan \theta = -\frac{4}{3}$ and $\sin \theta < 0$
43. $\sec \theta = -\frac{17}{15}$ and $\cot \theta > 0$
44. $\cot \theta = 2$ and $\cos \theta < 0$

In Exercises 45 through 48 insert the proper sign, $>$ or $<$ or $=$, between the given expressions.

45. $\sin 90°$ $2 \sin 45°$
46. $\cos 360°$ $2 \cos 180°$
47. $\tan 180°$ $\tan 0°$
48. $\sin 270°$ $3 \sin 90°$

In Exercises 49 through 52 use Equations (3-6) to express the following as functions of positive angles.

49. $\sin(-35°)$, $\tan(-70°)$
50. $\cos(-137°)$, $\csc(-214°)$
51. $\cot(-312°)$, $\sec(-241°)$
52. $-\cos(-187°)$, $-\tan(-96°)$

3-3 Radians

For many problems in which trigonometric functions are used, particularly those involving the solution of triangles, the degree measurement of angles is quite sufficient. However, in numerous other types of applications and in more theoretical discussions, another way of expressing the measure of angle is more meaningful and convenient. This unit of measurement is the *radian*. **A radian is the measure of an angle with its vertex at the center of a circle and with an intercepted arc on the circle equal in length to the radius of the circle.** See Figure 3-7.

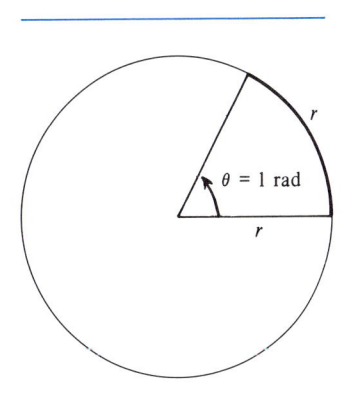

Figure 3-7

Since the circumference of any circle in terms of its radius is given by $c = 2\pi r$, the ratio of the circumference to the radius is 2π. This means that the radius may be laid off 2π (about 6.28) times along the circumference, regardless of the length of the radius. Therefore, we see that radian measure is independent of the radius of the circle. In Figure 3-8 the numbers on each of the radii indicate the number of radians in the angle measured in standard position. The circular arrow shows an angle of 6 rad.

Since the radius may be laid off 2π times along the circumference, it follows that there are 2π rad in one complete rotation. Also, there are 360° in one complete rotation. Therefore, 360° is equivalent to 2π rad. It then follows that the relation between degrees and radians is 2π rad = 360°. Thus:

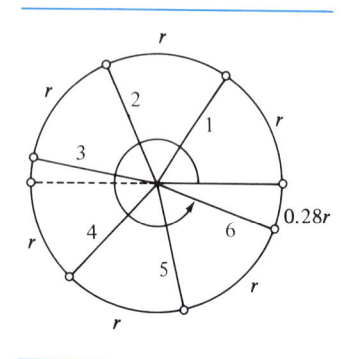

Figure 3-8

$$(3\text{-}7) \qquad \pi \text{ rad} = 180°$$

From this relation we find:

$$(3\text{-}8) \qquad 1° = \frac{\pi}{180} \text{ rad} = 0.01745 \text{ rad}$$

and

$$(3\text{-}9) \qquad 1 \text{ rad} = \frac{180°}{\pi} = 57.3°$$

We see from Equations (3-7) through (3-9) that:

1. To convert an angle measured in degrees to the same angle measured in radians, we multiply the number of degrees by $\pi/180$.
2. To convert an angle measured in radians to the same angle measured in degrees, we multiply the number of radians by $180/\pi$.

3-3 Radians

Example A

$$18° = \left(\frac{\pi}{180}\right)(18) = \frac{\pi}{10} = \frac{3.14}{10} = 0.314 \text{ rad}$$

$$120° = \left(\frac{\pi}{180}\right)(120) = \frac{2\pi}{3} = \frac{6.28}{3} = 2.09 \text{ rad}$$

$$0.4 \text{ rad} = \left(\frac{180°}{\pi}\right)(0.4) = \frac{72°}{3.14} = 22.9°$$

$$2 \text{ rad} = \left(\frac{180°}{\pi}\right)(2) = \frac{360°}{3.14} = 115°$$

Due to the nature of the definition of the radian, it is very common to express radians in terms of π. Expressing angles in terms of π is illustrated in the following example.

Example B

$$30° = \left(\frac{\pi}{180}\right)(30) = \frac{\pi}{6} \text{ rad}$$

$$45° = \left(\frac{\pi}{180}\right)(45) = \frac{\pi}{4} \text{ rad}$$

$$\frac{\pi}{2} \text{ rad} = \left(\frac{180°}{\pi}\right)\left(\frac{\pi}{2}\right) = 90°$$

$$\frac{3\pi}{4} \text{ rad} = \left(\frac{180°}{\pi}\right)\left(\frac{3\pi}{4}\right) = 135°$$

We wish now to make a very important point. Since π is a special way of writing the number (slightly greater than 3) that is the ratio of the circumference of a circle to its diameter, it is the ratio of one distance to another. Thus, radians really have no units and *radian measure amounts to measuring angles in terms of numbers.* It is this property of radians that makes them useful in many situations. Therefore, when radians are being used it is customary that no units are indicated for the angle. When no units are indicated, the radian is understood to be the unit of angle measurement.

Example C

$$60° = \left(\frac{\pi}{180}\right)(60) = \frac{\pi}{3} = 1.05$$

$$2.50 = \left(\frac{180°}{\pi}\right)(2.50) = \frac{450°}{3.14} = 143°$$

Since no units are indicated for 1.05 or 2.50 in this example, they are known to be radian measure.

Often when one first encounters radian measure, expressions such as sin 1 and sin θ = 1 are confused. The first is equivalent to sin 57.3°, since 57.3° = 1 (radian). The second means that θ is the angle for which the sine is 1. Since sin 90° = 1, we can say that θ = 90° or that θ = $\pi/2$. The following examples give additional illustrations of evaluating expressions involving radians.

Example D

$$\sin \frac{\pi}{3} = \frac{\sqrt{3}}{2}, \text{ since } \frac{\pi}{3} = 60°.$$

$$\sin 0.6021 = 0.5664, \text{ since } 0.6021 = 34°\, 30'.$$

$$\tan \theta = 1.709 \text{ means that } \theta = 59°\, 40'$$
(smallest positive θ).

Since 59° 40' = 1.041, we can state that tan 1.041 = 1.709.

Example E

Express θ in radians, such that cos θ = 0.8829 and $0 < \theta < 2\pi$.

We are to find θ in radians for the given value of the cos θ. Also, since θ is restricted to values between 0 and 2π, we must find a first-quadrant angle and a fourth-quadrant angle (cos θ is positive in the first and the fourth quadrants). From the table we see that

$$\cos 28° = 0.8829.$$

3-3 Radians

Therefore, for the fourth-quadrant angle, $\cos(360° - 28°) = \cos 332° = 0.8829$. Converting 28° and 332° to radians, we have

$$\theta = 0.489 \quad \text{or} \quad \theta = 5.79.$$

We can also use the table directly to find the function of an acute angle given in radians. The following examples illustrate this use of the table.

Example F

In determining the value of sin 0.4538, we locate 0.4538 in the column labeled radians and opposite it we note 0.4384 in the sine column. Therefore, sin 0.4538 = 0.4384. In the same manner we find

$$\tan 0.9977 = 1.550 \quad \cos 0.6807 = 0.7771$$
$$\sec 1.1374 = 2.381.$$

If we wish to find the value of a function of an angle greater than $\pi/2$, we must first determine which quadrant the angle is in and then find the reference angle. In this determination we should note the radian measure equivalents of 90°, 180°, 270°, and 360°. For 90° we have $\pi/2 = 1.571$, for 180° we have $\pi = 3.142$, for 270° we have $3\pi/2 = 4.712$, and for 360° we have $2\pi = 6.283$.

Example G

Find sin 3.402. Since 3.402 is greater than 3.142 but less than 4.712, we know that this angle is in the third quadrant and has a reference angle of 3.402 − 3.142 = 0.260. Thus,

$$\sin 3.402 = -\sin 0.260 = -0.2560$$

using the nearest value shown in the table.

Now find cos 5.210. Since 5.210 is between 4.712 and 6.283, we know that this angle is in the fourth quadrant and its reference angle is 6.283 − 5.210 = 1.073. Thus,

$$\cos 5.210 = \cos 1.073 = 0.4772$$

using the nearest value shown in the table.

Exercises 3-3

In Exercises 1 through 8 express the given angle measurements in terms of π.

1. 15°, 150°
2. 12°, 225°
3. 75°, 330°
4. 36°, 315°
5. 210°, 270°
6. 240°, 300°
7. 160°, 260°
8. -66°, 350°

In Exercises 9 through 16 the given numbers express angle measure. Express the measure of each angle in terms of degrees.

9. $\dfrac{2\pi}{5}, \dfrac{3\pi}{2}$
10. $\dfrac{3\pi}{10}, \dfrac{5\pi}{6}$
11. $\dfrac{\pi}{18}, \dfrac{7\pi}{4}$
12. $\dfrac{7\pi}{15}, \dfrac{4\pi}{3}$
13. $\dfrac{17\pi}{18}, \dfrac{5\pi}{3}$
14. $\dfrac{11\pi}{36}, \dfrac{5\pi}{4}$
15. $\dfrac{\pi}{12}, \dfrac{3\pi}{20}$
16. $\dfrac{7\pi}{30}, \dfrac{4\pi}{15}$

In Exercises 17 through 24 express the given angles in radian measure. (Use 3.14 as an *approximation* for π.)

17. 23°0′
18. 54°0′
19. 252°0′
20. 104°0′
21. 333°30′
22. 168°40′
23. 178.5°
24. 86.1°

In Exercises 25 through 32 the given numbers express angle measure. Express the measure of each angle in terms of degrees.

25. 0.750
26. 0.240
27. 3.00
28. 1.70
29. 2.45
30. 34.4
31. 16.4
32. 100

In Exercises 33 through 40 evaluate the given trigonometric functions by first changing the radian measure to degree measure. When using the table, choose the nearest value shown.

33. $\sin \dfrac{\pi}{4}$
34. $\cos \dfrac{\pi}{6}$
35. $\tan \dfrac{5\pi}{12}$

36. $\sin \frac{7\pi}{18}$ 37. $\cot \frac{5\pi}{6}$ 38. $\tan \frac{4\pi}{3}$
39. $\cos 4.59$ 40. $\sec 3.27$

In Exercises 41 through 48 evaluate the given trigonometric functions directly, without first changing the radian measure to degree measure. When using the table, choose the nearest value shown.

41. $\tan 0.7359$ 42. $\sec 0.9308$ 43. $\cot 4.239$
44. $\tan 3.471$ 45. $\cos 2.072$ 46. $\cot 2.340$
47. $\sin 4.861$ 48. $\csc 6.190$

In Exercises 49 through 56 find θ for $0 < \theta < 2\pi$. Interpolation is required in Exercises 53 through 56.

49. $\sin \theta = 0.3090$ 50. $\cos \theta = -0.9135$
51. $\tan \theta = -0.2126$ 52. $\sin \theta = -0.0436$
53. $\cos \theta = 0.6742$ 54. $\tan \theta = 1.860$
55. $\sec \theta = -1.306$ 56. $\csc \theta = 3.940$

3-4 Applications of the Use of Radian Measure

Radian measure has numerous applications in mathematics and other fields. In this section we illustrate the usefulness of radian measure in specific geometric and physical applications.

From geometry we know that the length of an arc on a circle is proportional to the central angle and that the length of the arc of a complete circle is the circumference. Letting s stand for the length of arc, we may state that $s = 2\pi r$ for a complete circle. Since 2π is the central angle (in radians) of the complete circle, we have

(3-10)
$$s = \theta r$$

for any circular arc with central angle θ. If we know the central angle in radians and the radius of a circle, we can find the length of a circular arc directly by using Equation (3-10). (See Figure 3-9.)

Example A
If $\theta = \pi/6$ and $r = 3.00$ in.,

$$s = \left(\frac{\pi}{6}\right)(3.00) = \frac{\pi}{2} = 1.57 \text{ in.}$$

If the arc length is 7.20 ft for a central angle of 150.0° on a certain circle, we may find the radius of the circle by

$$7.20 = (150.0)\left(\frac{\pi}{180}\right) \quad r = \frac{5\pi}{6} r \quad \text{or}$$

$$r = \frac{6(7.20)}{5(3.14)} = 2.75 \text{ ft.}$$

Another geometric application of radians is in finding the area of a sector of a circle. (See Figure 3-10.) We recall from geometry that areas of sectors of circles are proportional to their central angles. The area of a circle is given by $A = \pi r^2$. This can be written as $A = \frac{1}{2}(2\pi)r^2$. We now note that the angle for a complete circle is 2π. Therefore, the area of any sector of a circle in terms of the radius and the central angle is:

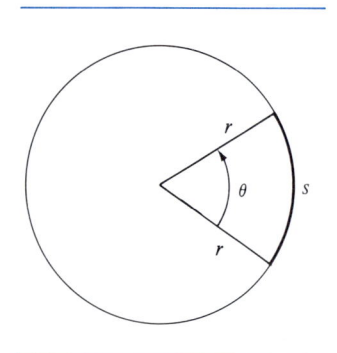

Figure 3-9

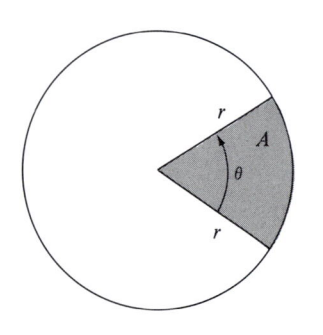

Figure 3-10

(3-11) $$A = \tfrac{1}{2} \theta r^2$$

Example B
The area of a sector of a circle with central angle 18.0° and radius 5.00 in. is

$$A = \frac{1}{2}(18.0)\left(\frac{\pi}{180}\right)(5.00)^2 = \frac{1}{2}\left(\frac{\pi}{10}\right)(25.0)$$

$$= 3.93 \text{ in}^2.$$

Given that the area of a sector is 75.5 ft² and the radius is 12.2 ft, we can find the central angle by

$$75.5 = \tfrac{1}{2}\theta(12.2)^2 \qquad \theta = \frac{2(75.5)}{(12.2)^2} = \frac{151}{149} = 1.01.$$

3-4 Applications of the Use of Radian Measure

This means that the central angle is 1.01 rad, or 57.9°.

The next illustration deals with velocity. We know that average velocity is defined by the equation $v = s/t$, where v is the average velocity, s is the distance traveled, and t is the elapsed time. If an object is moving around a circular path with constant speed, the actual distance traveled is the length of arc traversed. Therefore, if we divide both sides of Equation (3-10) by t, we obtain:

(3-12) $$\frac{s}{t} = \frac{\theta}{t} r \quad \text{or} \quad v = \omega r$$

Equation (3-12) expresses the relationship between the *linear velocity* v and the *angular velocity* ω of an object moving around a circle of radius r. The most convenient units for ω are radians per unit of time. In this way the formula can be used directly. In practice, however, ω is often given in revolutions per minute or some similar unit. In cases like these, it is necessary to convert the units of ω to radians per unit of time before substituting in Equation (3-12).

Example C

An object is moving about a circle of radius 6.00 in. with an angular velocity of 4.00 rad/sec. The linear velocity is

$$v = (6.00)(4.00) = 24.0 \text{ in./sec.}$$

(Remember that radians are numbers and are not included in the final set of units.) This means that the object is moving along the circumference of the circle at 24.0 in./sec.

Example D
A flywheel rotates with an angular velocity of 20.0 rev/min. If its radius is 18.0 in., find the linear velocity of a point on the rim.

Since there are 2π rad in each revolution,

$$20.0 \text{ rev/min} = 40.0\pi \text{ rad/min}.$$

Therefore,

$$v = (40.0)(3.14)(18.0) = 2260 \text{ in./min}.$$

This means that the linear velocity is 2260 in./min, which is equivalent to 188 ft/min, or 3.13 ft/sec.

Exercises 3-4

1. In a circle of radius 10.0 in., find the length of arc subtended on the circumference by a central angle of 60°0'.
2. In a circle of diameter 4.50 m, find the length of arc subtended on the circumference by a central angle of 42°0'.
3. Find the area of the circular sector indicated in Exercise 1.
4. Find the area of a sector of a circle, given that the central angle is 120°0' and the diameter is 56.0 cm.
5. Find the radian measure of an angle at the center of a circle of radius 5.00 in. which intercepts an arc length of 6.00 in.
6. Find the central angle of a circle which intercepts an arc length of 780 mm when the radius of the circle is 520 mm.
7. Two concentric (same center) circles have radii of 5.00 and 6.00 in. Find the portion of the area of the sector of the larger circle which is outside the smaller circle when the central angle is 30°0'.

3-4 Applications of the Use of Radian Measure 89

8. In a circle of radius 6.00 in., the length of arc of a sector is 10.0 in. What is the area of the sector?

9. A pendulum 3.00 ft long oscillates through an angle of 5°0'. Find the distance through which the end of the pendulum swings in going from one extreme position to the other.

10. The radius of the earth is about 3960 mi. What is the length, in miles, of an arc of the earth's equator for a central angle of 1°0'?

11. In turning, an airplane traveling at 540 km/hr moves through a circular arc for 2 min. What is the radius of the circle, given that the central angle is 8°0'?

12. A flywheel rotates at 300 rev/min. If the radius is 6.00 in., through what total distance does a point on the rim travel in 30.0 sec?

13. For the flywheel in Exercise 12, how far does a point halfway out along a radius move in 1 sec?

14. An automobile is traveling at 60.0 mi/hr (88.0 ft/sec). The tires are 28.0 in. in diameter. What is the angular velocity of the tires in radians per second?

15. Find the velocity, due to the rotation of the earth, of a point on the surface of the earth at the equator. (See Exercise 10.)

16. An astronaut in a spacecraft circles the moon once each 1.95 hr. If his altitude is constant at 70.0 mi, what is his velocity? The radius of the moon is 1080 mi.

17. What is the linear velocity of the point in Exercise 12?

18. A phonograph record 6.90 in. in diameter rotates 45.0 times per minute. What is the linear velocity of a point on the rim in feet per minute?

19. A circular sector whose central angle is 210° 0′ is cut from a circular piece of sheet metal of diameter 12.0 cm. A cone is then formed by bringing the two radii of the sector together. What is the lateral surface area of the cone?
20. The propeller of an airplane is 8.00 ft in diameter and rotates at 2200 rev/min. What is the linear velocity of a point on the tip of the propeller?
21. The moon is about 240,000 mi from the earth. It takes the moon about 28 days to make one revolution. What is its angular velocity about the earth in radians per hour?
22. Find the linear velocity of the moon in miles per hour. (See Exercise 21.)

3-5 Circular Functions

There is another method by which the trigonometric functions may be developed. In this section we demonstrate this approach to the trigonometric functions and show that it is consistent with our previous development. However, the material presented later in the text is not dependent on this discussion.

Let us consider the circle with its center at the origin and its radius equal to 1. This is called the *unit circle* and its equation is $x^2 + y^2 = 1$. If a point P starts at $(1, 0)$ and moves θ units around the circumference of the circle, counterclockwise if $\theta > 0$ and clockwise if $\theta < 0$, we can locate the exact position of P for any specific value of θ. See Figure 3-11.

3-5 Circular Functions

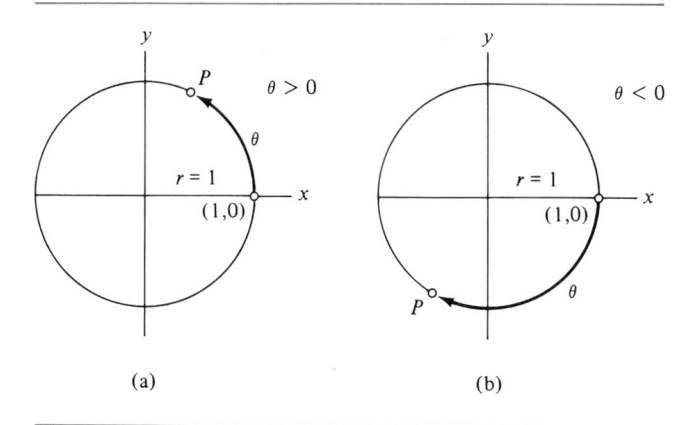

Figure 3-11

Example A
Find the coordinates of P if $\theta = \pi$.

Since the circumference of the circle is 2π, a distance of π is exactly one-half of the circle. Thus, P would have the coordinates of $(-1, 0)$. See Figure 3-12.

Now find the coordinates of P if $\theta = -5\pi/2$. Since θ is negative we measure in a clockwise direction from $(1, 0)$ a distance equal to $5\pi/2$. Since the circumference is 2π, point P must move 1¼ times around the circle, ending at the point $(0, -1)$. See Figure 3-13.

Example B
In which quadrant is P if $\theta = 2$?

If we measure a distance of 2 counterclockwise around the circle from the point $(1, 0)$, we locate P between the points $(0, 1)$ and $(-1, 0)$. The point $(0, 1)$ is located at a distance of one-fourth the circumference, or approximately 1.571. The point $(-1, 0)$ is located at a distance of one-half the circumference, or approximately 3.142. Since 2 is

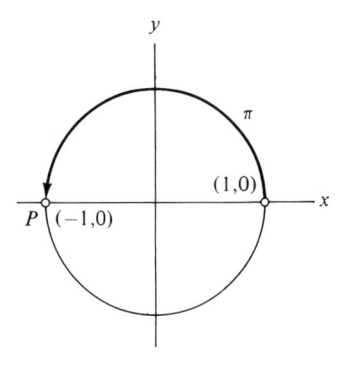

Figure 3-12

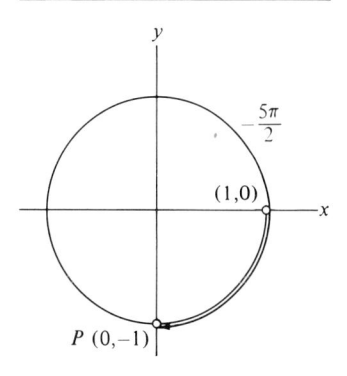

Figure 3-13

between 1.571 and 3.142, the point P is in the second quadrant. See Figure 3-14.

In which quadrant is P if $\theta = -5$?

Measuring in a clockwise direction from (1, 0) we locate P somewhere between three-fourths of the way around the circle (approximately a distance of 4.712) and all the way around the circle (approximately a distance of 6.283). Therefore, we find P is in the first quadrant. See Figure 3-15.

Now let us consider the coordinates (x, y) of point P. We define the x-coordinate of P as the *cosine* of θ and the y-coordinate of P as the *sine* of θ. Therefore, we have

$$x = \cos \theta \quad \text{and} \quad y = \sin \theta.$$

This also means that we may write the coordinates of P as $(\cos \theta, \sin \theta)$.

Another important function is the *tangent* function, which we define as the ratio of the y-coordinate of P to the x-coordinate of P. Thus,

$$\tan \theta = \frac{y}{x} \quad (x \neq 0).$$

These three functions are called *circular functions* since they are defined in terms of the coordinates of a point on the unit circle. The definitions of the circular functions are consistent with the definitions of the trigonometric functions (Equations 2-1), since in the unit circle $r = 1$. The arc length θ is the same length as the angle θ when the angle θ is measured in radians. Since $r = 1$, we have

$$\sin \theta = \frac{y}{r} = \frac{y}{1} = y \qquad \cos \theta = \frac{x}{r} = \frac{x}{1} = x$$

$$\tan \theta = \frac{y}{x}.$$

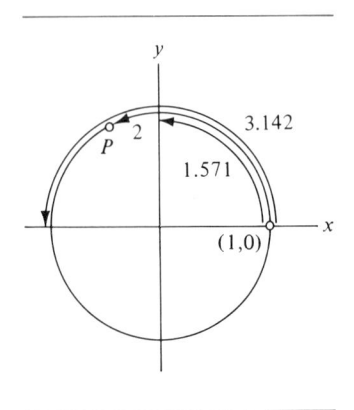

Figure 3-14

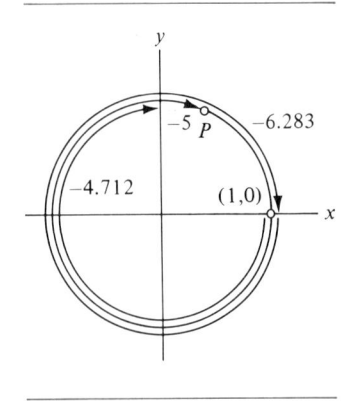

Figure 3-15

3-5 Circular Functions

By noting the quadrant in which point P lies and the signs of the coordinates x and y, we can determine in which quadrants the circular functions are positive or negative:

Quadrant in Which P Lies	$\sin \theta = y$	$\cos \theta = x$	$\tan \theta = \dfrac{y}{x}$
I	+	+	+
II	+	−	−
III	−	−	+
IV	−	+	−

Example C
To find the sign of $\sin 2\pi/3$, we note that a distance of $2\pi/3$ from $(1, 0)$ places P in the second quadrant. Thus, $\sin 2\pi/3$ is positive.

To find the sign of $\cos(-\pi/6)$, we note that a distance of $-\pi/6$ from $(1, 0)$ places P in the fourth quadrant. Thus, $\cos(-\pi/6)$ is positive.

To find the sign of $\tan 2$, we note that a distance of 2 from $(1, 0)$ places P in the second quadrant. Thus, $\tan 2$ is negative.

To find the sign of $\cos(-5)$, we note that a distance of -5 from $(1, 0)$ places P in the first quadrant. Thus, $\cos(-5)$ is positive.

A line from the origin $(0, 0)$ through any point in the plane will intersect the unit circle. Therefore, we can determine the circular functions of the number representing the arc length on the unit circle from $(1, 0)$ to the point of intersection. This is done by the use of similar triangles.

Example D
Let P lie on the line joining (0, 0) and (3, 4). Find $\sin \theta$, $\cos \theta$, and $\tan \theta$. See Figure 3-16.

To find the circular functions of θ we must determine the coordinates of P, which is on the unit circle and on the line joining (0, 0) and (3, 4). The distance from (0, 0) to (3, 4) is

$$r = \sqrt{3^2 + 4^2} = 5.$$

From the similar triangles we have

$$\frac{5}{1} = \frac{4}{y} = \frac{3}{x},$$

which gives us $x = \frac{3}{5}$ and $y = \frac{4}{5}$ for the coordinates of P. Therefore,

$$\sin \theta = \frac{4}{5} \quad \cos \theta = \frac{3}{5} \quad \tan \theta = \frac{\frac{4}{5}}{\frac{3}{5}} = \frac{4}{3}.$$

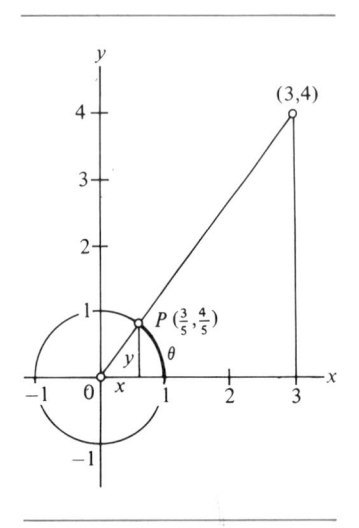

Figure 3-16

If we know one of the circular functions of a number θ, we can determine the other circular functions by use of the Pythagorean theorem. The following example illustrates the method.

Example E
Given that $\sin \theta = -\frac{5}{13}$ and $\tan \theta > 0$, find the values of $\cos \theta$ and $\tan \theta$.

Since $\sin \theta = -\frac{5}{13}$ is the y-coordinate of point P on the unit circle, we have

$$x^2 + \left(-\frac{5}{13}\right)^2 = 1$$

$$x^2 = 1 - \frac{25}{169} = \frac{144}{169}$$

$$x = \pm \frac{12}{13}.$$

However, since $\sin \theta < 0$ and $\tan \theta > 0$, point P is in the third quadrant. This means that $x = \cos \theta = -\frac{12}{13}$ and $\tan \theta = \frac{5}{12}$.

3-5 Circular Functions

Exercises 3-5

In Exercises 1 through 8 find the coordinates of point P for θ equal to the given distances on the unit circle from $(1, 0)$.

1. $\dfrac{\pi}{2}$ 2. $\dfrac{3\pi}{2}$ 3. $-\pi$ 4. $-\dfrac{\pi}{2}$

5. $\dfrac{11\pi}{2}$ 6. $\dfrac{9\pi}{2}$ 7. -8π 8. -5π

In Exercises 9 through 16 find which quadrant P is in, given that θ is equal to the given distances on the unit circle from $(1, 0)$.

9. 3 10. 1 11. -1.5 12. -2.7
13. 4 14. 1.8 15. -12 16. -7

In Exercises 17 through 32 find the signs of the given circular functions.

17. $\sin \dfrac{\pi}{3}$
18. $\cos \dfrac{\pi}{4}$
19. $\cos \dfrac{3\pi}{4}$
20. $\sin \dfrac{7\pi}{6}$
21. $\cos \left(-\dfrac{\pi}{6}\right)$
22. $\tan \left(-\dfrac{4\pi}{3}\right)$
23. $\tan \left(-\dfrac{3\pi}{5}\right)$
24. $\cos \dfrac{\pi}{7}$
25. $\sin 0.5$
26. $\tan 2$
27. $\tan 6$
28. $\sin 3.4$
29. $\cos(-4)$
30. $\sin(-6)$
31. $\tan(-1.7)$
32. $\cos(-0.6)$

In Exercises 33 through 48 the point P is on the line segment joining the origin and the given point. Find the circular function values of each real number θ.

33. $(5, 12)$ 34. $(8, 15)$
35. $(-3, 4)$ 36. $(-24, 7)$
37. $(-15, -8)$ 38. $(-4, -3)$

39. $(7, -24)$ 40. $(12, -5)$
41. $(8, 6)$ 42. $(30, 16)$
43. $(48, -14)$ 44. $(10, -24)$
45. $(-2, -5)$ 46. $(-1, -6)$
47. $(-3, 2)$ 48. $(-4, 2)$

In Exercises 49 through 56 find the values of the circular functions $\sin \theta$, $\cos \theta$, and $\tan \theta$ under the given conditions.

49. $\sin \theta = \frac{3}{5}$, $\cos \theta > 0$
50. $\cos \theta = \frac{7}{25}$, $\sin \theta > 0$
51. $\cos \theta = -\frac{5}{13}$, $\tan \theta > 0$
52. $\sin \theta = \frac{12}{13}$, $\cos \theta < 0$
53. $\tan \theta = -\frac{8}{15}$, $\cos \theta < 0$
54. $\tan \theta = \frac{4}{3}$, $\sin \theta < 0$
55. $\sin \theta = -\frac{24}{25}$, $\tan \theta < 0$
56. $\cos \theta = \frac{15}{17}$, $\tan \theta < 0$

Exercises for Chapter 3

In Exercises 1 through 8 find the trigonometric functions of θ given that the terminal side of θ passes through the given point.

1. $(6, 8)$ 2. $(5, 12)$ 3. $(-12, 5)$ 4. $(-4, 3)$
5. $(7, -2)$ 6. $(3, -1)$ 7. $(-2, -3)$ 8. $(-1, 5)$

In Exercises 9 through 12 express the given trigonometric functions in terms of the same function of a positive acute angle.

9. $\cos 132°$, $\tan 194°$ 10. $\sin 243°$, $\cot 318°$
11. $\sin 289°$, $\sec(-15°)$ 12. $\cos 103°$, $\csc(-100°)$

Exercises for Chapter 3

In Exercises 13 through 16 express the given trigonometric functions in terms of a positive acute angle less than 45°.

13. sin 62°, cos 117°
14. tan 83°, cot 123°
15. sec 247°, cos 295°
16. csc 236°, sec 286°

In Exercises 17 through 20 express the given angle measurements in terms of π.

17. 40°, 153°
18. 22.5°, 324°
19. 48°, 202.5°
20. 27°, 162°

In Exercises 21 through 28 the given numbers represent angle measure. Express the measure of each angle in terms of degrees.

21. $\dfrac{7\pi}{5}$, $\dfrac{13\pi}{18}$
22. $\dfrac{3\pi}{8}$, $\dfrac{7\pi}{20}$
23. $\dfrac{\pi}{15}$, $\dfrac{11\pi}{6}$
24. $\dfrac{17\pi}{10}$, $\dfrac{5\pi}{4}$
25. 0.560
26. 1.35
27. 3.60
28. 14.5

In Exercises 29 through 36 express the given angles in radians. (Do not answer in terms of π.)

29. 100°0′
30. 305°0′
31. 20°30′
32. 148°30′
33. 262°25′
34. 18°47′
35. 136.2°
36. 385.4°

In Exercises 37 through 44 determine the values of the given trigonometric functions.

37. cos 245°0′
38. sin 141°0′
39. tan 256°42′
40. cos 162°32′
41. sin $\dfrac{9\pi}{5}$
42. sec $\dfrac{5\pi}{8}$
43. csc 2.150
44. tan 0.7999

In Exercises 45 through 48 find θ for $0° < \theta < 360°$.

45. $\sin \theta = 0.2924$
46. $\cot \theta = -2.560$
47. $\cos \theta = 0.3297$
48. $\tan \theta = -0.7730$

In Exercises 49 through 52 find θ for $0 < \theta < 2\pi$.

49. $\cos \theta = 0.8387$
50. $\sin \theta = 0.1045$
51. $\sin \theta = -0.8646$
52. $\tan \theta = 2.850$

In Exercises 53 through 62 solve the given problems.

53. Determine the values of all the functions of θ given that $\sin \theta = -\frac{4}{5}$ and $\cos \theta > 0$.
54. Determine the values of all the functions of θ given that $\tan \theta = \frac{1}{3}$ and $\csc \theta < 0$.
55. A pendulum 5.00 ft long swings through an angle of 4.50°. Through what distance does the bob swing in going from one extreme position to the other?
56. In a circle of radius 4.00 cm, find the length of arc subtended on the circumference by a central angle of 82° 0'.
57. Find the area of a sector of a circle, given that the central angle is 230.0° and the diameter is 8.50 cm.
58. Find the area of a sector of a circle, given that the central angle is 80° 0' and the radius is 3.62 in.
59. If the apparatus shown in Figure 3-17 is rotating at 2.00 rev/sec, what is the linear velocity of the ball?
60. A lathe is to cut material at the rate of 350 ft/min. Calculate the radius of the cylindrical piece that is turned at the rate of 120 rev/min.
61. The chain on a child's swing is 8.25 ft long. What is the linear velocity of a child if she swings through an angle of 120° in 2.50 sec? See Figure 3-18.

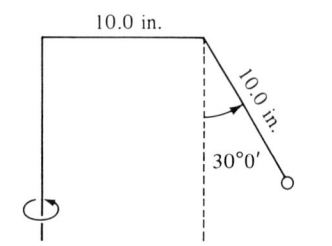

Figure 3-17

Figure 3-18

Exercises for Chapter 3

62. A thermometer needle passes through 55° 0' for a temperature change of 40.0°. If the needle is 5.00 cm long and the scale is circular, how long must the scale be for a maximum temperature change of 150°? See Figure 3-19.

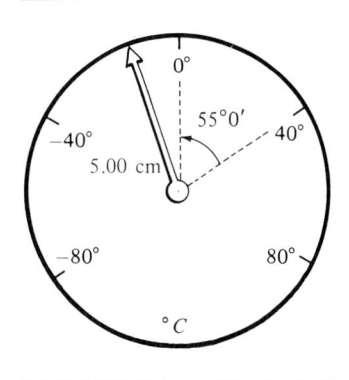

Figure 3-19

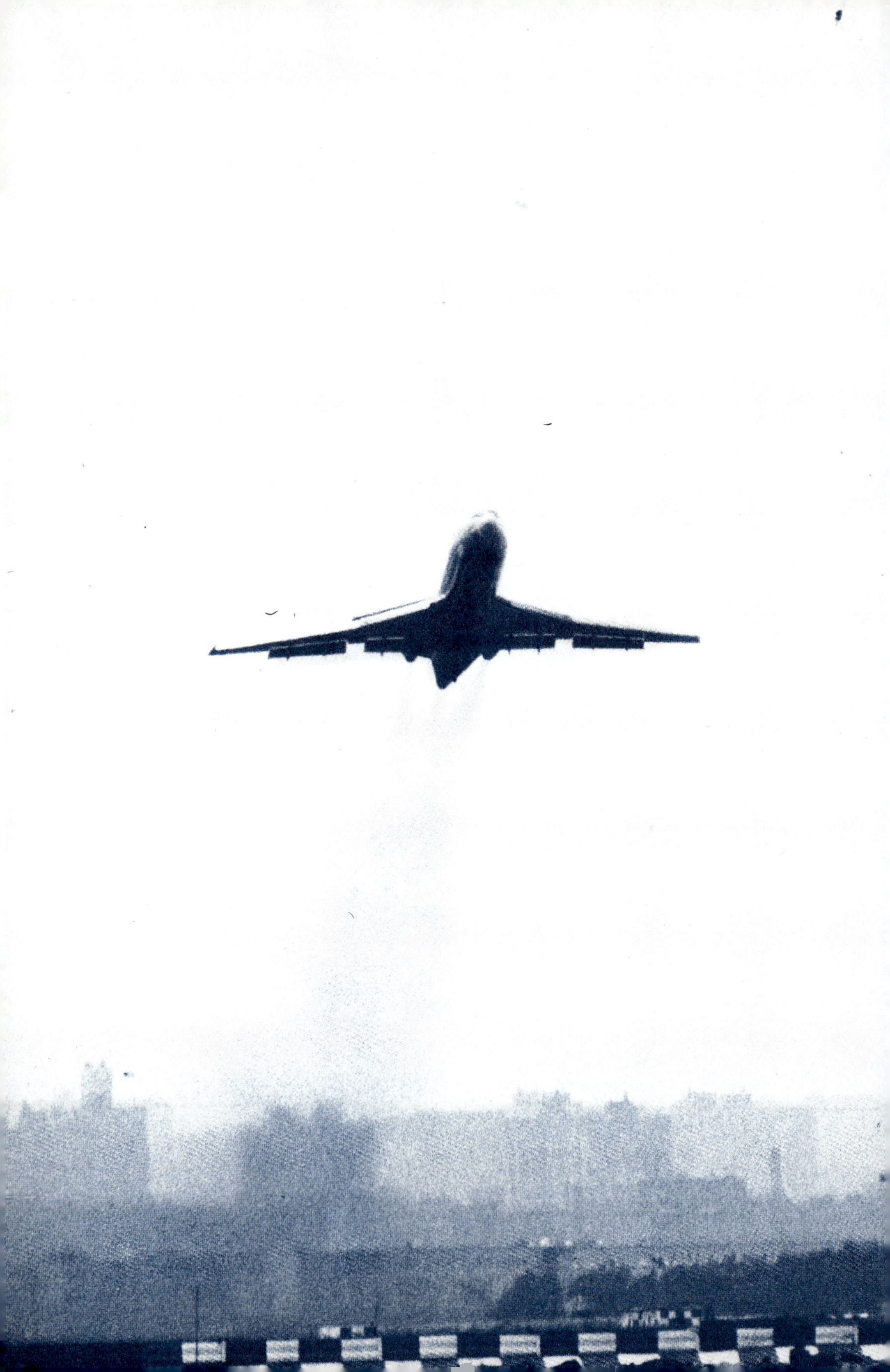

4

Vectors

and Oblique

Triangles

One of the important applications of trigonometry is in analyzing the motion of objects. The course of a jet, such as the one shown, is determined through the use of trigonometry by considering such factors as the force generated by the engines and the wind velocity.

4-1 Vectors

A great many quantities with which we deal may be described by specifying their magnitudes. Generally, one can describe lengths of objects, areas, time intervals, monetary amounts, temperatures, and numerous other quantities by specifying a number: the magnitude of the quantity. Such quantities are known as *scalar* quantities.

Many other quantities are fully described only when both their magnitude and direction are specified. Such quantities are known as *vectors*. Examples of vectors are velocity, force, and momentum. Vectors are of utmost importance in many fields of science and technology. The following example illustrates a vector quantity and the distinction between scalars and vectors.

Example A
A jet flies over a certain point traveling at 600 mi/hr. From this statement alone we know only the *speed* of the jet. Speed is a scalar quantity, and it designates only the magnitude of the rate.

If we were to add the phrase "in a direction 10° south of west" to the sentence above about the jet, we would be specifying the direction of travel as well as the speed. We then know the *velocity* of the jet; that is, we know the *direction* of travel as well as the *magnitude* of the rate at which the jet is traveling. Velocity is a vector quantity.

Let us analyze an example of the action of two vectors. Consider a boat moving in a stream. For purposes of this example, we shall assume that the boat is driven by a motor which can move it at 4 mi/hr in still water. We shall assume that the current is moving downstream at 3 mi/hr. We immediately see that the motion of the boat depends on the direction in which it is headed. If the boat heads downstream, it can travel at 7 mi/hr, for the current is going 3 mi/hr and the boat moves at

4 mi/hr with respect to the water (see Figure 4-1a). If the boat heads upstream, however, it progresses at the rate of only 1 mi/hr, since the action of the boat and that of the stream are counter to each other (see Figure 4-1b). If the boat heads across the stream, the point it reaches on the other side will not be directly opposite the point from which it started. We can see this is so because we know that as the boat heads across the stream, the stream is moving the boat downstream *at the same time* (see Figure 4-1c).

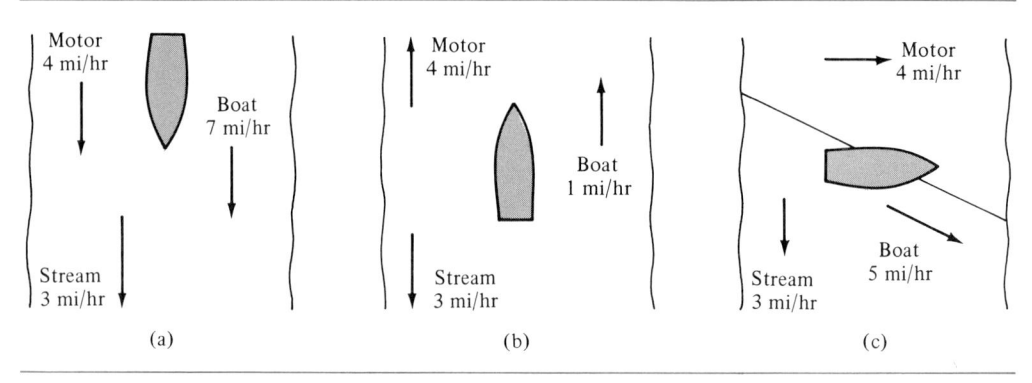

Figure 4-1

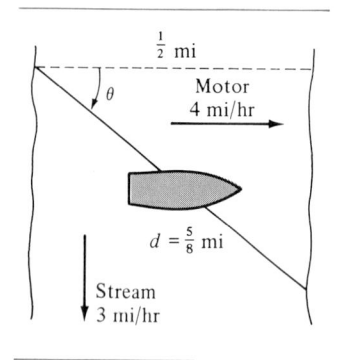

Figure 4-2

This last case should be investigated further. Let us assume that the stream is $\frac{1}{2}$ mi wide where the boat is crossing. It will then take the boat $\frac{1}{8}$ hr to cross. In $\frac{1}{8}$ hr the stream will carry the boat $\frac{3}{8}$ mi downstream. Therefore, when the boat reaches the other side it will be $\frac{3}{8}$ mi downstream. From the Pythagorean theorem, we find that the boat traveled $\frac{5}{8}$ mi from its starting point to its finishing point.

$$d^2 = \left(\frac{4}{8}\right)^2 + \left(\frac{3}{8}\right)^2 = \frac{16+9}{64} = \frac{25}{64} \qquad d = \frac{5}{8} \text{ mi}$$

Since this $\frac{5}{8}$ mi was traveled in $\frac{1}{8}$ hr, the magnitude of the velocity of the boat was actually

$$v = \frac{d}{t} = \frac{\frac{5}{8}}{\frac{1}{8}} = \frac{5}{8} \cdot \frac{8}{1} = 5 \text{ mi/hr.}$$

4-1 Vectors

Also, we see that the direction of this velocity can be represented along a line making an angle θ with the line directed across the stream (see Figure 4-2).

We have just seen two velocity vectors being *added*. Note that these vectors are not added the way numbers are added. We have to take into account their direction as well as their magnitude. Reasoning along these lines, let us now define the sum of two vectors. [Note: A vector quantity is represented by a letter printed in boldface type. In writing, one usually places an arrow over the letter. The same letter in italic (lightface) type represents the magnitude only.]

Let **A** and **B** represent vectors directed from O to P and from P to Q, respectively (see Figure 4-3). The resultant vector **R** from the *initial point* O to the *terminal point* Q represents the sum of vectors **A** and **B**. This is equivalent to letting the two vectors that are added be the sides of a parallelogram. The resultant is then the diagonal of the parallelogram, as shown in Figure 4-4. (When a parallelogram is used to find the resultant of two vectors, the vectors are generally placed with their endpoints together.)

Three or more vectors may be added by placing the initial point of the second at the terminal point of the first, the initial point of the third at the terminal point of the second, and so forth (Figure 4-5). The resultant is the vector from the initial point of the first to the terminal point of the last. **In general, a *resultant* is a single vector which can replace any number of other vectors and still produce the same physical effect.**

Vectors may be subtracted by reversing the direction of the vector being subtracted and then proceeding as in adding vectors.

Note that in Figures 4-3, 4-4 and 4-5 no attempt was made to place the vectors in particular positions, except with respect to direction and magnitude. **Any vector may be moved for purposes of adding and subtracting, so long as its magnitude and direction remain unchanged.**

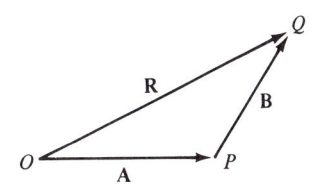

Figure 4-3

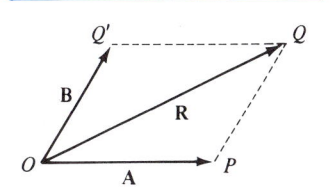

Figure 4-4

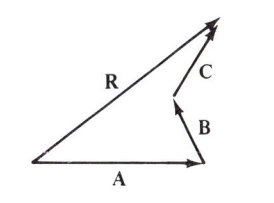

Figure 4-5

Example B

Given vectors **A** and **B** as shown in Figure 4-6(a), we move them for purposes of addition as shown in Figure 4-6(b). The parallelogram method of adding is indicated in Figure 4-6(c). The vector difference **A** − **B** is shown in Figure 4-6(d). Note that the direction of vector **B** is reversed for the subtraction, but the magnitude is the same.

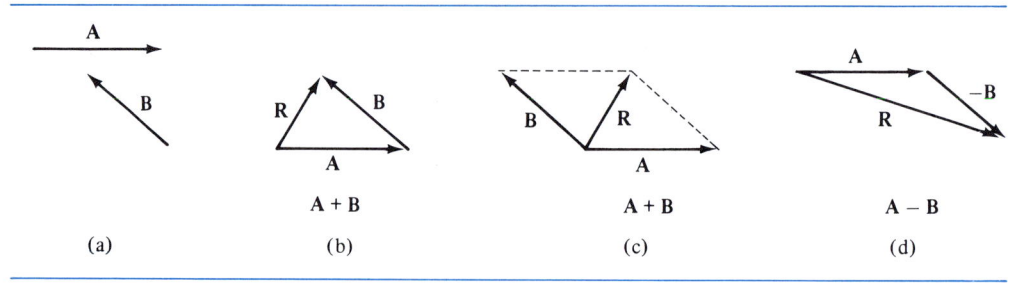

Figure 4-6

In addition to being able to add and subtract vectors, we often need to consider a given vector as the sum of two other vectors. **Two vectors which when added together give the original vector are called the *components* of the original vector.** In the example of the boat, the velocities of 4 mi/hr cross-stream and 3 mi/hr downstream are components of the 5 mi/hr vector directed at the angle θ.

In practice, certain components of a vector are of particular importance. If a vector is so placed that its initial point is at the origin of a rectangular coordinate system and its direction is indicated by an angle in standard position, we may find its x- and y-components. These components are vectors directed along the coordinate axes which, when added together, result in the given vector. **Finding these component vectors is called *resolving the vector into its components.***

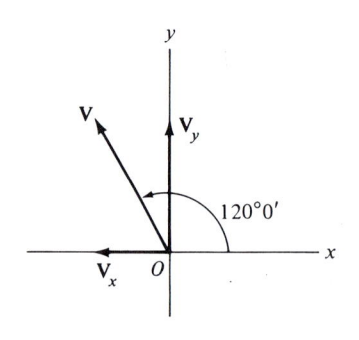

Figure 4-7

Example C

Resolve a vector 10.0 units long and directed at an angle of 120° 0′ into its x- and y-components (see Figure 4-7).

4-1 Vectors

Placing the initial point of the vector at the origin and putting the angle in standard position, we see that the vector directed along the x-axis, $\mathbf{V}_x$, is related to the vector $\mathbf{V}$ of magnitude V by

$$V_x = V \cos 120°0' = -V \cos 60°0'.$$

(The minus sign indicates that the x-component is directed in the negative direction; that is, to the left.) The vector directed along the y-axis, $\mathbf{V}_y$, is related to the vector $\mathbf{V}$ by

$$V_y = V \sin 120°0' = V \sin 60°0'.$$

Thus, the vectors $\mathbf{V}_x$ and $\mathbf{V}_y$ have the magnitudes

$$V_x = -10.0(0.5000) = -5.00$$
$$V_y = 10.0(0.8660) = 8.66.$$

Therefore, we have resolved the given vector into two components, one directed along the negative x-axis of magnitude 5.00 and the other directed along the positive y-axis of magnitude 8.66.

Adding vectors by diagrams gives only approximate results. By use of the trigonometric functions and the Pythagorean theorem it is possible to obtain accurate numerical results for the sum of vectors. In the following example we illustrate how this is done when the two given vectors are at right angles.

Example D

Add vectors **A** and **B**, with $A = 94.1$ and $B = 32.6$. The vectors are at right angles as shown in Figure 4-8.

We can find the magnitude R of the resultant vector **R** by use of the Pythagorean theorem. This leads to

$$R = \sqrt{A^2 + B^2} = \sqrt{(94.1)^2 + (32.6)^2}$$
$$= \sqrt{8855 + 1063} = \sqrt{9918}$$
$$= 99.6.$$

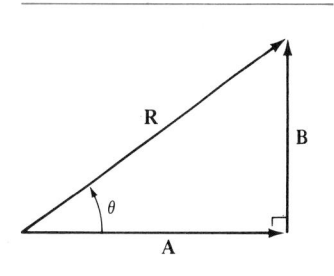

Figure 4-8

Note that the result has been rounded off to three significant digits.

We shall now determine the direction of **R** by specifying its direction as the angle θ, that is, the angle **R** makes with **A**. Therefore,

$$\tan \theta = \frac{B}{A} = \frac{32.6}{94.1} = 0.3464.$$

To the nearest 10', $\theta = 19°10'$. Thus, **R** is a vector with magnitude $R = 99.6$ and in a direction $19°10'$ from vector **A**.

The following two examples show how vectors which are not at right angles are added. The basic procedure is to resolve each into its x- and y-components. These x-components and y-components are added, thus giving the x- and y-components of the resultant. Then, by using the Pythagorean theorem and the tangent as in Example D, we find the magnitude and direction of the resultant.

Example E
Find the resultant **R** of the two vectors given in Figure 4-9(a)—**A** of magnitude 48.0 and direction $57°0'$ and **B** of magnitude 31.0 and direction $322°0'$.

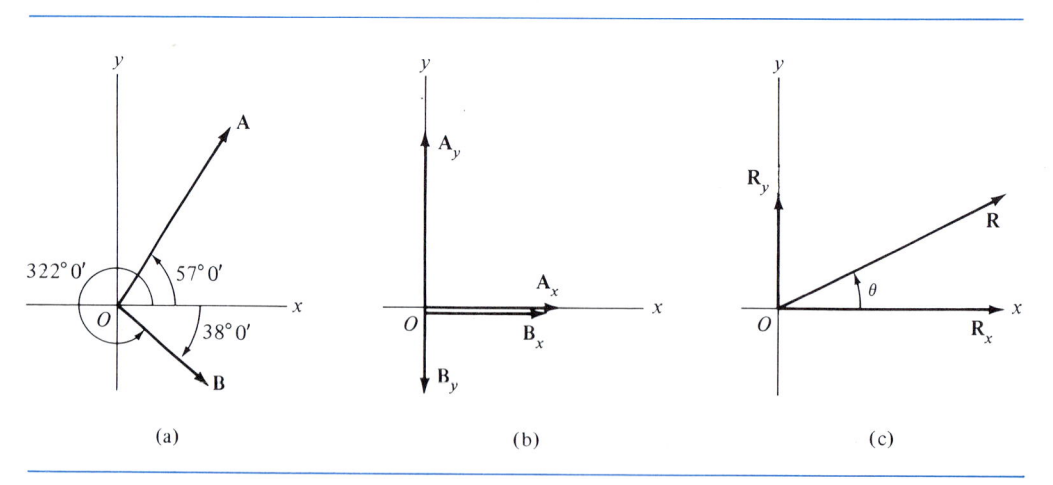

Figure 4-9

4-1 Vectors

$A_x = (48.0)(\cos 57° 0') = (48.0)(0.5446) = 26.1$

$A_y = (48.0)(\sin 57° 0') = (48.0)(0.8387) = 40.3$

$B_x = (31.0)(\cos 38° 0') = (31.0)(0.7880) = 24.4$

$B_y = -(31.0)(\sin 38° 0') = (31.0)(0.6157)$
$= -19.1$

$R_x = A_x + B_x = 26.1 + 24.4 = 50.5$

$R_y = A_y + B_y = 40.3 - 19.1 = 21.2$

$R = \sqrt{(50.5)^2 + (21.2)^2} = \sqrt{2550 + 449}$
$= \sqrt{2999} = 54.8$

$\tan \theta = \dfrac{R_y}{R_x} = \dfrac{21.2}{50.5} = 0.4198 \qquad \theta = 22°50'$

The resultant vector is 54.8 units long and is directed at an angle of 22°50', as shown in Figure 4-9(c).

Example F

Find the resultant of the two given vectors with $A = 322$ and $B = 210$, as shown in Figure 4-10(a).

$A_x = (322)(\cos 75°0') = (322)(0.2588) = 83.3$

$A_y = (322)(\sin 75°0') = (322)(0.9659) = 311$

(a)

(b)

(c)

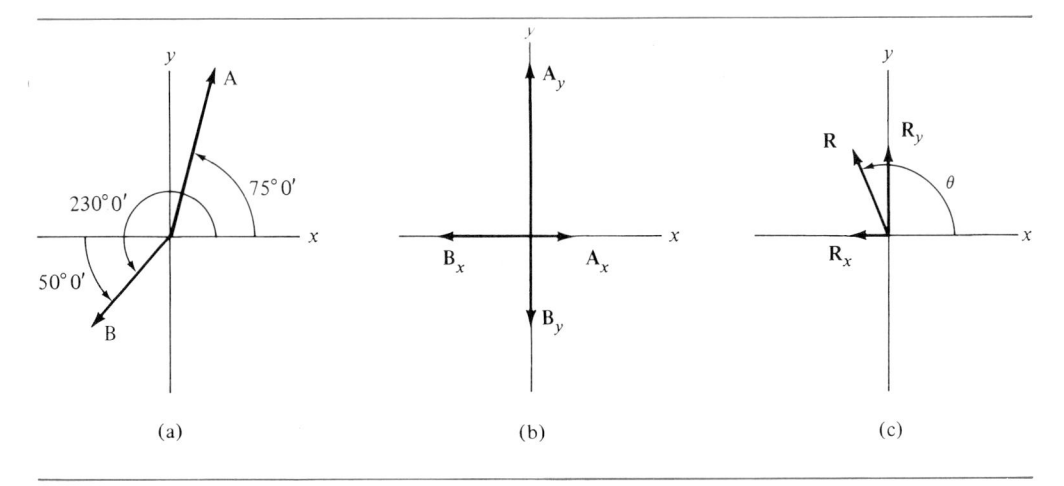

Figure 4-10

$B_x = (-210)(\cos 50°0') = (-210)(0.6428)$
$= -135$
$B_y = (-210)(\sin 50°0') = (-210)(0.7660)$
$= -161$
$R_x = 83.3 - 135 = -51.7$
$R_y = 311 - 161 = 150$
$R = \sqrt{(-51.7)^2 + (150)^2} = 159$
$\tan \theta = \dfrac{150}{-51.7} = -2.901 \quad \theta = 180°0' - 71°0'$
$= 109°0'$

The resultant vector is 159 units long and directed at an angle of 109°0', as shown in Figure 4-10(c).

Some general formulas can be derived from the previous examples. For a given vector **A** directed at an angle θ and of magnitude A, with components A_x and A_y, we have the following relations:

(4-1) $\qquad A_x = A \cos \theta \qquad A_y = A \sin \theta$

(4-2) $\qquad A = \sqrt{A_x^2 + A_y^2}$

(4-3) $\qquad \tan \theta = \dfrac{A_y}{A_x}$

Exercises 4-1

In Exercises 1 through 12, with the vectors indicated in Figure 4-11, find the vector sums by means of diagrams.

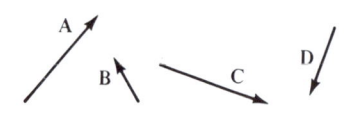

Figure 4-11

1. A + B
2. A + C
3. B + C
4. B + D
5. A + B + D
6. B − C
7. A + D − C
8. C − A

4-1 Vectors

9. $2A + B$ $(2A = A + A)$
10. $B - 2C + 3D$
11. $A + 2B + 3C - 4D$
12. $\frac{1}{2}A - C + 2B$

In Exercises 13 through 24 find the x- and y-components of the given vectors by use of the trigonometric functions. In Exercises 21 through 24 express angles to the nearest minute and other results to four significant digits.

13. Magnitude 8.60, $\theta = 68°0'$
14. Magnitude 9750, $\theta = 243°0'$
15. Magnitude 76.8, $\theta = 145°0'$
16. Magnitude 0.0998, $\theta = 476°0'$
17. Magnitude 9.04, $\theta = 283.5°$
18. Magnitude 16,000, $\theta = 156.5°$
19. Magnitude 15.9, $\theta = 122.4°$
20. Magnitude 153, $\theta = 267.8°$
21. Magnitude 1.000, $\theta = 57°29'$
22. Magnitude 81.82, $\theta = 178°46'$
23. Magnitude 67.81, $\theta = 337°27'$
24. Magnitude 2231, $\theta = 432°12'$

In Exercises 25 through 36, with the given sets of components, find R and θ. In Exercises 29 through 32 express angles to the nearest 0.1°. In Exercises 33 through 36 express angles to the nearest minute and other results to four significant digits.

25. $R_x = 5.18$, $R_y = 8.56$
26. $R_x = 89.6$, $R_y = -52.0$
27. $R_x = -0.982$, $R_y = 2.56$
28. $R_x = 729$, $R_y = -209$
29. $R_x = -646$, $R_y = 2030$
30. $R_x = -31.2$, $R_y = -41.2$
31. $R_x = 0.694$, $R_y = -1.24$
32. $R_x = 7.62$, $R_y = -6.35$
33. $R_x = 12.37$, $R_y = 51.20$
34. $R_x = -2134$, $R_y = -8715$
35. $R_x = 0.9701$, $R_y = -1.237$
36. $R_x = -176.4$, $R_y = 231.0$

112   Vectors and Oblique Triangles

In Exercises 37 through 48 add the given vectors by using the trigonometric functions and the Pythagorean theorem. In Exercises 45 through 48 express angles to the nearest minute and other results to four significant digits.

37. $A = 18.0$, $\theta_A = 0°0'$
    $B = 12.0$, $\theta_B = 27°0'$
38. $A = 150$, $\theta_A = 90°0'$
    $B = 128$, $\theta_B = 43°0'$
39. $A = 56.0$, $\theta_A = 76°0'$
    $B = 24.0$, $\theta_B = 200°0'$
40. $A = 6.89$, $\theta_A = 123°0'$
    $B = 29.0$, $\theta_B = 260°0'$
41. $A = 9.82$, $\theta_A = 34.0°$
    $B = 17.4$, $\theta_B = 752.5°$
42. $A = 1.65$, $\theta_A = 36.5°$
    $B = 0.980$, $\theta_B = 252.0°$
43. $A = 12.6$, $\theta_A = 98.4°$
    $B = 15.1$, $\theta_B = 332.2°$
44. $A = 121$, $\theta_A = 292.0°$
    $B = 112$, $\theta_B = 198.7°$
45. $A = 143.6$, $\theta_A = 57°28'$
    $B = 181.2$, $\theta_B = 112°32'$
46. $A = 2.150$, $\theta_A = 210°15'$
    $B = 6.014$, $\theta_B = 291°13'$
47. $A = 27.13$, $\theta_A = 108°9'$
    $B = 14.00$, $\theta_B = 212°58'$
48. $A = 5213$, $\theta_A = 381°51'$
    $B = 9231$, $\theta_B = 227°1'$

4-2 Application of Vectors

The following examples show how vectors are used in certain situations.

Example A
An object on a horizontal table is acted on by two horizontal forces. The two forces have magnitudes of 6.00 and 8.00 lb, and the angle between their

4-2 Application of Vectors

lines of action is 90°0′. What is the resultant of these forces?

By means of an appropriate diagram (Figure 4-12) we may better visualize the actual situation. We then note that a good choice of axes (unless specified, it is often convenient to choose the x- and y-axes to fit the problem) is to have the x-axis in the direction of the 6.00-lb force and the y-axis in the direction of the 8.00-lb force. (This is possible since the angle between them is 90°.) With this choice we note that the two given forces will be the x- and y-components of the resultant. Therefore, we arrive at the following results:

$$F_x = 6.00 \text{ lb} \quad F_y = 8.00 \text{ lb}$$

$$F = \sqrt{36.0 + 64.0} = 10.0 \text{ lb}$$

$$\tan \theta = \frac{F_y}{F_x} = \frac{8.00}{6.00} = 1.333 \quad \theta = 53°10'.$$

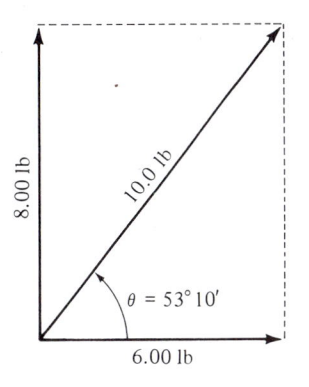

Figure 4-12

We would state that the resultant has a magnitude of 10.0 lb and acts at an angle of 53°10′ from the 6.00-lb force.

Example B

A ship sails 32.0 mi due east and then turns 40°0′ north of east. After sailing an additional 16.0 mi, where is it with reference to the starting point?

The distance an object moves and the direction in which it moves give the *displacement* of an object. Therefore, in this problem we are to determine the resultant displacement of the ship from the two given displacements. The problem is diagramed in Figure 4-13, where the first displacement has been labeled vector **A** and the second vector **B**.

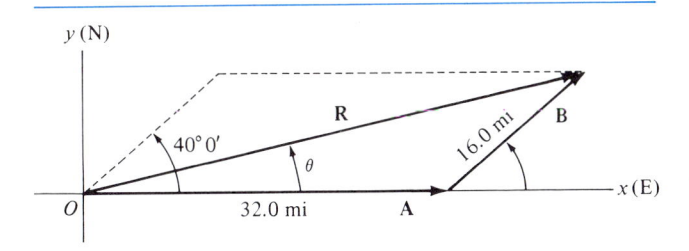

Figure 4-13

Since east corresponds to the positive x-direction, we see that the x-component of the resultant is $\mathbf{A} + \mathbf{B}_x$ and the y-component of the resultant is $\mathbf{B}_y$. Therefore, we have the following results.

$$R_x = A + B_x = 32.0 + 16.0(\cos 40°0')$$
$$= 32.0 + 16.0(0.7660) = 32.0 + 12.3$$
$$= 44.3 \text{ mi}$$
$$R_y = 16.0(\sin 40°0') = 16.0(0.6428)$$
$$= 10.3 \text{ mi}$$
$$R = \sqrt{(44.3)^2 + (10.3)^2} = \sqrt{2069}$$
$$= 45.5 \text{ mi}$$
$$\tan \theta = \frac{10.3}{44.3} = 0.2325 \qquad \theta = 13°10'$$

Therefore, the ship is 45.5 mi from the starting point, in a direction 13°10' north of east.

Example C

An airplane headed due east is in a wind blowing from the southeast. What is the resultant velocity of the plane with respect to the surface of the earth if the plane's velocity with respect to the air is 110 mi/hr and that of the wind is 40.0 mi/hr? (See Figure 4-14.)

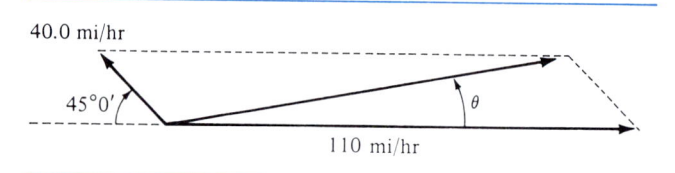

Figure 4-14

Let v_{px} be the velocity of the plane in the x-direction (east), v_{py} the velocity of the plane in the y-direction, v_{wx} the x-component of the velocity of the wind, v_{wy} the y-component of the velocity of the wind, and $\mathbf{v}_{pa}$ the velocity of the plane with respect to the air. Therefore,

$$v_{px} = v_{pa} + v_{wx} = 110 - 40.0(\cos 45°0')$$
$$= 110 - 28.3 = 81.7 \text{ mi/hr}$$
$$v_{py} = v_{wy} = 40.0(\sin 45°0') = 28.3 \text{ mi/hr}$$
$$v = \sqrt{(81.7)^2 + (28.3)^2} = \sqrt{6675 + 801}$$
$$= 86.5 \text{ mi/hr}$$
$$\tan \theta = \frac{28.3}{81.7} = 0.3464 \qquad \theta = 19°10'.$$

We have determined that the plane is traveling 86.5 mi/hr in a direction 19°10' north of east. From this we observe that a plane does not necessarily head in the direction of its destination.

Example D
A force of 12.0 kg is to be resolved into two forces, one of which is 7.00 kg and makes an angle of 90°0' with the 12.0-kg force. Find the magnitude of the other force and the angle it makes with the 12.0-kg force. See Figure 4-15.

Since the 12.0-kg force has been chosen to be horizontal, it can have no net vertical component. Thus, the vector **A** must have a downward vertical component of 7.00 kg. Also, we can see that the 12.0-kg force is the horizontal component of **A**. Thus, $A_x = 12.0$ kg and $A_y = -7.00$ kg. From the Pythagorean theorem we have

$$A = \sqrt{A_x^2 + A_y^2} = \sqrt{(12.0)^2 + (-7.00)^2} = \sqrt{193}$$
$$= 13.9 \text{ kg.}$$

To find the angle θ we have

$$\tan \theta = \frac{A_y}{A_x} = \frac{-7.00}{12.0} = -0.5833$$

and
$$\theta = 30°20'.$$

(The negative sign for $\tan \theta$ indicates a fourth-quadrant angle.) Thus, the force **A** is 13.9 kg at an angle of 30°20' from the 12.0-kg force.

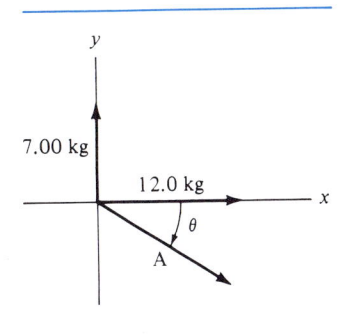

Figure 4-15

Exercises 4-2

1. Two forces, one of 45.0 lb and the other of 68.0 lb, act on the same object and at right angles to each other. Find the resultant of these forces.

2. Two forces act on an object at right angles to each other. If one force is 25.0 kg and the other is 14.0 kg, find the resultant of these forces.

3. Two forces, one of 150 kg and the other of 220 kg, pull on an object. The angle between these forces is 45°0′. What is the resultant of these forces?

4. The angle between two forces acting on an object is 63°0′. If the two forces are 86.0 lb and 103 lb, respectively, what is their resultant?

5. A jet travels 450 mi due west from a city. It then turns and travels 240 mi south. What is its displacement from the city?

6. A ship sails 78.3 km due north after leaving its pier. It then turns and sails 51.2 km east. What is the displacement of the ship from its pier?

7. Town B is 52.0 km southeast of town A. Town C is 45.0 km due west of town B. What is the displacement of town C from town A?

8. John is surveying and finds that Ann is 36.5 ft northeast of him. Ann knows that she is 20.0 ft north of a telephone pole. What is John's displacement from the telephone pole?

9. What are the horizontal and vertical components of the velocity of a stone thrown into the air with a velocity of 120 ft/sec at an angle of 48.0° with respect to the horizontal?

10. A rocket is traveling at an angle of 74.5° with respect to the horizontal at a speed of 3500 km/hr. What are the horizontal and vertical components of the velocity?

11. A stone is thrown horizontally from a plane traveling at 100 m/sec. If the stone is thrown at 50.0 m/sec in a direction perpendicular to the direction of the plane, what is the velocity of the stone just after it is released?
12. A bird is flying east with a velocity of 25.0 mi/hr with respect to the air. If the wind is from the southwest at 10.0 mi/hr, what is the resultant speed of the bird? In what direction is it traveling?
13. The resultant of two forces acting at right angles to each other is 40.0 lb. If one of the forces is 30.0 lb, find the other force and the angle it makes with the resultant.
14. A force of 150 g (grams) is to be resolved into two forces, one of which is 50.0 g and makes an angle of 90° with the 150-g force. Find the other force and the angle it makes with the 150-g force.
15. A boat travels across a river, reaching the opposite bank at a point directly opposite that from which it left. If the boat travels 6.00 mi/hr in still water and the current of river flows at 3.00 mi/hr, what is the velocity of the boat in the water?
16. A plane is flying west at a rate of 105 mi/hr. This resultant course is due to a 20.0 mi/hr wind from the south and the velocity of the plane relative to the air. Find the velocity of the plane relative to the air.

4-3 Oblique Triangles, the Law of Sines

To this point we have limited our study of triangle solution to right triangles. However, many triangles which require solution do not contain a right angle. Such a triangle is termed an *oblique triangle*. Let us now discuss the solutions of oblique triangles.

Vectors and Oblique Triangles

In Section 2-4 we stated that we need three parts, at least one of them a side, to solve any triangle. With this in mind we may determine that there are four possible combinations of parts from which we may solve a triangle. These combinations are:

Case 1: two angles and one side
Case 2: two sides and the angle opposite one of them
Case 3: two sides and the included angle
Case 4: three sides

There are several ways in which oblique triangles may be solved, but we shall restrict our attention to the two most useful methods, the *Law of Sines* and the *Law of Cosines*. In this section we discuss the Law of Sines and show that it may be used to solve Case 1 and Case 2.

Let ABC be an oblique triangle with sides a, b, and c opposite angles A, B, and C, respectively. By drawing a perpendicular h from B to side b or its extension, we see from Figure 4-16(a) that:

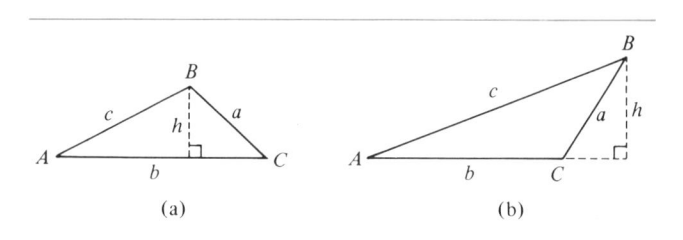

(4-4) $\qquad h = c \sin A \quad$ or $\quad h = a \sin C$

Figure 4-16

And from Figure 4-16(b) we see that:

(4-5) $\quad h = c \sin A \quad$ or $\quad h = a \sin (180° - C) = a \sin C$

4-3 Oblique Triangles, the Law of Sines

We note that the results are precisely the same in Equations (4-4) and (4-5). Setting the results for h equal to each other, we have:

(4-6) $\quad c \sin A = a \sin C \quad$ or $\quad \dfrac{a}{\sin A} = \dfrac{c}{\sin C}$

By dropping a perpendicular from A to a we also derive the result:

(4-7) $\quad c \sin B = b \sin C \quad$ or $\quad \dfrac{b}{\sin B} = \dfrac{c}{\sin C}$

Combining Equations (4-6) and (4-7) we have the **Law of Sines:**

(4-8) $\qquad \dfrac{a}{\sin A} = \dfrac{b}{\sin B} = \dfrac{c}{\sin C}$

The Law of Sines is a statement of proportionality between the sides of a triangle and the sines of the angles opposite them.

Now we may see how the Law of Sines is applied to the solution of a triangle in which two angles and one side are known (Case 1). If two angles are known, the third may be found from the fact that the sum of the angles in a triangle is 180°. At this point we must be able to determine the ratio between the given side and the sine of the angle opposite it. Then, by use of the Law of Sines, we may find the other sides.

Example A
Given that $c = 6.00$, $A = 60°0'$, and $B = 40°0'$, find a, b, and C.

First, we can see that
$$C = 180°0' - (60°0' + 40°0') = 80°0'.$$
Thus,
$$\frac{a}{\sin 60°0'} = \frac{6.00}{\sin 80°0'} \quad \text{or} \quad a = \frac{6.00(0.8660)}{0.9848}$$
$$= 5.28$$
$$\frac{b}{\sin 40°0'} = \frac{6.00}{\sin 80°0'} \quad \text{or} \quad b = \frac{6.00(0.6428)}{0.9848}$$
$$= 3.92.$$
Thus, $a = 5.28$, $b = 3.92$, and $C = 80°0'$.

Example B
Solve the triangle with the following given parts: $a = 63.7$, $A = 56°0'$, and $B = 97°0'$.

We may immediately determine that $C = 27°0'$. Thus,
$$\frac{b}{\sin 97°0'} = \frac{63.7}{\sin 56°0'} \quad \text{or}$$
$$b = \frac{63.7(0.9925)}{0.8290} = 76.3$$
and
$$\frac{c}{\sin 27°0'} = \frac{63.7}{\sin 56°0'} \quad \text{or}$$
$$c = \frac{63.7(0.4540)}{0.8290} = 34.9$$
Thus, $b = 76.3$, $c = 34.9$, and $C = 27°0'$.

Example C
Solve the triangle with the following given parts: $b = 5.06$, $A = 42.0°$, and $C = 28.5°$.

We determine that $B = 109.5°$. Thus,
$$\frac{a}{\sin 42.0°} = \frac{5.06}{\sin 109.5°} \quad \text{or}$$
$$a = \frac{5.06(0.6691)}{0.9426} = 3.59$$

4-3 Oblique Triangles, the Law of Sines

and

$$\frac{c}{\sin 28.5°} = \frac{5.06}{\sin 109.5°} \quad \text{or}$$

$$c = \frac{5.06(0.4772)}{0.9426} = 2.56.$$

Thus, $a = 3.59$, $c = 2.56$, and $B = 109.5°$.

If the given information is appropriate, the Law of Sines may be used to solve applied problems. The following example illustrates the use of the Law of Sines in such a problem.

Example D

A plane traveling at 650 mi/hr with respect to the air is headed 30°0′ east of north. The wind is blowing from the south, which causes the actual course to be 27°0′ east of north. Find the velocity of the wind and the velocity of the plane with respect to the ground.

From the given information the angles are determined as shown in Figure 4-17. Then, applying the Law of Sines, we have the relations

$$\frac{v_w}{\sin 3°0′} = \frac{v_{pg}}{\sin 150°0′} = \frac{650}{\sin 27°0′}$$

where v_w is the magnitude of the velocity of the wind and v_{pg} is the magnitude of the velocity of the plane with respect to the ground. Thus,

$$v_w = \frac{650(0.0523)}{0.4540} = 74.9 \text{ mi/hr}$$

and

$$v_{pg} = \frac{650(0.5000)}{0.4540} = 716 \text{ mi/hr}.$$

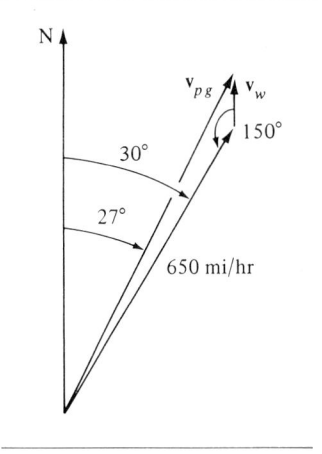

Figure 4-17

If we have information equivalent to Case 2 (two sides and the angle opposite one of them), we may find that there are *two* triangles which satisfy the given information. The following example illustrates this point.

Example E
Solve the triangle with the following given parts: $a = 60.0$, $b = 40.0$, and $B = 30°0'$.

By making a good scale drawing (Figure 4-18) we note that the angle opposite a may be either at position A or A'. Both positions of this angle satisfy the given parts. Therefore, there are two triangles which result. Using the Law of Sines, we have

$$\frac{60.0}{\sin A} = \frac{40.0}{\sin 30°0'} \quad \text{or}$$

$$\sin A = \frac{60.0(0.5000)}{40.0} = 0.7500.$$

Therefore, $A = 48°40'$ and $C = 101°20'$. Using the Law of Sines again to find c, we have

$$\frac{c}{\sin 101°20'} = \frac{40.0}{\sin 30°0'} \quad \text{or}$$

$$c = \frac{40.0(0.9805)}{0.5000} = 78.4.$$

Therefore, one solution is $A = 48°40'$, $C = 101°20'$, and $c = 78.4$.

The other solution is found by interpreting $\sin A = 0.7500$ to mean $A' = 131°20'$. For this case we have C' (the angle opposite c when $A' = 131°20'$) as $18°40'$. Using the Law of Sines to find c', we have

$$\frac{c'}{\sin 18°40'} = \frac{40.0}{\sin 30°0'} \quad \text{or}$$

$$c' = \frac{40.0(0.3201)}{0.5000} = 25.6.$$

This means that the second solution is $A' = 131°20'$, $C' = 18°40'$, and $c' = 25.6$.

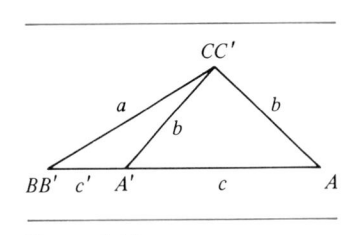

Figure 4-18

Example F
In Example E, if $b > 60.0$ only one solution would result. In this case, side b would intercept side c at A. It also intercepts the extension of side c, but this would require that angle B not be included in the triangle (see Figure 4-19). Thus only one solution may result if $b > a$.

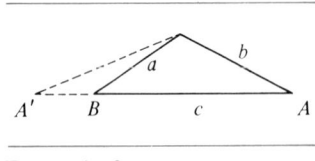

Figure 4-19

4-3 Oblique Triangles, the Law of Sines

In Example E, there would be no solution if side b were not at least 30.0. For if this were the case, side b would not be sufficiently long to even touch side c. It can be seen that b must at least equal $a \sin B$. If it is just equal to $a \sin B$, there is one solution, a right triangle (Figure 4-20).

Summarizing the results for Case 2 as illustrated in Examples E and F, we make the following conclusions. Given sides a and b and angle A (assuming here that a and A ($A < 90°$) are the given corresponding parts), we have:

1. No solution if $a < b \sin A$
2. A right triangle solution if $a = b \sin A$
3. Two solutions if $b \sin A < a < b$
4. One solution if $a > b$

Since two solutions may result from it, Case 2 is often referred to as the "ambiguous case." If we attempt to use the Law of Sines for the solution of Case 3 or Case 4, we find that we do not have sufficient information to complete one of the ratios. These cases can, however, be solved by the Law of Cosines, which we consider in the next section.

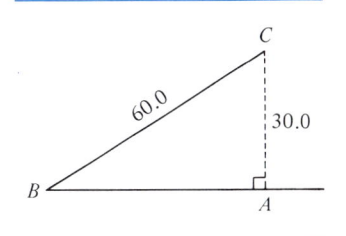

Figure 4-20

Example G
Given the three sides $a = 5.00$, $b = 6.00$, and $c = 7.00$, we would set up the ratios

$$\frac{5.00}{\sin A} = \frac{6.00}{\sin B} = \frac{7.00}{\sin C}.$$

However, since there is no way to determine a complete ratio from these equations, we cannot find the solution of the triangle in this manner.

Exercises 4-3

In Exercises 1 through 16 solve the triangles with the given parts.

1. $a = 45.7$, $A = 65°0'$, $B = 49°0'$
2. $b = 3.07$, $A = 26°0'$, $C = 120°0'$

3. $c = 4380$, $A = 37°0'$, $B = 34°0'$
4. $a = 93.2$, $B = 17°50'$, $C = 82°40'$
5. $a = 4.60$, $b = 3.10$, $A = 18°0'$
6. $b = 3.62$, $c = 2.94$, $B = 69°20'$
7. $b = 0.0742$, $B = 51°0'$, $C = 3°30'$
8. $c = 729$, $B = 121°0'$, $C = 44°10'$
9. $a = 63.8$, $B = 58.5°$, $C = 22.0°$
10. $a = 13.0$, $A = 55.0°$, $B = 67.5°$
11. $b = 438$, $B = 47.3°$, $C = 64.5°$
12. $b = 283$, $B = 12.7°$, $C = 76.4°$
13. $a = 5.24$, $b = 4.44$, $B = 48°10'$
14. $a = 89.4$, $c = 37.3$, $C = 15°40'$
15. $a = 45.0$, $b = 126$, $A = 64°0'$
16. $a = 10.0$, $c = 5.00$, $C = 30°0'$

In Exercises 17 through 22 use the Law of Sines to solve the given problems.

17. Determine the lengths of the two steel supports of the sign shown in Figure 4-21.
18. Determine the unknown sides of the four-sided piece of land shown in Figure 4-22.
19. The angles of elevation of an airplane, measured from points A and B, 7540 m apart on the same side of the airplane (the airplane and points A and B are in the same vertical plane), are $32°0'$ and $44°0'$. How far is point A from the airplane?
20. Resolve vector **A** ($A = 160$) into two components in the directions u and v as shown in Figure 4-23.
21. A ship leaves a port and travels due west. At a certain point it turns $30°0'$ north of west and travels an additional 42.0 mi to a point 63.0 mi from the port. How far from the port is the point where the ship turned?

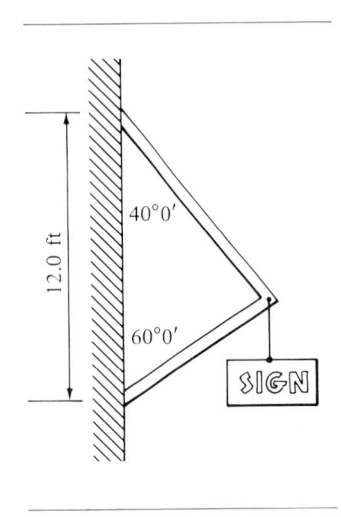

Figure 4-21

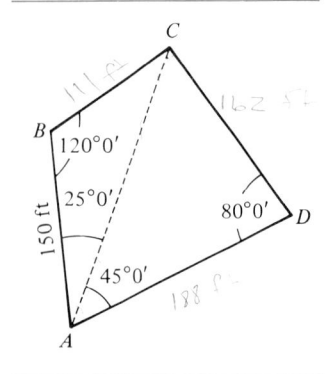

Figure 4-22

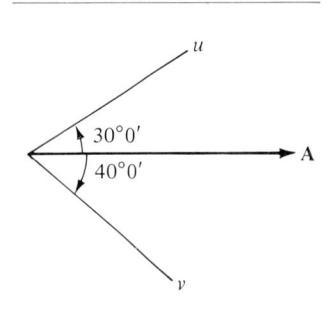

Figure 4-23

22. City B is $40°0'$ south of east of city A. A pilot wishes to know what direction he should head the plane in flying from A to B if the wind is from the west at 40.0 km/hr and his velocity with respect to the air is 300 km/hr. What should his heading be?

4-4 The Law of Cosines

As we noted at the end of the preceding section, the Law of Sines cannot be used if the only information given is that of Case 3 or Case 4. Therefore, it is necessary to develop a method of finding at least one more part of the triangle. Here we can use the Law of Cosines. After obtaining another part by the Law of Cosines, we can then use the Law of Sines to complete the solution. We do this because the Law of Sines generally provides a simpler method of solution than the Law of Cosines.

Consider any oblique triangle, for example either of the ones in Figure 4-24. For each we obtain $h = b \sin A$. By using the Pythagorean theorem we obtain $a^2 = h^2 + x^2$ for each. Thus:

(4-9) $$a^2 = b^2 \sin^2 A + x^2$$

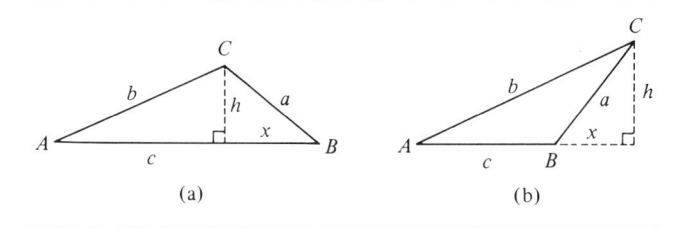

Figure 4-24

In Figure 4-24(a), we have $c - x = b \cos A$, or $x = c - b \cos A$. In Figure 4-24(b), we have $c + x = b \cos A$, or $x = b \cos A - c$. Substituting these relations into Equation (4-9), we obtain:

(4-10)

$$a^2 = b^2 \sin^2 A + (c - b \cos A)^2$$

and

$$a^2 = b^2 \sin^2 A + (b \cos A - c)^2$$

Each of these, when expanded, gives

$$a^2 = b^2 \sin^2 A + b^2 \cos^2 A + c^2 - 2bc \cos A$$

or

(4-11)

$$a^2 = b^2(\sin^2 A + \cos^2 A) + c^2 - 2bc \cos A$$

Recalling the definitions of the trigonometric functions, we know that $\sin \theta = y/r$ and $\cos \theta = x/r$. Therefore, $\sin^2 \theta + \cos^2 \theta = (y^2 + x^2)/r^2$. However, $x^2 + y^2 = r^2$, which means that:

(4-12) $\sin^2 \theta + \cos^2 \theta = 1$

This equation holds for any angle θ, since we made no assumptions as to the properties of θ. By substituting Equation (4-12) into Equation (4-11), we arrive at the **Law of Cosines**:

(4-13)

$$a^2 = b^2 + c^2 - 2bc \cos A$$
$$b^2 = a^2 + c^2 - 2ac \cos B$$
$$c^2 = a^2 + b^2 - 2ab \cos C$$

4-4 The Law of Cosines

The second and third forms in Equations (4-13) are derived in the same manner as the first form. Therefore, if we know two sides and the included angle (Case 3) we may directly solve for the side opposite the given angle. Then, by using the Law of Sines, we may complete the solution. If we are given all three sides (Case 4), we may solve for the angle opposite one of these sides by use of the Law of Cosines. Again we use the Law of Sines to complete the solution.

Example A
Solve the triangle with $a = 45.0$, $b = 67.0$, and $C = 35°0'$.

Using the Law of Cosines, we have

$$c^2 = (45.0)^2 + (67.0)^2 - 2(45.0)(67.0)(0.8192)$$
$$= 2025 + 4489 - 4940 = 1574$$
$$c = 39.7.$$

From the Law of Sines, we now have

$$\frac{45.0}{\sin A} = \frac{67.0}{\sin B} = \frac{39.7}{0.5736}$$

which leads to

$$\sin A = 0.6502 \quad \text{or} \quad A = 40°30'$$

and

$$B = 104°30'.$$

Example B
If, in Example A, $C = 145°0'$, we have

$$c^2 = (45.0)^2 + (67.0)^2$$
$$\quad - 2(45.0)(67.0)(-0.8192)$$
$$= 2025 + 4489 + 4940 = 11{,}450$$
$$c = 107.$$

The Law of Sines then gives $A = 14°0'$ and $B = 21°0'$.

Example C
Solve the triangle for which $a = 49.3$, $b = 21.6$, and $c = 42.6$.

$$\cos A = \frac{b^2 + c^2 - a^2}{2bc}$$

$$= \frac{(21.6)^2 + (42.6)^2 - (49.3)^2}{2(21.6)(42.6)}$$

$$= \frac{466.6 + 1815 - 2430}{1840} = -0.0807$$

Since the value of $\cos A$ is negative, we know that A is between $90°0'$ and $180°0'$. Thus,

$$A = 180°0' - 85°20' = 94°40'.$$

We then find that $B = 25°50'$ and $C = 59°30'$ from the Law of Sines.

Example D
Find the resultant **R** of vector **A** of magnitude 42.3 and direction 25.5° and vector **B** of magnitude 21.4 and direction 96.1°. See Figure 4-25.

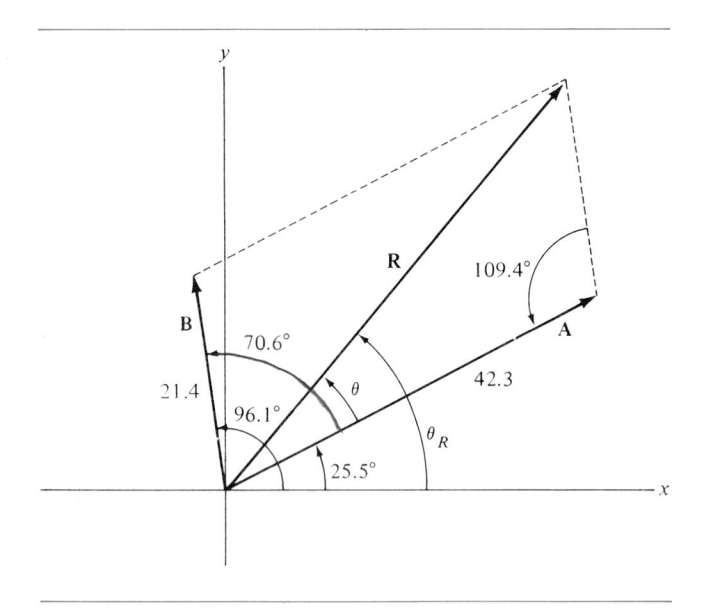

Figure 4-25

4-4 The Law of Cosines

From the figure we see that an angle of 109.4° is between the sides of 42.3 and 21.4, with **R** opposite the 109.4° angle. Thus, from the Law of Cosines, the magnitude of **R** is found as follows:

$$R^2 = (42.3)^2 + (21.4)^2$$
$$\quad - 2(42.3)(21.4)(\cos 109.4°)$$
$$= 1789 + 458.0 - 1810(-0.3322)$$
$$= 1789 + 458.0 + 601.3 = 2848$$
$$R = 53.4.$$

We now use the Law of Sines to find the angle between **R** and **A**.

$$\frac{21.4}{\sin \theta} = \frac{53.4}{\sin 109.4°} \quad \text{or}$$

$$\sin \theta = \frac{(21.4)(0.9432)}{53.4} = 0.3780$$

Thus, $\theta = 22.2°$. This means that

$$\theta_R = 25.5° + 22.2° = 47.7°.$$

The resultant has a magnitude of 53.4 and a direction of 47.7°.

Example E

Find the resultant of two vectors having magnitudes of 78.0 and 45.0 and directed toward the east and 15°0' east of north, respectively. See Figure 4-26. The magnitude of the resultant is given by

$$R = \sqrt{(78.0)^2 + (45.0)^2 - 2(78.0)(45.0)(\cos 105°0')}$$
$$= \sqrt{6084 + 2025 + 1817} = 99.6.$$

Also, $\theta = 25°50'$ is found from the Law of Sines.

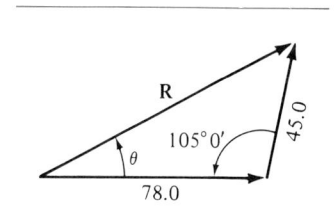

Figure 4-26

Example F

A vertical radio antenna is to be built on a hill which makes an angle of 6°0′ with the horizontal. If guy wires are to be attached at a point 150 ft up on the antenna and at points 100 ft from the base of the antenna, what will be the lengths of guy wires positioned directly up and directly down the hill?

Making an appropriate figure such as Figure 4-27, we are able to establish the equations necessary for the solution:

$$L_u^2 = (100)^2 + (150)^2 - 2(100)(150) \cos 84°0'$$

$$L_d^2 = (100)^2 + (150)^2 - 2(100)(150) \cos 96°0'$$

$$L_u^2 = 10,000 + 22,500 - 30,000(0.1045)$$

$$= 32,500 - 3135 = 29,370$$

$$L_u = 171 \text{ ft}$$

$$L_d^2 = 32,500 + 3135 = 35,640$$

$$L_d = 189 \text{ ft.}$$

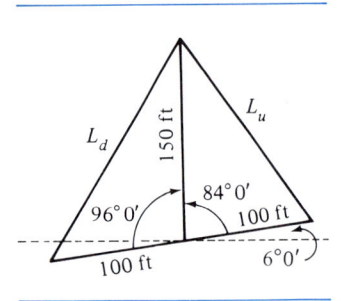

Figure 4-27

Exercises 4-4

In Exercises 1 through 16 solve the triangles with the given parts.

1. $a = 6.00$, $b = 7.56$, $C = 54°0'$
2. $b = 87.3$, $c = 34.0$, $A = 130°0'$
3. $a = 4530$, $b = 924$, $C = 98°0'$
4. $a = 0.0845$, $c = 0.116$, $B = 85°0'$
5. $a = 39.5$, $b = 45.2$, $c = 67.1$
6. $a = 23.3$, $b = 27.2$, $c = 29.1$
7. $a = 385$, $b = 467$, $c = 800$
8. $a = 0.243$, $b = 0.263$, $c = 0.153$
9. $a = 320$, $b = 847$, $C = 158.0°$

4-4 The Law of Cosines

10. $b = 18.3$, $c = 27.1$, $A = 58.7°$
11. $a = 21.4$, $c = 4.28$, $B = 86.3°$
12. $a = 11.3$, $b = 5.10$, $C = 77.6°$
13. $a = 103$, $c = 159$, $C = 104°30'$
14. $a = 49.3$, $b = 54.5$, $B = 114°0'$
15. $a = 0.493$, $b = 0.595$, $c = 0.639$
16. $a = 69.7$, $b = 49.3$, $c = 56.2$

In Exercises 17 through 20 use the Law of Cosines and Law of Sines to find the resultant of the given vectors.

17. $A = 93.8$, $\theta_A = 21°20'$
    $B = 17.4$, $\theta_B = 145°50'$
18. $A = 84.7$, $\theta_A = 159°20'$
    $B = 62.3$, $\theta_B = 32°50'$
19. $A = 243$, $\theta_A = 27°10'$
    $B = 521$, $\theta_B = 84°30'$
20. $A = 43.2$, $\theta_A = 62°40'$
    $B = 21.4$, $\theta_B = 101°30'$

In Exercises 21 through 28 use the Law of Cosines to solve the given problems.

21. To measure the distance AC, a man walks 500 ft from A to B, then turns $30°0'$ to face C, and walks 680 ft to C. What is the distance AC?

22. An airplane traveling at 700 km/hr leaves the airport at noon going due east. At 2 P.M. the pilot turns $10.0°$ north of east. How far is the plane from the airport at 3 P.M.?

23. Two forces, one of 56.0 kg and the other of 67.0 kg, are applied to the same object. The resultant force is 82.0 kg. What is the angle between the two forces?

24. A triangular metal frame has sides of 8.00 ft, 12.0 ft, and 16.0 ft. What is the largest angle between parts of the frame?

25. A boat which can travel 6.00 mi/hr in still water heads downstream at an angle of $20°0'$ with the bank. If the stream is flowing at the rate of 3.00 mi/hr, how fast is the boat traveling and in what direction?

132 Vectors and Oblique Triangles

26. An airplane's velocity with respect to the air is 520 mi/hr, and the plane is headed 24°0' north of west. The wind is from the due southwest and has a velocity of 55.0 mi/hr. What is the true direction of the plane? What is its velocity with respect to the ground?

27. One end of a 13.1-m pole is 15.7 m from an observer's eyes and the other end is 19.3 m from her eyes. Through what angle does the observer see the pole?

28. A triangular piece of land is bounded by 234 ft of stone wall, 205 ft of road frontage, and 147 ft of fencing. What angle does the fence make with the road?

4-5 Area of a Triangle

There are times when we wish to find the area of a triangle and do not know the height of the triangle. In this case it is possible to find the area in terms of its sides and angles. In this section we develop the necessary formulas.

To find the area of triangle ABC, we first draw in the height h from B to side b or its extension, as shown in Figure 4-28. We see from Figure 4-28(a) that

$$h = a \sin C.$$

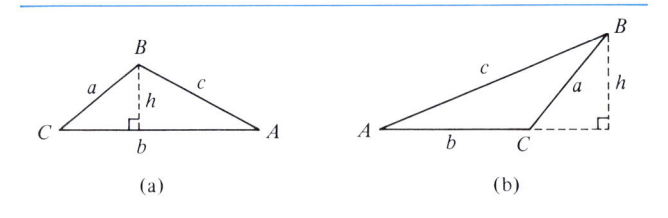

Figure 4-28

Also, from Figure 4-28(b) we see that

$$h = a \sin (180°0' - C) = a \sin C.$$

4-5 Area of a Triangle

By substituting the expression $h = a \sin C$ into the formula for the area K of triangle ABC, we have

$$K = \tfrac{1}{2} bh = \tfrac{1}{2} b(a \sin C)$$

or

(4-14)

$$K = \tfrac{1}{2} ab \sin C$$
$$K = \tfrac{1}{2} bc \sin A$$
$$K = \tfrac{1}{2} ac \sin B$$

The second and third forms in Equations 4-14 are derived in the same manner as the first form.

Example A
Find the area of the triangle for which $a = 16.0$, $b = 19.0$, and $C = 27°0'$.
    In this case we have

$$K = \tfrac{1}{2}(16.0)(19.0)(\sin 27°0')$$
$$= 152(0.4540) = 69.0.$$

Thus, the area is 69.0 square units.

Example B
Find the area of the triangle for which $b = 5.00$, $c = 17.0$, and $A = 130°0'$.
    Since $\sin 130°0' = \sin(180°0' - 130°0') = \sin 50°0'$, we have

$$K = \tfrac{1}{2}(5.00)(17.0) \sin 50°0' = (42.5)(0.7660)$$
$$= 32.6.$$

Therefore, the area of the triangle is 32.6 square units.

In the cases we have considered to this point, two sides and the included angle are known. For other types of known information, it may be necessary

134 Vectors and Oblique Triangles

to use the Law of Sines or the Law of Cosines before using Equations (4-14). Such cases are illustrated in the following examples.

Example C
Find the area of the triangle for which $a = 32.0$, $B = 68.0°$, and $C = 53.0°$. See Figure 4-29.

To find the area of this triangle we must find a second side by use of the Law of Sines. First, we see that

$$A = 180.0° - (68.0° + 53.0°) = 59.0°.$$

From the Law of Sines we have

$$\frac{c}{\sin 53.0°} = \frac{32.0}{\sin 59.0°} \quad \text{or}$$

$$c = \frac{32.0(0.7986)}{0.8572} = 29.8.$$

Now, from the third form of Equations (4-14), we have

$$K = \tfrac{1}{2}(32.0)(29.8)(0.9272) = 442.$$

The area of the triangle is 442 square units.

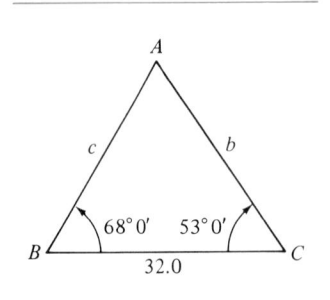

Figure 4-29

Example D
Find the area of the triangle for which $a = 278$, $b = 341$, and $c = 143$. See Figure 4-30.

First, we find B by use of the Law of Cosines:

$$\cos B = \frac{a^2 + c^2 - b^2}{2ac} = \frac{278^2 + 143^2 - 341^2}{2(278)(143)}$$

$$= \frac{77{,}280 + 20{,}450 - 116{,}300}{79{,}510}$$

$$= -0.2336$$

Thus, $B = 180°0' - 76°30' = 103°30'$. Now

$$K = \tfrac{1}{2}(143)(278)(0.9724) = 19{,}300.$$

The area of the triangle is 19,300 square units.

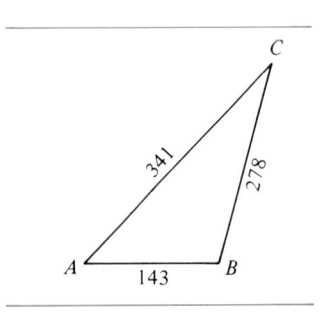

Figure 4-30

4-5 Area of a Triangle

Example E
Find the area of the triangle whose longest and shortest sides are 62.4 cm and 31.5 cm, respectively, and whose smallest angle is 18°50′.

We note that we are given two sides and the angle opposite one of them (the smallest angle is opposite the shortest side) so that we may find there are two triangles which satisfy the given information. See Figure 4-31.

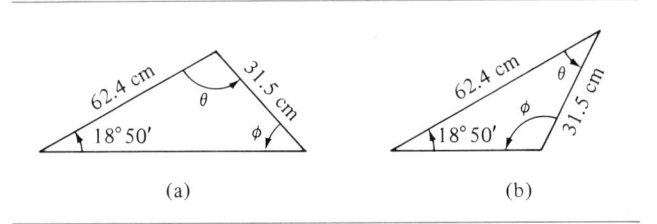

Figure 4-31

Before we find the area of the triangle, we must find the angle included between the known sides. By using the Law of Sines we can find the angle opposite the 62.4 cm side:

$$\sin \phi = \frac{62.4(\sin 18°50′)}{31.5} = 0.6395.$$

If $\phi = 39°50′$, then $\theta = 180°0′ - (39°50′ + 18°50′) = 121°20′$. This cannot be the case since the longest side is 62.4 cm, which must be opposite the largest angle (Figure 4-31a). Therefore, $\phi = 180°0′ - 39°50′ = 140°10′$ and $\theta = 180°0′ - (140°10′ + 18°50′) = 21°0′$ (Figure 4-31b).

Now that we have the two sides and the included angle, we may find the area of the triangle:

$$K = \tfrac{1}{2}(62.4)(31.5)(\sin 21°0′)$$
$$= 352 \text{ cm}^2.$$

Exercises 4-5

In Exercises 1 through 24 find the areas of the triangles with the given parts.

1. $a = 5.00$, $b = 7.00$, $C = 31°0'$
2. $a = 8.00$, $b = 11.00$, $C = 27°0'$
3. $b = 12.0$, $c = 9.00$, $A = 57°0'$
4. $b = 21.0$, $c = 15.0$, $A = 72°0'$
5. $a = 18.4$, $c = 21.3$, $B = 32°10'$
6. $a = 34.1$, $c = 27.2$, $B = 15°50'$
7. $b = 8.20$, $c = 4.30$, $A = 51°20'$
8. $a = 3.79$, $b = 5.12$, $C = 61°40'$
9. $a = 137$, $b = 242$, $C = 111.0°$
10. $b = 801$, $c = 207$, $A = 128.0°$
11. $a = 432$, $c = 278$, $B = 152.3°$
12. $a = 721$, $c = 634$, $B = 141.5°$
13. $b = 12.0$, $A = 27°0'$, $C = 82°0'$
14. $a = 28.0$, $B = 44°0'$, $C = 51°0'$
15. $c = 420$, $A = 31°0'$, $B = 98°30'$
16. $b = 385$, $A = 121°40'$, $C = 42°0'$
17. $c = 81.0$, $B = 62°0'$, $C = 73°20'$
18. $b = 14.0$, $A = 81°30'$, $B = 58°0'$
19. $a = 324$, $A = 105.2°$, $C = 19.8°$
20. $c = 500$, $B = 34.7°$, $C = 127.1°$
21. $a = 17.0$, $b = 20.0$, $c = 15.0$
22. $a = 150$, $b = 190$, $c = 220$
23. $a = 680$, $b = 430$, $c = 920$
24. $a = 82.0$, $b = 130$, $c = 71.0$

In Exercises 25 through 32 solve the given problems.

25. Two sides of a triangular piece of cardboard are 21.0 in. and 32.0 in. The angle between these sides is $52°0'$. Find the area of the cardboard.

26. Two sides of a triangular lot are 28.2 m and 17.8 m, and the angle between these sides is $25.0°$. Find the area of the lot.

Exercises for Chapter 4

27. Two angles of a triangular shelf are 102.0° and 27.0°, and the length of the side between these angles is 270 cm. Find the area of the shelf.
28. A triangular piece of carpeting has sides of 9.25 ft, 12.5 ft, and 14.0 ft. Find the area of the carpeting.
29. A triangle has angles of 22°0' and 58°0'. If the shortest side is 28.0 yd, find the area of the triangle.
30. Two angles of a triangle are 138°0' and 19°0'. What is the area of the triangle if the longest side is 173 mm?
31. The smallest angle of a triangle is 28°0'. If the shortest side of this triangle is 642 cm and the longest side is 1030 cm, what is the area of the triangle?
32. The longest and shortest sides of a triangle are 51.4 ft and 24.1 ft, respectively. If the smallest angle is 21.1°, find the area of the triangle.

Exercises for Chapter 4

In Exercises 1 through 4 find the x- and y-components of the given vectors by use of the trigonometric functions.

1. $A = 65.0$, $\theta_A = 28°0'$
2. $A = 8.05$, $\theta_A = 149°0'$
3. $A = 0.920$, $\theta_A = 215.0°$
4. $A = 657$, $\theta_A = 343.0°$

Vectors and Oblique Triangles

In Exercises 5 through 12 add the given vectors by use of the trigonometric functions and other equations shown in this chapter.

5. $A = 780$, $\theta_A = 28°0'$
   $B = 346$, $\theta_B = 320°0'$
6. $A = 0.0120$, $\theta_A = 10°30'$
   $B = 0.0078$, $\theta_B = 260°0'$
7. $A = 22.5$, $\theta_A = 130°10'$
   $B = 7.60$, $\theta_B = 200°0'$
8. $A = 18,700$ $\theta_A = 110°40'$
   $B = 4830$, $\theta_B = 350°20'$
9. $A = 52.1$, $\theta_A = 80.5°$
   $B = 27.0$, $\theta_B = 160.0°$
10. $A = 28.5$, $\theta_A = 192.5°$
    $B = 13.0$, $\theta_B = 289.0°$
11. $A = 51.3$, $\theta_A = 12.2°$
    $B = 42.6$, $\theta_B = 291.7°$
12. $A = 70.3$, $\theta_A = 122.5°$
    $B = 30.2$, $\theta_B = 214.8°$

In Exercises 13 through 28 solve the triangles with the given parts. Then find the area of each.

13. $A = 48°0'$, $B = 68°0'$, $a = 14.5$
14. $A = 132°0'$, $b = 7.50$, $C = 32°0'$
15. $a = 22.8$, $B = 33°30'$, $C = 125°20'$
16. $A = 71°0'$, $B = 48°30'$, $c = 8.42$
17. $b = 76.0$, $c = 40.5$, $B = 110°0'$
18. $A = 77°0'$, $a = 12.0$, $c = 5.00$
19. $b = 14.5$, $c = 13.0$, $C = 56°40'$
20. $B = 40°0'$, $b = 7.00$, $c = 18.0$
21. $a = 186$, $B = 130.0°$, $c = 106$
22. $b = 750$, $c = 1100$, $A = 56.0°$
23. $a = 7.86$, $b = 2.45$, $C = 22.0°$
24. $a = 0.208$, $c = 0.697$, $B = 105.0°$
25. $a = 17.0$, $b = 12.0$, $c = 25.0$
26. $a = 900$, $b = 995$, $c = 1100$
27. $a = 5.30$, $b = 8.75$, $c = 12.5$
28. $a = 47.4$, $b = 40.0$, $c = 45.5$

Exercises for Chapter 4

In Exercises 29 through 40 solve the given problems.

29. A bullet is fired into the air at 2000 mi/hr at an angle of 25°0' with the horizontal. What is the vertical component of the velocity of the bullet?

30. A jet climbs at an angle of 35°0' while traveling 600 km/hr. How long will it take to climb to an altitude of 10,000 m? (*Note:* 1 km = 1000 m.)

31. A balloon is rising at the rate of 15.0 ft/sec and at the same time is being blown horizontally by the wind at the rate of 22.5 ft/sec. Find the resultant velocity.

32. A motorboat which travels at 8.00 mi/hr in still water heads directly across a stream which flows at 3.00 mi/hr. What is the resultant velocity of the boat?

33. To find the distance between points A and B on opposite sides of a river, a distance AC is measured off as 1000 m, where point C is on the same side of the river as A. Angle BAC is measured to be 102.0° and angle ACB is 33.0°. What is the distance between A and B?

34. A 22.0-ft ladder leans against a wall, making an angle of 29°0' with the wall. If the foot of the ladder is 10.7 ft from the foot of the wall, find the angle of inclination of the wall to the ground.

35. Two points on opposite sides of an obstruction are respectively 117 ft and 88.0 ft from a third point. The lines joining the first two points and the third point intersect at an angle of 115.0° at the third point. How far apart are the original two points?

36. Determine the angles of the structure indicated in Figure 4-32.

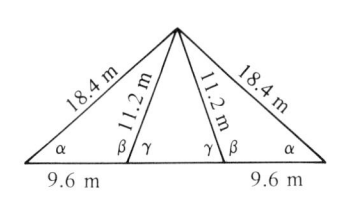

Figure 4-32

37. A piece of property is triangular in shape. It extends 920 ft along one road and 650 ft along a second road. If these roads meet at an angle of 72.0°, what is the area of the property?
38. A triangular patio has sides of 40.0 ft, 32.0 ft, and 28.0 ft. What is the area of the patio?
39. An engineer on a hill in the middle of a plain notes that the angles of depression of two objects on the plain below (on directly opposite sides of the hill) are 30°0' and 40°0'. He knows that the objects are 15,800 ft apart. How far is he from the closest object?
40. A velocity vector is the resultant of two other vectors. If the given velocity is 450 mi/hr and makes angles of 34°0' and 76°0' with the two components, find the magnitudes of the components.
41. A communications satellite is directly above the extension of a line between receiving towers A and B. It is determined from radio signals that the angle of elevation of the satellite from tower A is 89.2°, and the angle of elevation from tower B is 86.5°. If A and B are 1290 km apart, how far is the satellite from A? (Neglect the curvature of the earth.)
42. Two cars are at the intersection of two straight roads. One car travels 5.20 mi on one road, and the other travels 3.75 mi on the other road. The drivers contact each other on CB radio and find they are at points known to be 4.50 mi apart. What angle do the roads make at the intersection?

Exercises for Chapter 4

43. Atlanta is 290 mi and 51.0° south of east from Nashville. The pilot of an airplane due north of Atlanta radios Nashville and finds her plane is on a line 10.5° south of east from Nashville. How far is the plane from Nashville?

44. Boston is 650 km and 21.0° south of west of Halifax, Nova Scotia. Radio signals locate a ship 10.5° east of south of Halifax, and 5.6° north of east of Boston. How far is the ship from each city?

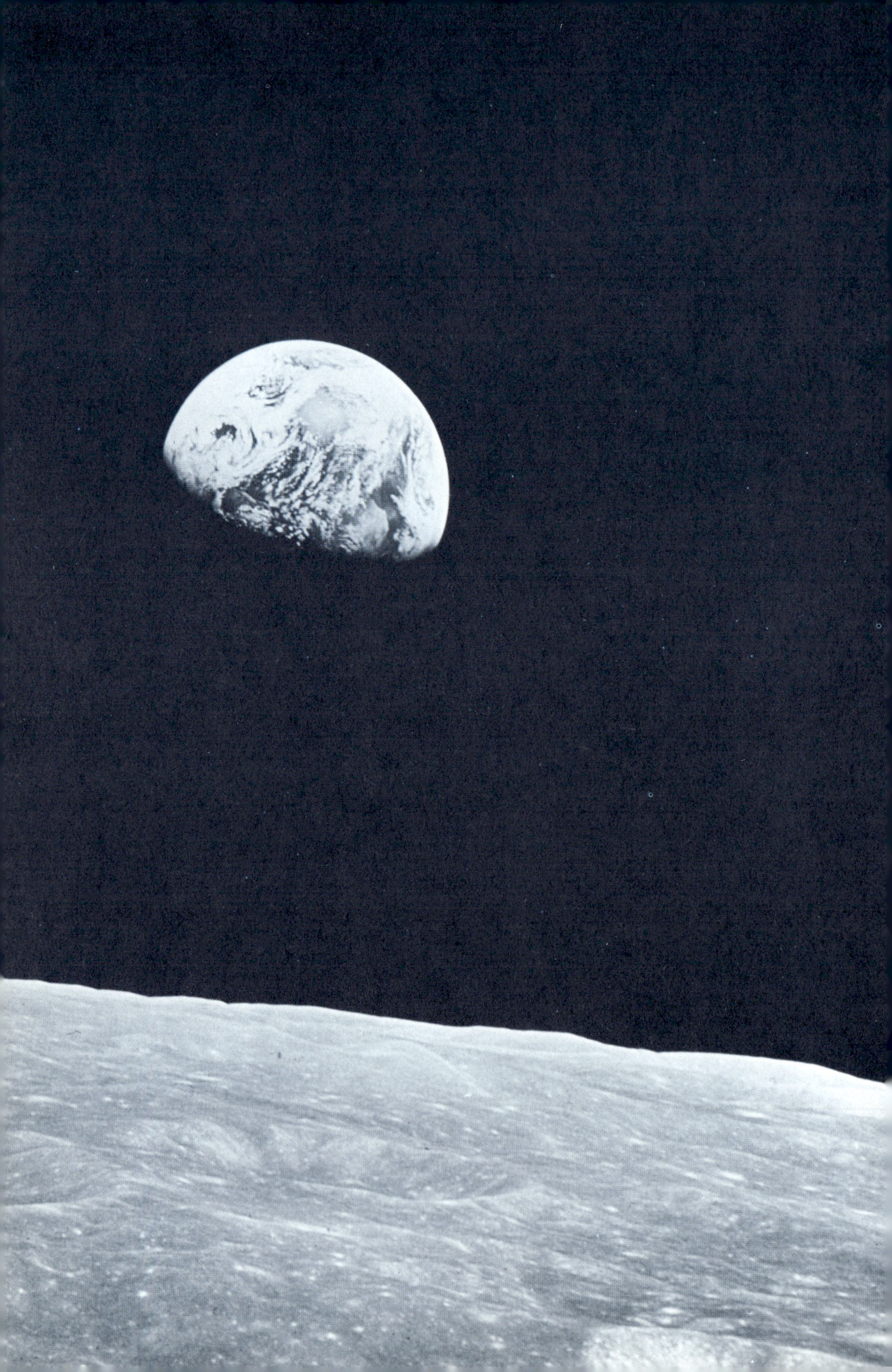

Logarithms

5

Logarithms were first developed to help perform complicated calculations in navigation and astronomy. Today's space travel and navigation, as was required to take this photograph of the earth from the moon, involves innumerable such calculations performed by various methods developed over the years.

5-1 Definition of a Logarithm

We have found that there is a great deal of computational work in solving problems in trigonometry. In most of these problems we have rounded off results to three significant digits. These computations may be completed in a number of ways, including electronic calculators and the slide rule. To be able to perform these calculations without a calculator, we shall develop a basic method which gives an accuracy of four significant digits.

The method developed in this chapter employs *logarithms.* Historically, logarithms were developed for purposes of computation, but they also are used extensively in mathematics for theoretical purposes. With the increased use of electronic calculators, logarithms are used less for computation, but their usefulness in mathematics remains of great importance.

Logarithms and their properties are based on exponents. Those who do not recall the uses and laws of exponents should review Appendix A before proceeding.

In developing algebra we deal with exponents in expressions of the form x^n, where n may be any rational number. Here we shall deal with expressions of the form b^x, where x is any real number. When we look at these expressions, we note the primary difference is that in the second expression the exponent is variable. We have not previously dealt with variable exponents. Thus let us define the *exponential function* to be:

(5-1) $$y = b^x$$

In Equation (5-1), x is called the logarithm of the number y to the base b. In our work with logarithms we shall restrict all numbers to the real number system. This leads us to choose the base as a positive number other than 1. We know that 1 raised to any power will result in 1, which would make y a constant regardless of the value of x. Negative numbers for b would result in imaginary values for y if x were any fractional exponent with an even integer for its denominator.

Example A
The equation $y = 2^x$ is an exponential function, where x is the logarithm of y to the base 2. This means that 2 raised to a given power gives us the corresponding value of y.

If $x = 2$, $y = 2^2 = 4$; this means that 2 is the logarithm of 4 to the base 2.

If $x = 4$, $y = 2^4 = 16$; this means that 4 is the logarithm of 16 to the base 2.

If $x = \frac{1}{2}$, $y = 2^{1/2} = 1.41$; this means that $\frac{1}{2}$ is the logarithm of 1.41 to the base 2.

Using the definition of a logarithm, Equation (5-1) may be solved for x. It is written in the form:

(5-2) $$x = \log_b y$$

This equation is read in accordance with the definition of x in Equation (5-1): **x equals the logarithm of y to the base b.** This means that x is the power to which the base b must be raised in order to equal the number y; that is, x is a logarithm, and a logarithm is an exponent. Note that Equations (5-1) and (5-2) state the same relationship but in a different manner. **Equation (5-1) is the exponential form and Equation (5-2) is the logarithmic form.**

Example B
The equation $y = 2^x$ would be written as $x = \log_2 y$ if we put it in logarithmic form. When we choose values of y to find the corresponding values of x from this equation, we ask ourselves: "2 raised to what power gives y?" Hence if $y = 4$, we know that 2^2 is 4 and x would be 2. If $y = 8$, $2^3 = 8$, or $x = 3$.

Example C

Note that $3^2 = 9$ in logarithmic form is $2 = \log_3 9$; $4^{-1} = \frac{1}{4}$ in logarithmic form is $-1 = \log_4 (\frac{1}{4})$. Remember, the exponent may be negative. The base must be positive.

Example D

$(64)^{1/3} = 4$ in logarithmic form is $\frac{1}{3} = \log_{64} 4$

$(32)^{3/5} = 8$ in logarithmic form is $\frac{3}{5} = \log_{32} 8$

Example E

$\log_2 32 = 5$ in exponential form is $32 = 2^5$

$\log_6 (\frac{1}{36}) = -2$ in exponential form is $\frac{1}{36} = 6^{-2}$

Example F

Find b, given that $-4 = \log_b (\frac{1}{81})$.
Writing this in exponential form, we have $\frac{1}{81} = b^{-4}$. Thus $b = 3$, since $3^4 = 81$.

Example G

Find y, given that $\log_4 y = \frac{1}{2}$.
In exponential form we have $y = 4^{1/2}$ or $y = 2$.

We see that exponential form is very useful for determining values written in logarithmic form. For this reason it is important that you learn to transform readily from one form to the other.

Exercises 5-1

In Exercises 1 through 12 express the given equations in logarithmic form.

1. $3^3 = 27$
2. $5^2 = 25$
3. $4^4 = 256$
4. $8^2 = 64$
5. $4^{-2} = \frac{1}{16}$
6. $3^{-2} = \frac{1}{9}$
7. $2^{-6} = \frac{1}{64}$
8. $(12)^0 = 1$
9. $8^{1/3} = 2$
10. $(81)^{3/4} = 27$
11. $(\frac{1}{4})^2 = \frac{1}{16}$
12. $(\frac{1}{2})^{-2} = 4$

In Exercises 13 through 24 express the given equa- in exponential form.

13. $\log_3 81 = 4$
14. $\log_{11} 121 = 2$
15. $\log_9 9 = 1$
16. $\log_{15} 1 = 0$
17. $\log_{25} 5 = \frac{1}{2}$
18. $\log_8 16 = \frac{4}{3}$
19. $\log_{243} 3 = \frac{1}{5}$
20. $\log_{1/32}(\frac{1}{8}) = \frac{3}{5}$
21. $\log_{10}(0.01) = -2$
22. $\log_7(\frac{1}{49}) = -2$
23. $\log_{0.5} 16 = -4$
24. $\log_{1/3} 3 = -1$

In Exercises 25 through 40 determine the value of the unknown.

25. $\log_4 16 = x$
26. $\log_5 125 = x$
27. $\log_{10} 0.01 = x$
28. $\log_{16}(\frac{1}{4}) = x$
29. $\log_7 y = 3$
30. $\log_8 N = 3$
31. $\log_8 y = -\frac{2}{3}$
32. $\log_7 y = -2$
33. $\log_b 81 = 2$
34. $\log_b 625 = 4$
35. $\log_b 4 = -\frac{1}{3}$
36. $\log_b 4 = \frac{2}{3}$
37. $\log_{10} 10^{0.2} = x$
38. $\log_5 5^{1.3} = x$
39. $\log_3 27^{-1} = x$
40. $\log_b(\frac{1}{4}) = -\frac{1}{2}$

5-2 Properties of Logarithms

When we are working with functions, we must keep in mind that a function is defined by the operation being performed on the independent variable, and not by the letter chosen to represent it. For consistency, however, it is normal practice to let y represent the dependent variable and x represent the independent variable. Therefore, the logarithmic function is:

(5-3) $$y = \log_b x$$

Equations (5-2) and (5-3) express the same *function,* the logarithmic function. They do not represent different functions, due to the difference in location of the variables, since they represent the same operation on the independent variable. Equation (5-3) simply expresses the function with the usual choice of variables.

Graphical representation of functions is often valuable when we wish to demonstrate their properties. We shall now plot the graphs of the exponential function (Equation 5-1) and the logarithmic function (Equation 5-3).

Example A

Plot the graph of $y = 2^x$.

Assuming values for x and then finding the corresponding values for y, we obtain the table shown.

x	-3	-2	-1	0	1	2	3	4
y	$\frac{1}{8}$	$\frac{1}{4}$	$\frac{1}{2}$	1	2	4	8	16

From these values we plot the curve shown in Figure 5-1.

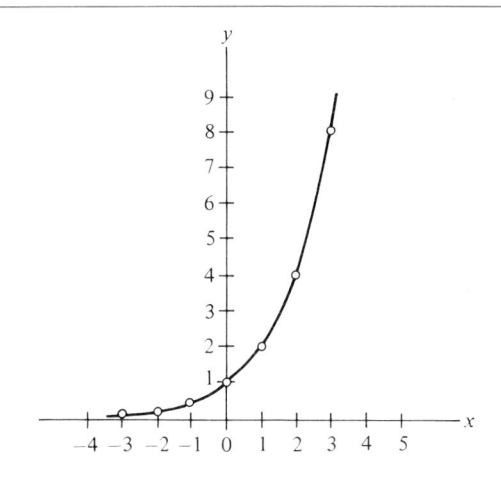

Figure 5-1

Example B

Plot the graph of $y = \log_3 x$.

We can find the points for this graph more easily if we first put the equation in exponential form: $x = 3^y$. By assuming values for y, we can find the corresponding values for x.

x	$\frac{1}{9}$	$\frac{1}{3}$	1	3	9	27
y	-2	-1	0	1	2	3

Using these values, we construct the graph seen in Figure 5-2.

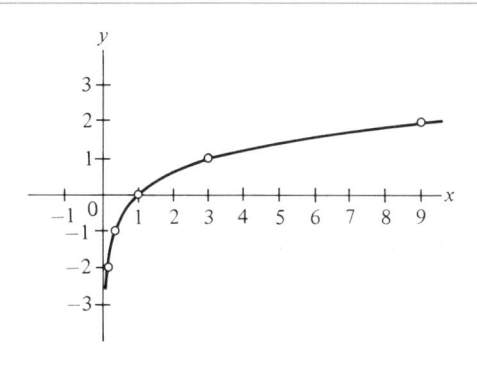

Figure 5-2

Any exponential or logarithmic curve, where $b > 1$, will be similar in shape to those of Examples A and B. From these curves we can draw certain conclusions:

1. If $0 < x < 1$, $\log_b x < 0$; if $x = 1$, $\log_b 1 = 0$; if $x > 1$, $\log_b x > 0$.
2. If $x > 1$, x increases more rapidly than $\log_b x$.
3. For all values of x, $b^x > 0$.
4. If $x > 1$, b^x increases more rapidly than x.

5-2 Properties of Logarithms

Since a logarithm is an exponent, it must follow the laws of exponents. The laws of greatest importance at this time are listed here for reference.

(5-4) $$b^u \cdot b^v = b^{u+v}$$

(5-5) $$\frac{b^u}{b^v} = b^{u-v}$$

(5-6) $$(b^u)^n = b^{nu}$$

We shall now show how these laws for exponents give certain useful properties to logarithms.

If we let $u = \log_b x$ and $v = \log_b y$ and write these equations in exponential form, we have $x = b^u$ and $y = b^v$. Therefore, forming the product of x and y, we obtain

$$xy = b^u b^v = b^{u+v} \quad \text{or} \quad xy = b^{u+v}.$$

Writing this last equation in logarithmic form yields

$$u + v = \log_b xy$$

or

(5-7) $$\log_b x + \log_b y = \log_b xy$$

Equation (5-7) states the property that **the logarithm of the product of two numbers is equal to the sum of the logarithms of the numbers.**

Using the same definitions of u and v to form the quotient of x and y, we then have

$$\frac{x}{y} = \frac{b^u}{b^v} = b^{u-v} \quad \text{or} \quad \frac{x}{y} = b^{u-v}.$$

Writing this last equation in logarithmic form, we have

$$u - v = \log_b\left(\frac{x}{y}\right)$$

or

$$(5\text{-}8) \qquad \log_b x - \log_b y = \log_b\left(\frac{x}{y}\right)$$

Equation (5-8) states the property that *the logarithm of the quotient of two numbers is equal to the logarithm of the numerator minus the logarithm of the denominator.*

If we again let $u = \log_b x$ and write this in exponential form, we have $x = b^u$. To find the nth power of x, we write

$$x^n = (b^u)^n = b^{nu}.$$

Expressing this equation in logarithmic form yields

$$nu = \log_b (x^n)$$

or

$$(5\text{-}9) \qquad n \log_b x = \log_b (x^n)$$

This last equation states that *the logarithm of the nth power of a number is equal to n times the logarithm of the number.* The exponent n may be integral or fractional and, therefore, we may use Equation (5-9) for finding powers and roots of numbers.

Example C

$$\log_4 15 = \log_4(3 \cdot 5) = \log_4 3 + \log_4 5$$
$$\log_4 \left(\tfrac{5}{3}\right) = \log_4 5 - \log_4 3$$
$$\log_4 9 = \log_4 (3^2) = 2 \log_4 3$$

Example D

$$\log_2 6 = \log_2(2 \cdot 3) = \log_2 2 + \log_2 3$$

This may be simplified further if we use the definition of a logarithm. Therefore, we have

$$\log_b b = 1$$

which is a very important property of logarithms. When written in exponential form, it is nothing more than a statement of $b = b^1$. Thus, $\log_2 2 = 1$, and we have

$$\log_2 6 = 1 + \log_2 3.$$

Example E

$$\log_3 \left(\tfrac{2}{9}\right) = \log_3 2 - \log_3 9 = \log_3 2 - \log_3(3^2)$$
$$= \log_3 2 - 2 \log_3 3 = \log_3 2 - 2(1)$$
$$= \log_3 2 - 2 = -2 + \log_3 2$$

Example F

$$\log_2 16 = \log_2(2^4) = 4 \log_2 2 = 4(1) = 4$$

This expression could have been evaluated directly from the definition of a logarithm. However, by use of Equation (5-9) we see that the same result is obtained. Therefore, any proper method can be followed to obtain the result.

Example G

$$\log_{10} \sqrt{7} = \log_{10}(7^{1/2}) = \tfrac{1}{2} \log_{10} 7$$

This demonstrates the property which is especially useful for finding roots of numbers. We see that we need merely multiply the logarithm of the number by the fractional exponent representing the root to obtain the logarithm of the root. Similarly,

$$\log_{10} \sqrt[5]{6} = \log_{10}(6^{1/5}) = \tfrac{1}{5} \log_{10} 6.$$

Example H
Use the basic properties of logarithms to solve for y in terms of x:

$$\log_b y = 2 \log_b x + \log_b a.$$

Using Equation (5-9) and then Equation (5-7), we have

$$\log_b y = \log_b (x^2) + \log_b a = \log_b (ax^2).$$

Now, since we have the logarithm to the base b of different expressions on each side of the resulting equation, the expressions must be equal. Therefore, $y = ax^2$.

Exercises 5-2

In Exercises 1 through 4 plot graphs of the given functions for $-3 \leq x \leq 3$.

1. $y = 3^x$
2. $y = 4^x$
3. $y = (\frac{1}{2})^x$
4. $y = (\frac{1}{3})^x$

In Exercises 5 through 8 plot graphs of the given functions for $-3 \leq y \leq 3$.

5. $y = \log_2 x$
6. $y = \log_3 x$
7. $y = \log_{1/2} x$
8. $y = \log_{1/3} x$

In Exercises 9 through 24 express each as a sum, difference, or multiple of logarithms. Wherever possible, evaluate logarithms of the result.

9. $\log_5 xy$
10. $\log_6 abc$
11. $\log_3 \left(\frac{r}{s}\right)$
12. $\log_2 \frac{xy}{z}$
13. $\log_2 (a^3)$
14. $\log_8 (n^5)$
15. $\log_3 18$
16. $\log_5 75$
17. $\log_2 (\frac{1}{6})$
18. $\log_{10} (0.05)$
19. $\log_3 \sqrt{6}$
20. $\log_2 \sqrt[3]{24}$
21. $\log_2 (4^2 \cdot 3^3)$
22. $\log_5 (\frac{4}{125})$
23. $\log_{10} 3000$
24. $\log_{10} (40)^2$

In Exercises 25 through 32 express each as the logarithm of a single quantity.

25. $\log_b a + \log_b c$
26. $\log_2 3 + \log_2 x$
27. $\log_5 9 - \log_5 3$
28. $\log_8 6 - \log_8 a$
29. $\log_b x^2 - \log_b \sqrt{x}$
30. $\log_4 3^3 + \log_4 9$
31. $2 \log_b 2 + 3 \log_b n$
32. $\frac{1}{2} \log_b a - 2 \log_b 5$

In Exercises 33 through 36 solve for y in terms of x.

33. $\log_b y = \log_b 2 + \log_b x$
34. $\log_b y = \log_b 6 + \log_b x - \log_b \sqrt{x}$
35. $\log_{10} y = 2 \log_{10} 7 - 3 \log_{10} x$
36. $4 \log_2 x - 2 \log_2 y = \log_2 9$

In Exercises 37 through 40, using $\log_{10} 2 = 0.301$, evaluate each of the given expressions.

37. $\log_{10} 4$
38. $\log_{10} 20$
39. $\log_{10} (0.5)$
40. $\log_{10} 8000$

5-3 Logarithms to the Base 10

Any number may be expressed in *scientific notation* as the product of a number between 1 and 10 and a power of 10. (An explanation of scientific notation is given in Appendix A.) Writing this as $N = P \times 10^k$ and taking logarithms of both sides, we have

$$\log_b N = \log_b (P \times 10^k) = \log_b P + \log_b 10^k$$
$$= \log_b P + k \log_b 10.$$

If we let $b = 10$, then $k \log_{10} 10 = k$, and this equation becomes:

(5-10) $\qquad \log_{10} N = k + \log_{10} P$

Equation (5-10) shows that if we have a method for finding logarithms to the base 10 of numbers between 1 and 10, then we can find the logarithm of *any* number to base 10. The value of k can be found by writing the number N in scientific notation, and P is a number between 1 and 10. Logarithms to the base 10 have been tabulated, and tables of these *common logarithms* may be found in Appendix D. Logarithms may be calculated for any base, but for purposes of computation, logarithms to the base 10 are the most convenient. From now on we shall not write the number 10 to indicate the base, and log N will be assumed to be to the base 10. Thus:

(5-11) $$\log N = k + \log P$$

In Equation (5-11), k is called the *characteristic* and log P is known as the *mantissa*. Remember, k is the power of 10 of the number when it is written in scientific notation and the term log P is the logarithm of the number between 1 and 10.

Example A
For $N = 3600 = 3.6 \times 10^3$, we see that the characteristic $k = 3$ and the mantissa log P = log 3.6. Therefore, log 3600 = 3 + log 3.6.

For $N = 80.9 = 8.09 \times 10^1$, we see that $k = 1$ and log P = log 8.09. Therefore, log 80.9 = 1 + log 8.09.

Example B
For $N = 0.00543 = 5.43 \times 10^{-3}$, we see that $k = -3$ and log P = log 5.43. Therefore, log 0.00543 = −3 + log 5.43.

For $N = 0.741 = 7.41 \times 10^{-1}$, we see that $k = -1$ and log P = log 7.41. Therefore, log 0.741 = −1 + log 7.41.

To find log P we use Table 2 in Appendix D. The following examples illustrate how to use this table.

Example C
Find log 572.

We first write the number in scientific notation as 5.72×10^2. The characteristic is 2, and we must now find log 5.72. We look in the column headed N and find 57 (the first two significant digits). Then, to the right of this, we look under the column headed 2 (the third significant digit) and we find 7574. All numbers between 1 and 10 will have common logarithms between 0 and 1 (log 1 = 0 and log 10 = 1). Therefore log 5.72 = 0.7574, and the logarithm of 572 = 2 + 0.7574. We then write this in the usual form of 2.7574.

Example D
Find log 0.00485.

When we write this number in scientific notation, we have 4.85×10^{-3}, and we see that $k = -3$. From the tables we find that log 4.85 = 0.6857. Thus log 0.00485 = −3 + 0.6857. We do *not* write this as −3.6857, for this would say that the mantissa was also negative, which it is not. To avoid confusion, we shall write it in the form 7.6857 − 10. We shall follow this policy whenever the characteristic is negative. That is, we shall write a negative characteristic as the appropriate positive number with 10 subtracted. For example, for a characteristic of −6, we write 4 before the mantissa and −10 after it.

Example E
Other examples of logarithms are as follows.

$89{,}000 = 8.9 \times 10^4$: $k = 4$, log 8.9 = 0.9494; log 89000 = 4.9494

$0.307 = 3.07 \times 10^{-1}$: $k = -1$, log 3.07 = 0.4871; log 0.307 = 9.4871 − 10

$0.00629 = 6.29 \times 10^{-3}$: $k = -3$, log 6.29 = 0.7987; log 0.00629 = 7.7987 − 10

158   Logarithms

Table 2 in Appendix D is a four-place table, which means that we can obtain accuracy to four significant digits. However, only three digits may be read directly, and the fourth place is found by interpolation. We discussed this in Section 2-3 in reference to finding values from trigonometric tables. The method for finding values from logarithmic tables is the same. It is illustrated in the following examples.

Example F
Find log 686300.
  In scientific notation, $686300 = 6.863 \times 10^5$. This means that the characteristic is 5. To find the mantissa from the table, we must interpolate, finding the value $\frac{3}{10}$ (since the fourth digit is 3) of the way between the log 6.86 and log 6.87. These two values are 0.8363 and 0.8370. The tabular difference is 7, and $(\frac{3}{10})(7) = 2$ (to one significant digit). Adding this to the mantissa 0.8363 gives 0.8365. Hence, log 686300 = 5.8365.

Example G
Find log 0.02178.
  In scientific notation, $0.02178 = 2.178 \times 10^{-2}$. Therefore, the characteristic is -2. To find the mantissa, we must interpolate, finding the value $\frac{8}{10}$ of the way between log 2.17 and log 2.18. These two values are 0.3365 and 0.3385. The tabular difference is 20, and $(\frac{8}{10})(20) = 16$. Adding this to 0.3365, we find the mantissa to be 0.3381. Therefore, log 0.02178 = 8.3381 − 10.

We may also use Table 2 to find N if we know log N. In this case we may refer to N as the *antilogarithm* of log N. The following examples illustrate the determination of antilogarithms.

Example H
Given $\log N = 1.5263$, find N.

Direct observation of the given logarithm tells us that the characteristic is 1 and that $N = P \times 10^1$. In Table 2 we find 5263 opposite 33 and under 6. Thus $P = 3.360$. This means that $N = 3.360 \times 10^1$, or in ordinary notation, $N = 33.60$.

Example I
Given $\log N = 8.2611 - 10$, find N.

Using the method described in Example D, we determine that the characteristic is $8 - 10 = -2$. We look for 0.2611 in the tables, and find that it is between 2601 (log 1.82) and 2625 (log 1.83). These two values have a tabular difference of 24, and the difference between 2611 and 2601 is 10. Thus, the number we want is $\frac{10}{24}$, or 0.4 of the way between 1.82 and 1.83. Hence,

$$N = 1.824 \times 10^{-2} = 0.01824.$$

The basic properties of logarithms allow us to find the logarithms of products, quotients, and roots. The following example illustrates the method.

Example J
Find $\log \sqrt{0.846}$.

From the properties of logarithms we write $\log \sqrt{0.846} = \frac{1}{2} \log 0.846$. From the tables we find that $\log 8.46 = 0.9274$. Therefore, $\log 0.846 = 9.9274 - 10$. To obtain the desired logarithm we must divide this by 2. This will result in a 5 to be subtracted. To ensure that our answer is in the usual form of a negative characteristic, we shall write $\log 0.846$ as $19.9274 - 20$. Thus, $\log \sqrt{0.846} = 9.9637 - 10$ when we divide through by 2. In this type of problem we choose that part of the characteristic which is to be subtracted so that 10 will result after division. This is done by *adding* the proper multiple of 10 to each part of the characteristic.

Logarithms

Exercises 5-3

In Exercises 1 through 16 find the common logarithm of each of the given numbers.

1. 567
2. 60.5
3. 0.0640
4. 0.000566
5. 9.24×10^6
6. 3.19×10^{15}
7. 1.17×10^{-4}
8. 8.04×10^{-8}
9. 1.053
10. 73.27
11. 0.2384
12. 0.004309
13. 7.331×10^8
14. 1.656×10^{-5}
15. $\sqrt{0.002006}$
16. $\sqrt[3]{38310000}$

In Exercises 17 through 32 find the antilogarithm N from the given logarithms.

17. 4.4378
18. 0.9294
19. $8.6955 - 10$
20. $3.0212 - 10$
21. 3.3010
22. 8.8241
23. $9.8597 - 10$
24. $7.4409 - 10$
25. 1.9495
26. 2.4367
27. $6.6090 - 10$
28. $9.3755 - 10$
29. 0.1543
30. 10.2750
31. $17.7625 - 20$
32. $35.6641 - 40$

In Exercises 33 through 36 find the logarithms of the given numbers.

33. A certain radar signal has a frequency of 1.15×10^9 cycles/sec.
34. The earth travels about 595,000,000 mi in one year.
35. A certain bank charges 7.5 percent interest on loans.
36. The radius of a red blood corpuscle is 0.00038 cm.

5-4 Computations Using Logarithms

As we mentioned in the introduction to this chapter, logarithms allow us to do calculations more easily than by ordinary arithmetic methods. Using logarithms we can reduce a difficult problem to a series of additions and subtractions. Also, using logarithms we can do the necessary calculations in a trigonometry problem, and can determine the answer to such problems as $\sqrt[5]{7.6}$ or $(89)^{0.3}$.

The tables in Appendix D can give us accuracy to four significant digits. If greater accuracy is required, electronic calculators or more extensive tables may be used. Such tables are available in many references, and are used in the same manner as the four-place tables. The following examples illustrate the use of logarithms in computations.

Example A
By the use of logarithms, calculate the value of $(5.670)(21.50)$.

Equation (5-7) tells us that $\log xy = \log x + \log y$. If we find $\log 5.670$ and $\log 21.50$ and add them, we have the logarithm of the product. Using this result, we look up the antilogarithm, which is the desired product.

$$\begin{array}{r} \log 5.670 = 0.7536 \\ \underline{\log 21.50 = 1.3324} \\ \log (5.670)(21.50) = 2.0860 \\ \log 121.9 = 2.0860 \end{array}$$

Thus $(5.670)(21.50) = 121.9$.

Example B
By the use of logarithms, calculate the value of
$$\frac{8.640}{45.55}.$$

From Equation (5-8), we know that $\log x/y = \log x - \log y$. Therefore, by subtracting log 45.55 from log 8.640, we have the logarithm of the quotient. The antilogarithm gives the desired result.

$$\begin{array}{rl} \log 8.640 = & 10.9365 - 10 \\ \log 45.55 = & \underline{1.6585} \\ \log(8.640/45.55) = & 9.2780 - 10 \end{array}$$

We wrote the characteristic of log 8.640 as 10 − 10 so that when we subtracted, the part of the result containing the mantissa would be positive although the characteristic was negative. The antilogarithm of 9.2780 − 10 is 0.1897. Therefore,

$$\frac{8.640}{45.55} = 0.1897.$$

Example C
By the use of logarithms, calculate the value of $\sqrt[5]{0.03760}$.

From Equation (5-9), we know that $\log x^n = n \log x$. Therefore, by writing $\sqrt[5]{0.03760} = (0.03760)^{1/5}$, we know that we can find the logarithm of the result by multiplying log 0.03760 by $\frac{1}{5}$. Now, log 0.03760 = 8.5752 − 10. Since we wish to multiply this by $\frac{1}{5}$, we write this logarithm as

$$\log 0.03760 = 48.5752 - 50$$

by adding and subtracting 40. Multiplying by $\frac{1}{5}$, we have

$$\frac{1}{5}\log 0.03760 = \frac{1}{5}(48.5752 - 50)$$
$$= 9.7150 - 10.$$

The antilogarithm of 9.7150 − 10 is 0.5188. Therefore, $\sqrt[5]{0.03760} = 0.5188$.

The following examples illustrate calculations which involve the use of a combination of the basic properties of logarithms.

5-4 Computations Using Logarithms

Example D
Calculate the value of

$$N = \frac{6.875 \sqrt{98.66}}{7.880}.$$

The calculations are as follows:

$\log N = \log 6.875 + \frac{1}{2} \log 98.66 - \log 7.880$

$\begin{aligned}\log 6.875 &= 0.8373 \\ \tfrac{1}{2}\log 98.66 &= \underline{0.9970} \\ &\;1.8343 \\ \log 7.880 &= \underline{0.8965} \\ \log N &= 0.9378 \end{aligned}$ $\qquad \log 98.66 = 1.9941$

$N = 8.666.$

Example E
Calculate the value of

$$N = \left[\frac{(0.05325)\sqrt{0.8884}}{\sqrt[3]{895.3}}\right]^{0.3}$$

The calculations are as follows:

$\log N = 0.3(\log 0.05325 + \frac{1}{2} \log 0.8884$
$\qquad\qquad\qquad - \frac{1}{3} \log 895.3)$

$\begin{aligned}\log 0.05325 &= 8.7263 - 10 \\ \tfrac{1}{2}\log 0.8884 &= \underline{9.9743 - 10} \\ &\;18.7006 - 20 \\ \tfrac{1}{3}\log 895.3 &= \underline{0.9840} \\ &\;17.7166 - 20\end{aligned}$ $\qquad \begin{aligned}\log 0.8884 &= \\ 19.9486 &- 20\end{aligned}$

$\qquad\qquad\qquad\qquad\qquad\qquad \log 895.3 = 2.9520$

$0.3(97.7166 - 100) = 29.3150 - 30 \qquad N = 0.2065.$

Exercises 5-4

In the following exercises use logarithms to perform the indicated calculations.

1. $(5.980)(10.80)$
2. $(0.7640)(200.0)$
3. $(0.8256)(0.04532)$
4. $(0.0008080)(2623)$

5. $\dfrac{790.0}{8.000}$
6. $\dfrac{31.60}{0.4500}$
7. $\dfrac{76.98}{43.82}$
8. $\dfrac{0.008670}{0.6521}$
9. $(6.760)^3$
10. $(0.04030)^{0.6}$
11. $(0.9042)^5$
12. $(9065)^8$
13. $\sqrt[7]{7.090}$
14. $\sqrt[3]{95.40}$
15. $\sqrt{641.6}$
16. $\sqrt[8]{308.7}$
17. $\dfrac{(4510)(0.6120)}{738.0}$
18. $\dfrac{87.42}{(11.54)(0.9316)}$
19. $\dfrac{\sqrt{0.07530}}{86.02}$
20. $(\sqrt{5.270})(\sqrt[3]{42.19})$
21. $\dfrac{89.42\sqrt[3]{0.1142}}{0.04290}$
22. $\left(\dfrac{75.19}{900.5\sqrt{15.00}}\right)^{0.1}$
23. $(8.723)^{9.742}$ (Find log log 8.723.)
24. $(4.072)^{-10}$ (Be careful, especially if your method of solution leads to a negative "mantissa.")

5-5 Logarithms of Trigonometric Functions

When we are solving problems in trigonometry, we may also use logarithms of the trigonometric functions in order to perform necessary calculations. In such cases we could look up the function of the desired angle and then find the logarithm of this number to perform the required operation on it.

Example A
Find log sin 23°20′. (Use Tables 2 and 3 of Appendix D.)

sin 23°20′ = 0.3961   log 0.3961 = 9.5978 − 10.

5-5 Logarithms of Trigonometric Functions

To facilitate work when using trigonometric functions, we can use tables of logarithms of the trigonometric functions, which allow us to find these logarithms in one step.

Example B
Find log sin 23°20'. (Use Table 4.)
 By reading directly from the table we find that

sin 23°20' = 9.5978 − 10

Like similar tables, these tables enable us to find by interpolation values not directly listed. We may also find an angle directly if we know the logarithm of some function of that angle.

Example C
Find log tan 57°34'.
 We find that log tan 57°30' = 0.1958 and that log tan 57°40' = 0.1986. The tabular difference is 28, and 0.4(28) = 11 (to two digits). Thus,

log tan 57°34' = 0.1969.

Example D
Given that log cos θ = 9.9049 − 10, find θ.
 From Table 4 we find that

log cos 36°30' = 9.9052 − 10

and

log cos 36°40' = 9.9042 − 10.

We find that the tabular difference between listed values is 10 and the tabular difference between 9.9049 and 9.9052 is 3. Therefore,

θ = 36°33'.

(Be careful—the values of the cosine and its logarithm *decrease* as θ increases.)

Example E

Solve the following oblique triangle by using logarithms for your calculations: $a = 34.12$, $A = 31°20'$, $B = 52°43'$.

We first determine that the solution may be completed by the Law of Sines. Next we find $C = 95°57'$, and then we can find sides b and c. From the Law of Sines we have

$$b = \frac{a \sin B}{\sin A} \quad \text{and} \quad c = \frac{a \sin C}{\sin A}.$$

By using a and $\sin A$ in both calculations, we reduce the amount of information required from the tables.

$\log b = \log 34.12 + \log \sin 52°43' - \log \sin 31°20'$

$$\begin{array}{rl}
\log 34.12 = & 1.5330 \\
\log \sin 52°43' = & 9.9007 - 10 \\
& 11.4337 - 10 \\
\log \sin 31°20' = & 9.7160 - 10 \\
& 1.7177
\end{array}$$

$b = 52.20$

$\log c = \log 34.12 + \log \sin 95°57' - \log \sin 31°20'$

$$\begin{array}{rl}
\log 34.12 = & 1.5330 \\
\log \sin 95°57' = & 9.9976 - 10 \text{ (log sin } 84°3'\text{)} \\
& 11.5306 - 10 \\
\log \sin 31°20' = & 9.7160 - 10 \\
& 1.8146
\end{array}$$

$c = 65.26$

Example F

A ship passes a certain point at noon going north at 12.35 mi/hr. A second ship passes the same point at 1 P.M. going east at 16.42 mi/hr. How far apart are the ships at 3 P.M.?

5-5 Logarithms of Trigonometric Functions

At 3 P.M. the first ship is 37.05 mi from the point and the second ship is 32.84 mi from it (Figure 5-3). Since the angle between their directions is 90°, the distance d between them can be found from the Pythagorean theorem. Also, by finding the angle α from $\tan \alpha = 37.05/32.84$, we can then solve for d by using $d = 32.84/\cos \alpha$. This second method has the advantage that once we determine α, we can find $\log \cos \alpha$ immediately from the table by shifting from the log tan column to the log cos column:

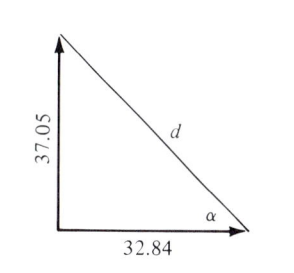

Figure 5-3

$\log \tan \alpha = \log 37.05 - \log 32.84$

$\log 37.05 = 1.5688$
$\log 32.84 = \underline{1.5164}$
$\log \tan \alpha = 0.0524$

$\alpha = 48°27'$

$\log d = \log 32.84 - \log \cos \alpha$

$\log 32.84 = 11.5164 - 10$
$\log \cos \alpha = \underline{9.8217 - 10}$
$\log d = 1.6947$

$d = 49.51$ mi.

Exercises 5-5

In Exercises 1 through 12 use Table 4 in Appendix D to find the values of the indicated logarithms.

1. $\log \sin 22°10'$
2. $\log \cos 31°40'$
3. $\log \sec 52°0'$
4. $\log \csc 61°50'$
5. $\log \sin 38°14'$
6. $\log \cos 12°7'$
7. $\log \tan 56°45'$
8. $\log \sin 75°42'$
9. $\log \cos 322°17'$
10. $\log \cot 228°12'$
11. $\log \cos 79°6'$
12. $\log \tan 85°52'$

168　Logarithms

In Exercises 13 through 20 use Table 4 to find the smallest positive θ.

13. $\log \sin \theta = 9.6740 - 10$
14. $\log \sec \theta = 0.0748$
15. $\log \cos \theta = 9.8056 - 10$
16. $\log \tan \theta = 9.9140 - 10$
17. $\log \tan \theta = 0.0599$
18. $\log \sin \theta = 8.9150 - 10$
19. $\log \cos \theta = 9.9998 - 10$
20. $\log \csc \theta = 0.0075$

In Exercises 21 through 28 solve the given triangles by logarithms.

21. $A = 82°5'$, $C = 90°0'$, $c = 86.17$
22. $B = 54°10'$, $C = 90°0'$, $b = 15.70$
23. $A = 65°40'$, $B = 72°10'$, $a = 9100$
24. $A = 63°14'$, $C = 18°16'$, $c = 0.5320$
25. $A = 67°10'$, $B = 44°42'$, $b = 9.328$
26. $A = 47°36'$, $a = 17.45$, $b = 10.29$
27. $a = 298.5$, $b = 382.6$, $C = 90°0'$
28. $a = 7392$, $b = 4218$, $c = 4005$

In Exercises 29 through 32 solve the given problems by logarithms.

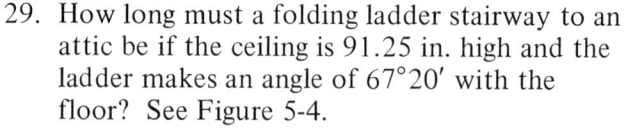

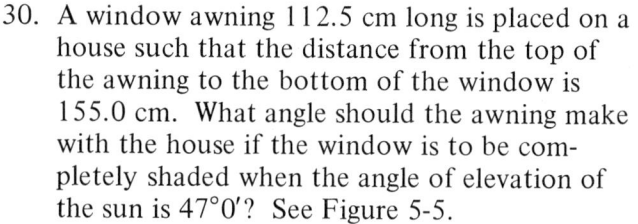

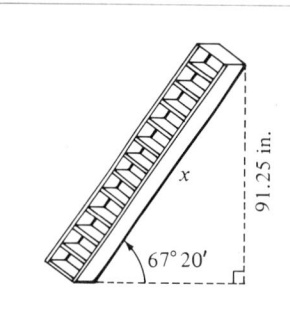

29. How long must a folding ladder stairway to an attic be if the ceiling is 91.25 in. high and the ladder makes an angle of 67°20' with the floor? See Figure 5-4.

30. A window awning 112.5 cm long is placed on a house such that the distance from the top of the awning to the bottom of the window is 155.0 cm. What angle should the awning make with the house if the window is to be completely shaded when the angle of elevation of the sun is 47°0'? See Figure 5-5.

Figure 5-4

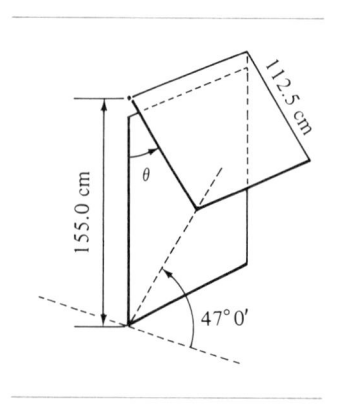

Figure 5-5

31. A surveyor finds one side of a rectangular piece of land to be 137.8 m and the angle between this side and the diagonal to be 36°17'. What is the area of the piece of land?
32. An airplane headed south has a speed with respect to the air of 418.5 mi/hr. The speed with respect to the ground is 425.0 mi/hr in a direction 3°16' east of south. Find the direction and speed of the wind.

5-6 Logarithms to Bases Other Than 10

Another number which is important as a base of logarithms is the number e, an irrational number equal to about 2.718. Logarithms to the base e are called *natural logarithms*. Just as the notation $\log x$ refers to logarithms to the base 10, the notation $\ln x$ is used to denote logarithms to the base e.

Since more than one base is important, there are times when it is useful to be able to change a logarithm in one base to another base. If $u = \log_b x$, then $b^u = x$. Taking logarithms of both sides of this last expression to the base a, we have

$$\log_a b^u = \log_a x$$

$$u \log_a b = \log_a x.$$

Solving this last equation for u, we have

$$u = \frac{\log_a x}{\log_a b}.$$

However, $u = \log_b x$, which means that:

(5-12) $$\log_b x = \frac{\log_a x}{\log_a b}$$

Equation (5-12) allows us to change a logarithm in one base to a logarithm in another base. The following examples illustrate the method of performing this operation.

Example A
Change log 20 = 1.3010 to a logarithm with base e; that is, find ln 20.

In Equation (5-12), if we let $b = e$ and $a = 10$, we have

$$\log_e x = \frac{\log_{10} x}{\log_{10} e}$$

or

$$\ln x = \frac{\log x}{\log e}.$$

In this example, $x = 20$. Therefore,

$$\ln 20 = \frac{\log 20}{\log e} = \frac{\log 20}{\log 2.718} = \frac{1.3010}{0.4343}.$$

This indicated division can itself be found by logarithms. Therefore,

$$\log 1.301 = 10.1143 - 10$$
$$\log 0.4343 = \underline{9.6378 - 10}$$
$$0.4765$$

The antilogarithm of 0.4765 is 2.996. Therefore, ln 20 = 2.996.

Example B
Find $\log_5 560$.

In Equation (5-12), if we let $b = 5$ and $a = 10$, we have

$$\log_5 x = \frac{\log x}{\log 5}.$$

In this example, $x = 560$. Therefore, we have

$$\log_5 560 = \frac{\log 560}{\log 5} = \frac{2.7482}{0.6990}.$$

Performing this division by logarithms, we have

$$\log 2.748 = 10.4390 - 10$$
$$\log 0.6990 = \underline{9.8445 - 10}$$
$$0.5945$$

(Note that 2.7482 was rounded off to four digits and written as 2.748, because we are using our four-place table.) The antilogarithm of 0.5945 is 3.931. Therefore, $\log_5 560 = 3.931$.

Example C
Given that $\ln 80 = 4.382$ and $\ln 10 = 2.303$, find $\log 80$.

In Equation (5-12), if we choose $b = 10$ and $a = e$, we have

$$\log x = \frac{\ln x}{\ln 10}.$$

In this example, $x = 80$. Therefore,

$$\log 80 = \frac{\ln 80}{\ln 10} = \frac{4.382}{2.303}.$$

Performing this division by logarithms, we have

$\log 4.382 = 0.6417$
$\log 2.303 = \underline{0.3623}$
0.2794

The antilogarithm of 0.2794 is 1.903. Therefore, from this calculation $\log 80 = 1.903$. From the table we find that $\log 80 = 1.9031$. It can be seen that these two values agree to three decimal places.

Example D
Find ln 0.811.
Following the procedure used in Example A, we have

$$\ln x = \frac{\log x}{\log e}.$$

In this case $x = 0.811$. Here we shall not express the characteristic of log 0.811 as $9 - 10$. We are not using this logarithm for calculation but as part of a calculation, and we need its value in its explicit form. Therefore, $\log 0.811 = 9.9090 - 10 = -0.0910$. Thus,

$$\ln 0.811 = \frac{\log 0.811}{\log 2.718}$$

$$= \frac{-0.0910}{0.4343} = -0.2095.$$

The numerical value of this quotient also can be determined by using logarithms.

Exercises 5-6

In Exercises 1 through 12 use logarithms to the base 10 to find the natural logarithms of the given numbers.

1. 26.0
2. 51.4
3. 631
4. 293
5. 1.56
6. 1.39
7. 45.7
8. 65.6
9. 0.501
10. 0.991
11. 0.052
12. 0.0020

In Exercises 13 through 16 use logarithms to the base 10 to find the indicated logarithms.

13. $\log_7 42$
14. $\log_{12} 122$
15. $\log_2 86$
16. $\log_{20} 86$

In Exercises 17 through 20 use ln 10 = 2.303 and the indicated natural logarithm to find the indicated logarithms.

17. log 40 (ln 40 = 3.689)
18. log 150 (ln 150 = 5.011)
19. log 2.02 (ln 2.02 = 0.7031)
20. log 4.95 (ln 4.95 = 1.5994)

Exercises for Chapter 5

In Exercises 1 through 12 determine the value of x.

1. $\log_{10} x = 4$
2. $\log_9 x = 3$
3. $\log_5 x = -1$
4. $\log_4 x = -\frac{1}{2}$
5. $\log_2 64 = x$
6. $\log_{12} 144 = x$
7. $\log_8 32 = x$
8. $\log_9 27 = x$
9. $\log_x 36 = 2$
10. $\log_x 243 = 5$
11. $\log_x 10 = \frac{1}{2}$
12. $\log_x 8 = \frac{3}{4}$

In Exercises 13 through 16 express each as a sum, difference, or multiple of logarithms. Wherever possible, evaluate logarithms of the result.

13. $\log_2 28$
14. $\log_6 \left(\frac{5}{36}\right)$
15. $\log_4 \sqrt{48}$
16. $\log_3 (9^2 \cdot 6^3)$

In Exercises 17 through 20 solve for y in terms of x.

17. $\log_6 y = \log_6 4 - \log_6 x$
18. $\log_3 y = \frac{1}{2} \log_3 7 + \frac{1}{2} \log_3 x$
19. $\log_2 y + \log_2 x = 3$
20. $6 \log_4 y = 8 \log_4 4 - 3 \log_4 x$

Logarithms

In Exercises 21 through 32 use logarithms to perform the indicated calculations.

21. $(13.60)(0.6930)$

22. $(0.07255)(4320)$

23. $\dfrac{9.826}{0.08004}$

24. $\dfrac{87.64}{108.2}$

25. $(5.670)^{20}$

26. $(0.9823)^{10}$

27. $\sqrt[4]{17.22}$

28. $(0.006247)^{0.2}$

29. $\dfrac{\sqrt{8645}}{19.49}$

30. $[(9.060)(13.45)]^{1/3}$

31. $\dfrac{\sqrt[5]{22.46}\,(14.98)}{\sqrt[3]{0.8664}}$

32. $(12.66)^{1.096}$

In Exercises 33 through 36 use logarithms to solve the given triangles.

33. $A = 45°0'$, $B = 67°10'$, $a = 76.50$
34. $B = 123°0'$, $C = 15°43'$, $a = 0.9122$
35. $b = 50.00$, $c = 30.00$, $A = 115°15'$
36. $a = 3.200$, $b = 4.520$, $c = 5.615$

In Exercises 37 through 40 use logarithms to the base 10 to find the natural logarithms of the given numbers.

37. 8.86

38. 33.0

39. 2.07

40. 0.542

In Exercises 41 through 42 use logarithms to make any indicated calculations.

41. If P dollars are invested at an interest rate r which is compounded n times a year, the value A of the investment t years later is given by the formula

$$A = P\left(1 + \frac{r}{n}\right)^{nt}$$

What is the value after 5 years of \$5636 invested at $4\frac{1}{2}$ percent and compounded quarterly?

42. Points A and B are on opposite sides of a lake. A third point C is found such that $AC = 402.5$ ft and $BC = 317.9$ ft. What is the distance AB if the angle BAC is $41°18'$?

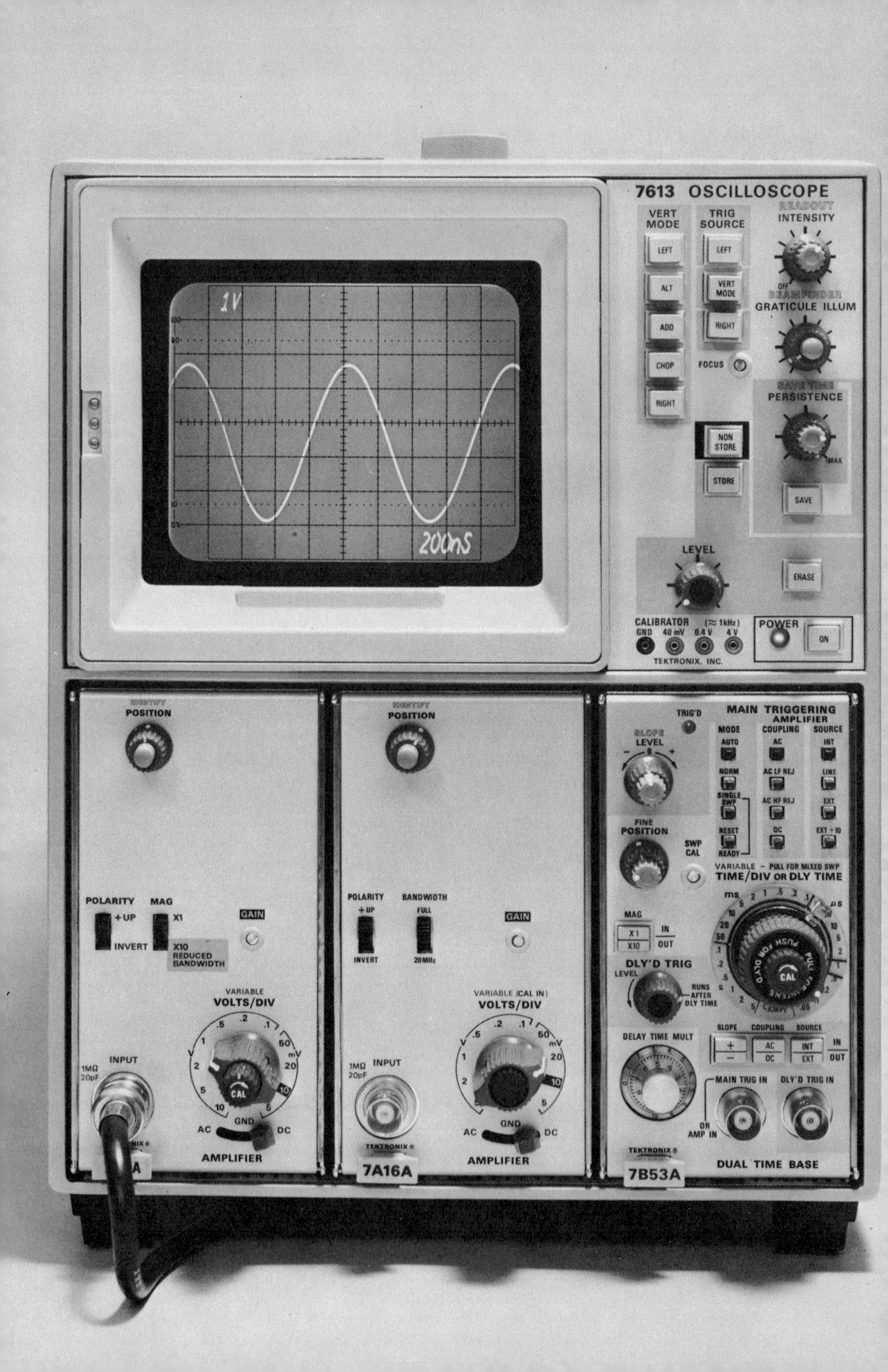
7613 OSCILLOSCOPE
VERT MODE
LEFT
ALT
ADD
CHOP
RIGHT
TRIG SOURCE
LEFT
VERT MODE
RIGHT
FOCUS
READOUT
INTENSITY
OFF
BEAMFINDER
GRATICULE ILLUM
SAVE TIME
PERSISTENCE
MAX
NON STORE
STORE
SAVE
LEVEL
ERASE
CALIBRATOR (≈ 1kHz)
GND 40 mV 0.4 V 4 V
TEKTRONIX, INC.
POWER
ON
1V
200nS
IDENTIFY
POSITION
POLARITY
+UP
INVERT
MAG
X1
X10
REDUCED BANDWIDTH
GAIN
VARIABLE
VOLTS/DIV
CAL
INPUT
1MΩ
20pF
AC
GND
DC
AMPLIFIER
IDENTIFY
POSITION
POLARITY
+UP
INVERT
BANDWIDTH
FULL
20MHz
GAIN
VARIABLE (CAL IN)
VOLTS/DIV
1MΩ
20pF
INPUT
TEKTRONIX®
7A16A
AC
GND
DC
AMPLIFIER
TRIG'D
MAIN TRIGGERING
AMPLIFIER
SLOPE
LEVEL
MODE
AUTO
NORM
SINGLE SWP
RESET
READY
COUPLING
AC
AC LF REJ
AC HF REJ
DC
SOURCE
INT
LINE
EXT
EXT ÷10
FINE
POSITION
SWP CAL
VARIABLE – PULL FOR MIXED SWP
TIME/DIV OR DLY TIME
ms
µs
PUSH FOR DLY'D
CAL
MAG
X1
X10
IN
OUT
DLY'D TRIG
LEVEL
RUNS AFTER DLY TIME
DELAY TIME MULT
SLOPE
COUPLING
SOURCE
AC
DC
INT
EXT
IN
OUT
MAIN TRIG IN
OR AMP IN
DLY'D TRIG IN
TEKTRONIX®
7B53A
DUAL TIME BASE

6

Graphs of the Trigonometric Functions

The graphs of the trigonometric functions are often useful in analyzing phenomena such as sound and electricity. This oscilloscope shows two wave patterns being compared.

6-1 Graphs of $y = a \sin x$ and $y = a \cos x$

One of the clearest ways to demonstrate the variation of the trigonometric functions is by means of their graphs. The graphs are useful for analyzing properties of the trigonometric functions and in themselves are valuable in many applications. Some of these applications are indicated in the exercises.

The graphs are constructed on the rectangular coordinate system. In plotting the trigonometric functions, it is normal to express the angle in radians. In this way x and the function of x are expressed as *numbers*, and these numbers may have any desired unit of measurement. Therefore, to determine the graphs one must be able to use angles expressed in radians. If necessary, Section 3-3 should be reviewed for this purpose.

In this section the graphs of the sine and cosine functions are demonstrated. We begin by constructing a table of values of x and y for the function $y = \sin x$:

x	0	$\frac{\pi}{6}$	$\frac{\pi}{3}$	$\frac{\pi}{2}$	$\frac{2\pi}{3}$	$\frac{5\pi}{6}$	π	$\frac{7\pi}{6}$	$\frac{4\pi}{3}$	$\frac{3\pi}{2}$	$\frac{5\pi}{3}$	$\frac{11\pi}{6}$	2π
y	0	0.5	0.87	1	0.87	0.5	0	−0.5	−0.87	−1	−0.87	−0.5	0

Plotting these values, we obtain the graph shown in Figure 6-1.

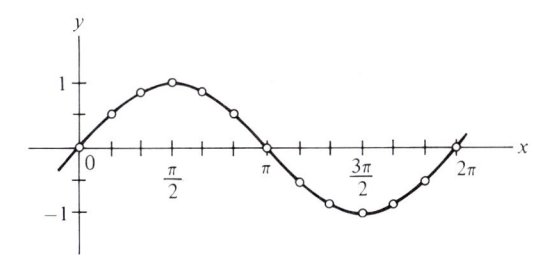

Figure 6-1

The graph of $y = \cos x$ may be constructed in the same manner. The following table gives the proper values for the graph of $y = \cos x$. The graph is shown in Figure 6-2.

x	0	$\frac{\pi}{6}$	$\frac{\pi}{3}$	$\frac{\pi}{2}$	$\frac{2\pi}{3}$	$\frac{5\pi}{6}$	π	$\frac{7\pi}{6}$	$\frac{4\pi}{3}$	$\frac{3\pi}{2}$	$\frac{5\pi}{3}$	$\frac{11\pi}{6}$	2π
y	1	0.87	0.5	0	−0.5	−0.87	−1	−0.87	−0.5	0	0.5	0.87	1

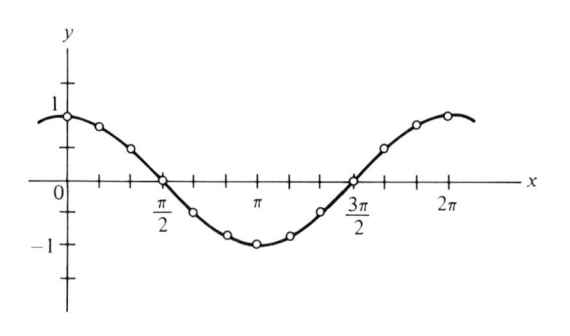

Figure 6-2

The graphs are continued beyond the values shown in the table to indicate that they continue indefinitely in each direction. From the values and the graphs, it can be seen that the two graphs are of exactly the same shape, with the cosine curve displaced $\pi/2$ units to the left of the sine curve. The shape of these curves should be recognized readily, with special note as to the points at which they cross the axes. This information will be especially valuable in sketching similar curves, since the basic shape always remains the same. We shall find it unnecessary to plot numerous points every time we wish to sketch such a curve.

From the graph of the sine function, we see that its domain (see Section 1-1) consists of all real numbers. The range of the sine function consists of all real numbers greater than or equal to −1 and less than or equal to +1. The domain and range of the cosine function are the same as those of the sine function.

We can also establish the graphs of $y = \sin x$ by following a point on the terminal side of an angle,

6-1 Graphs of y = a sin x and y = a cos x

one unit from the origin, as the terminal side rotates through the four quadrants. The sine of the angle is equal to the y-coordinate divided by the radius vector, and the radius vector in this case is equal to 1. Thus, the sine of the angle is represented by the y-coordinate (compare with Section 3-5). See Figure 6-3.

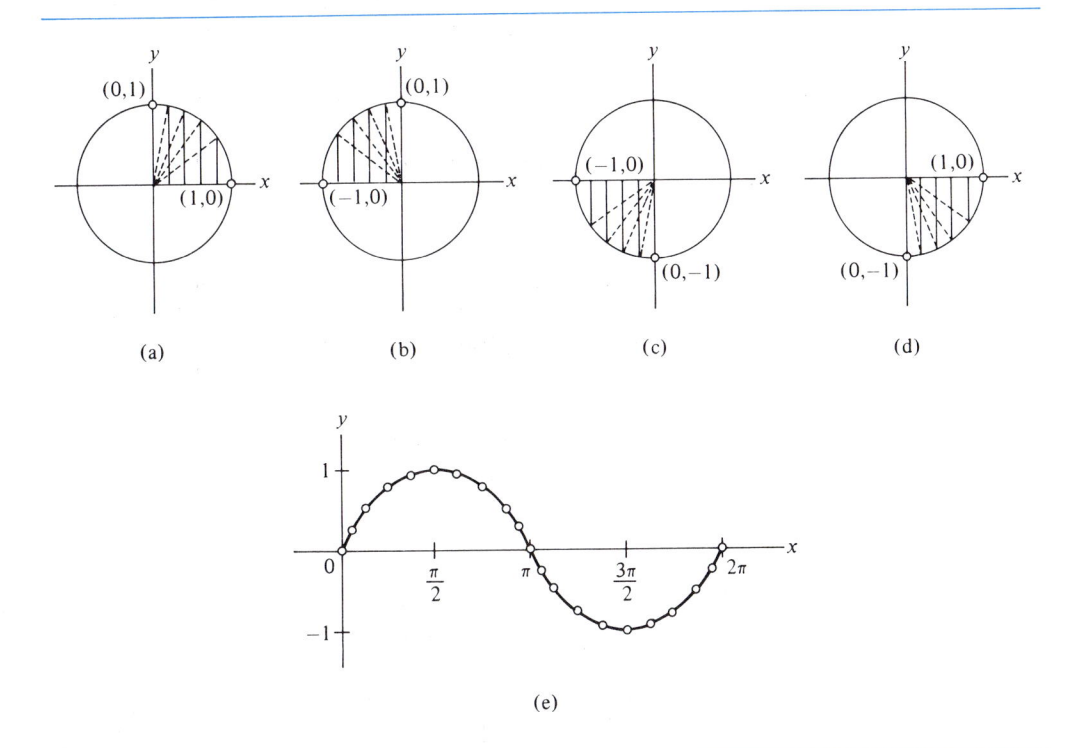

Figure 6-3

In Figure 6-3(a) the angle is shown increasing from 0 to $\pi/2$ rad. We can see that the y-coordinate is increasing from 0 to 1. At $\pi/2$ the y-coordinate is equal to 1. As the angle increases through the second quadrant (Figure 6-3b), the y-coordinate decreases from 1 to 0. When the angle equals π, the y-coordinate is zero. As the terminal side enters the third quadrant (Figure 6-3c), we see that the y-coordinate of the point on the terminal side is negative. It becomes more negative until, at $3\pi/2$, it reaches the value -1. In the fourth quadrant (Figure 6-3d), the value of the y-coordinate increases until, at 2π, it returns to the value of zero. As the terminal side continues to rotate, the y-coordinate would take on the same values as in the first rotation. The y-coordinate values obtained are shown in Figure 6-3(e).

Similarly, we can obtain the values of the cosine of the angle if we observe the x-coordinate. Since $\cos \theta = x/r$ and we let $r = 1$, we see that the cosine of the angle is represented by the x-coordinate. As the angle increases from 0 to 2π, we can see that the value of the x-coordinate starts with a value of 1, decreases to zero and then to -1, and then increases to zero and then to 1. As the terminal side of the angle continues to rotate, the cycle is repeated.

To obtain the graph of $y = a \sin x$, we note that all the y-values obtained for the graph of $y = \sin x$ are to be multiplied by the number a. In this case the greatest value of the sine function is a, instead of 1. **The number a is called the *amplitude* of the curve and represents the greatest y-value of the curve.** This is also true for $y = a \cos x$.

Example A

Plot the curve of $y = 2 \sin x$.

The table of values to be used is as follows; Figure 6-4 shows the graph.

x	0	$\frac{\pi}{6}$	$\frac{\pi}{3}$	$\frac{\pi}{2}$	$\frac{2\pi}{3}$	$\frac{5\pi}{6}$	π	$\frac{7\pi}{6}$	$\frac{4\pi}{3}$	$\frac{3\pi}{2}$	$\frac{5\pi}{3}$	$\frac{11\pi}{6}$	2π
y	0	1	1.73	2	1.73	1	0	-1	-1.73	-2	-1.73	-1	0

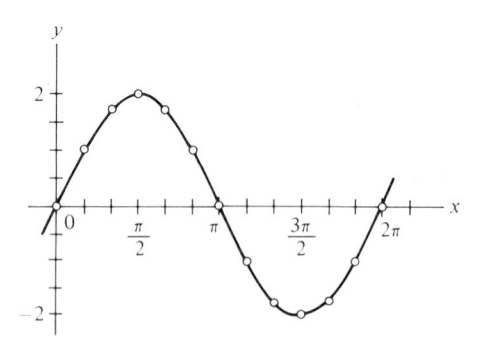

Figure 6-4

Example B

Plot the curve of $y = -3 \cos x$.

The table of values to be used is as follows; Figure 6-5 shows the graph.

x	0	$\frac{\pi}{6}$	$\frac{\pi}{3}$	$\frac{\pi}{2}$	$\frac{2\pi}{3}$	$\frac{5\pi}{6}$	π	$\frac{7\pi}{6}$	$\frac{4\pi}{3}$	$\frac{3\pi}{2}$	$\frac{5\pi}{3}$	$\frac{11\pi}{6}$	2π
y	-3	-2.6	-1.5	0	1.5	2.6	3	2.6	1.5	0	-1.5	-2.6	-3

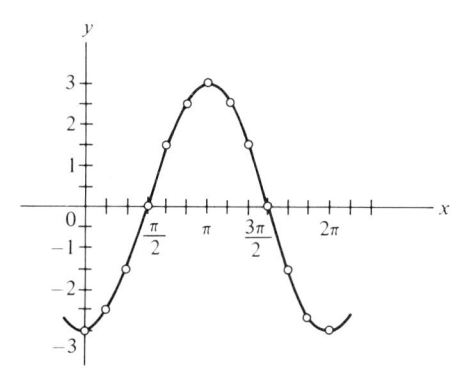

Figure 6-5

Note from Example B that the effect of the minus sign before the number a is to invert the curve. The effect of the number a can also be seen readily from Examples A and B.

By knowing the general shape of the sine curve, where it crosses the axes, and the amplitude, we can rapidly sketch curves of the form $y = a \sin x$ and $y = a \cos x$. There is generally no need to plot any more points than those corresponding to the values of the amplitude and those where the curve crosses the axes.

Example C

Sketch the graph of $y = 4 \cos x$.

First we set up a table of values for the points where the curve crosses the x-axis and for the highest and lowest points on the curve.

x	0	$\frac{\pi}{2}$	π	$\frac{3\pi}{2}$	2π
y	4	0	-4	0	4

Now we plot these points and join them, knowing the basic shape of the curve. See Figure 6-6.

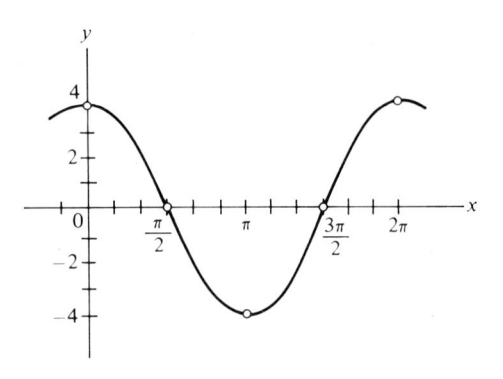

Figure 6-6

Example D

Sketch the curve of $y = -2 \sin x$.

We list here the important values associated with this curve.

x	0	$\frac{\pi}{2}$	π	$\frac{3\pi}{2}$	2π
y	0	-2	0	2	0

Since we know the general shape of the sine curve, we can now sketch the graph as shown in Figure 6-7. Note the inversion of the curve due to the minus sign.

6-1 Graphs of y = a sin x and y = a cos x

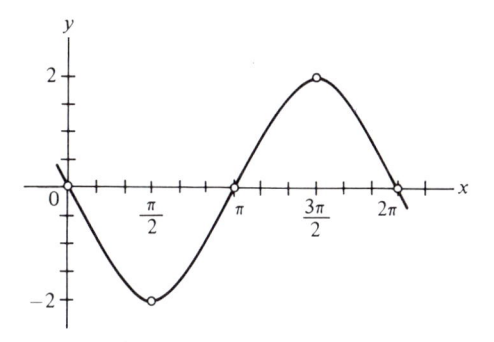

Figure 6-7

Exercises 6-1

In Exercises 1 through 4 complete the following table for the given functions and then plot the resulting graph.

x	$-\pi$	$-\dfrac{3\pi}{4}$	$-\dfrac{\pi}{2}$	$-\dfrac{\pi}{4}$	0	$\dfrac{\pi}{4}$	$\dfrac{\pi}{2}$	$\dfrac{3\pi}{4}$	π	$\dfrac{5\pi}{4}$	$\dfrac{3\pi}{2}$	$\dfrac{7\pi}{4}$	2π	$\dfrac{9\pi}{4}$	$\dfrac{5\pi}{2}$	$\dfrac{11\pi}{4}$	3π
y																	

1. $y = \sin x$
2. $y = \cos x$
3. $y = 10 \cos x$
4. $y = 10 \sin x$

In Exercises 5 through 16 sketch the curves of the indicated functions.

5. $y = 3 \sin x$
6. $y = 5 \sin x$
7. $y = \frac{5}{2} \sin x$
8. $y = 0.5 \sin x$
9. $y = 2 \cos x$
10. $y = 3 \cos x$
11. $y = 0.8 \cos x$
12. $y = \frac{3}{2} \cos x$
13. $y = -\sin x$
14. $y = -3 \sin x$
15. $y = -\cos x$
16. $y = -8 \cos x$

Although units of π are often convenient, we must remember that π is really only a number. Numbers which are not multiples of π may be used as well. In Exercises 17 through 20 plot the indicated graphs by finding the values of y corresponding to the values of 0, 1, 2, 3, 4, 5, 6, and 7 for x. (Remember, the numbers 0, 1, 2, and so forth represent radian measure.)

17. $y = \sin x$
18. $y = 3 \sin x$
19. $y = \cos x$
20. $y = 2 \cos x$

6-2 Graphs of $y = a \sin bx$ and $y = a \cos bx$

In graphing the curve of $y = \sin x$ we note that the values of y start repeating every 2π units of x. This is because $\sin x = \sin(x + 2\pi) = \sin(x + 4\pi)$ and so forth. **For any trigonometric function F, we say that it has a *period P* if $F(x) = F(x + P)$.** For functions which are periodic, such as the sine and cosine, the period refers to the x-distance between any point and the next corresponding point for which the values of y start repeating.

Let us now plot the curve $y = \sin 2x$. This means that we choose a value for x, multiply this value by 2, and find the sine of the result. This leads to the following table of values for this function.

x	0	$\frac{\pi}{8}$	$\frac{\pi}{4}$	$\frac{3\pi}{8}$	$\frac{\pi}{2}$	$\frac{5\pi}{8}$	$\frac{3\pi}{4}$	$\frac{7\pi}{8}$	π	$\frac{9\pi}{8}$	$\frac{5\pi}{4}$
$2x$	0	$\frac{\pi}{4}$	$\frac{\pi}{2}$	$\frac{3\pi}{4}$	π	$\frac{5\pi}{4}$	$\frac{3\pi}{2}$	$\frac{7\pi}{4}$	2π	$\frac{9\pi}{4}$	$\frac{5\pi}{2}$
y	0	0.7	1	0.7	0	-0.7	-1	-0.7	0	0.7	1

Plotting these values, we have the curve shown in Figure 6-8.

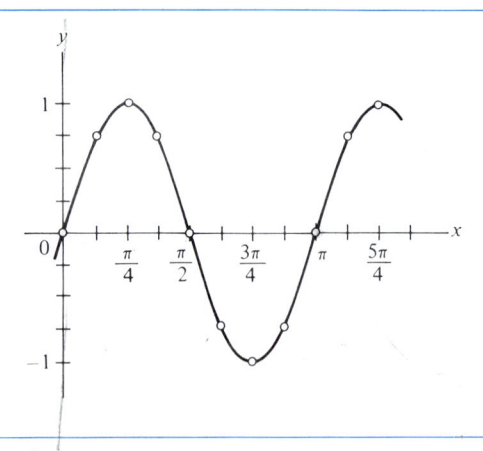

Figure 6-8

From the table and Figure 6-8, we note that the function $y = \sin 2x$ starts repeating after π units of x. The effect of the 2 before the x has been to make the period of this curve half the period of the curve of $\sin x$. This leads us to the following conclusion: **If the period of the trigonometric function $F(x)$ is P, then the period of $F(bx)$ is P/b.** Since each of the functions $\sin x$ and $\cos x$ has a period of 2π, each of the functions $\sin bx$ and $\cos bx$ has a period of $2\pi/b$.

Example A
The period of $\sin 3x$ is $2\pi/3$. That is, the curve of the function $y = \sin 3x$ will repeat every $2\pi/3$ (approximately 2.09) units of x.

The period of $\cos 4x$ is $2\pi/4 = \pi/2$. The period of $\sin \frac{1}{2}x$ is $2\pi/\frac{1}{2} = 4\pi$.

Example B
The period of $\sin \pi x$ is $2\pi/\pi = 2$. That is, the curve of the function $\sin \pi x$ will repeat every 2 units. Note that the periods of $\sin 3x$ and $\sin \pi x$ are nearly equal. This is to be expected since π is only slightly greater than 3.

The period of $\cos 3\pi x$ is $2\pi/3\pi = \frac{2}{3}$.

Combining the result for the period with the results of Section 6-1, we conclude that **each of the functions $y = a \sin bx$ and $y = a \cos bx$ has an amplitude of $|a|$ and a period of $2\pi/b$.** These properties are very useful in sketching these functions, as is shown in the following examples.

Example C
Sketch the graph of $y = 3 \sin 4x$ for $0 \leq x \leq \pi$.

We immediately conclude that the amplitude is 3 and the period is $2\pi/4 = \pi/2$. Therefore, we know that $y = 0$ when $x = 0$ and $y = 0$ when $x = \pi/2$. Also, we recall that the sine function is zero halfway between these values, which means that $y = 0$ when $x = \pi/4$. The function reaches its amplitude values halfway between these points where the value of the function is zero. Therefore, $y = 3$ for $x = \pi/8$ and $y = -3$ for $x = 3\pi/8$. A table for these important values of the function $y = 3 \sin 4x$ is shown on the next page.

x	0	$\frac{\pi}{8}$	$\frac{\pi}{4}$	$\frac{3\pi}{8}$	$\frac{\pi}{2}$	$\frac{5\pi}{8}$	$\frac{3\pi}{4}$	$\frac{7\pi}{8}$	π
y	0	3	0	−3	0	3	0	−3	0

Using this table and the knowledge of the form of the sine curve, we sketch the function (Figure 6-9).

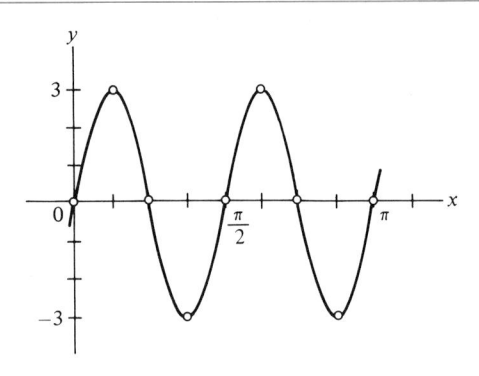

Figure 6-9

Example D

Sketch the graph of $y = -2 \cos 3x$ for $0 \leq x \leq 2\pi$.

We note that the amplitude is 2 and the period is $2\pi/3$. Since the cosine curve is at its amplitude value for $x = 0$, we have $y = -2$ for $x = 0$ (the negative value is due to the minus sign before the function) and $y = -2$ for $x = 2\pi/3$. The cosine function also reaches its amplitude value halfway between these values, or $y = 2$ for $x = \pi/3$. The cosine function is zero halfway between the x-values listed for the amplitude; that is, $y = 0$ for $x = \pi/6$ and for $x = \pi/2$. A table of the important values follows.

x	0	$\frac{\pi}{6}$	$\frac{\pi}{3}$	$\frac{\pi}{2}$	$\frac{2\pi}{3}$	$\frac{5\pi}{6}$	π	$\frac{7\pi}{6}$	$\frac{4\pi}{3}$	$\frac{3\pi}{2}$	$\frac{5\pi}{3}$	$\frac{11\pi}{6}$	2π
y	−2	0	2	0	−2	0	2	0	−2	0	2	0	−2

Using this table and the knowledge of the form of the cosine curve, we sketch the function as shown in Figure 6-10.

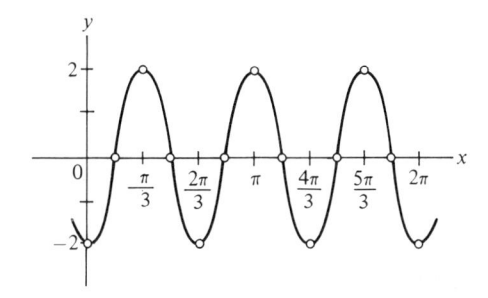

Figure 6-10

Example E

Sketch the function $y = \cos \pi x$ for $0 \leq x \leq \pi$.

For this function the amplitude is 1; the period is $2\pi/\pi = 2$. Since the value of the period is not in terms of π, it is more convenient to use regular decimal units for x when sketching than to use units in terms of π as in the previous graphs. Therefore, we have the following table.

x	0	0.5	1	1.5	2	2.5	3
y	1	0	−1	0	1	0	−1

The graph of this function is shown in Figure 6-11.

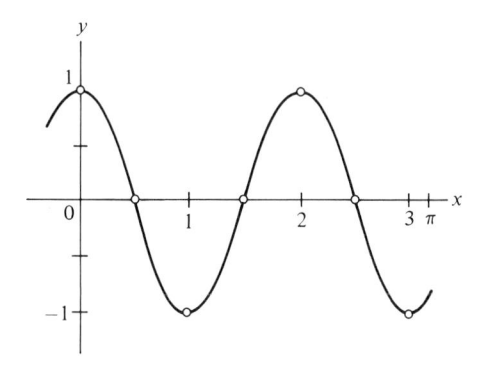

Figure 6-11

Exercises 6-2

In Exercises 1 through 20 find the period of each of the given functions.

1. $y = 2 \sin 6x$
2. $y = 4 \sin 2x$
3. $y = 3 \cos 8x$
4. $y = \cos 10x$
5. $y = -2 \sin 12x$
6. $y = -\sin 5x$
7. $y = -\cos 16x$
8. $y = -4 \cos 2x$
9. $y = 5 \sin 2\pi x$
10. $y = 2 \sin 3\pi x$
11. $y = 3 \cos 4\pi x$
12. $y = 4 \cos 10\pi x$
13. $y = 3 \sin \frac{1}{3}x$
14. $y = -2 \sin \frac{2}{5}x$
15. $y = -\frac{1}{2} \cos \frac{2}{3}x$
16. $y = \frac{1}{3} \cos \frac{1}{4}x$
17. $y = 0.4 \sin \frac{2\pi x}{3}$
18. $y = 1.5 \cos \frac{\pi x}{10}$
19. $y = 3.3 \cos \pi^2 x$
20. $y = 2.5 \sin \frac{2x}{\pi}$

In Exercises 21 through 40 sketch the graphs of the functions given for Exercises 1 through 20.

6-3 Graphs of $y = a \sin (bx + c)$ and $y = a \cos (bx + c)$

There is one more important quantity to be discussed in relation to graphing the sine and cosine functions. This quantity is the *phase angle* of the function. **In the function $y = a \sin (bx + c)$, c represents this phase angle.** Its meaning is illustrated in the following example.

Example A
Sketch the graph of $y = \sin (2x + \pi/4)$.
   Note that $c = \pi/4$. This means that to obtain the values for the table we must assume a value of x, multiply it by 2, add $\pi/4$ to this value, and then find the sine of this result. In this manner we arrive at the following table.

6-3 Graphs of y = a sin (bx + c) and y = a cos (bx + c)

x	$-\frac{\pi}{8}$	0	$\frac{\pi}{8}$	$\frac{\pi}{4}$	$\frac{3\pi}{8}$	$\frac{\pi}{2}$	$\frac{5\pi}{8}$	$\frac{3\pi}{4}$	$\frac{7\pi}{8}$	π
$2x + \frac{\pi}{4}$	0	$\frac{\pi}{4}$	$\frac{\pi}{2}$	$\frac{3\pi}{4}$	π	$\frac{5\pi}{4}$	$\frac{3\pi}{2}$	$\frac{7\pi}{4}$	2π	$\frac{9\pi}{4}$
y	0	0.7	1	0.7	0	-0.7	-1	-0.7	0	0.7

We use the value of $x = -\pi/8$ in the table, for we note that it corresponds to finding sin 0. Now using the values listed in the table, we plot the graph of $y = \sin(2x + \pi/4)$. See Figure 6-12.

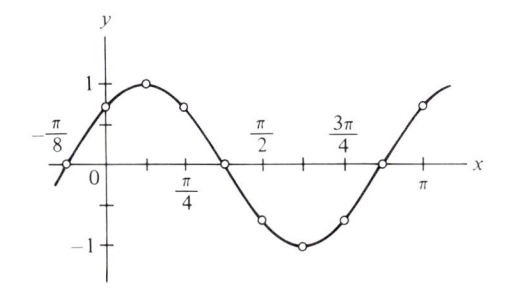

Figure 6-12

We can see from the table and from the graph that the curve of

$$y = \sin\left(2x + \frac{\pi}{4}\right)$$

is precisely the same as that of $y = \sin 2x$, except that it is shifted $\pi/8$ units to the left. The effect of c in the equation $y = a \sin(bx + c)$ is to shift the curve of $y = a \sin bx$ to the left if $c > 0$ and to shift the curve to the right if $c < 0$. The amount of this shift is given by c/b. Due to its importance in sketching curves, **the quantity c/b is called the displacement**.

Therefore, the results above combined with the results of Section 6-2 may be used to sketch curves of the functions $y = a \sin(bx + c)$ and $y = a \cos(bx + c)$. These are the important quantities to determine:

1. The amplitude (equal to $|a|$)
2. The period (equal to $2\pi/b$)
3. The displacement (equal to c/b)

By use of these quantities, the curves of these sine and cosine functions can be readily sketched.

Example B

Sketch the graph of $y = 2 \sin(3x - \pi)$ for $0 \leq x \leq \pi$.

First we note that $a = 2$, $b = 3$, and $c = -\pi$. Therefore, the amplitude is 2, the period is $2\pi/3$, and the displacement is $\pi/3$ to the right.

With this information we can tell that the curve starts at $x = \pi/3$ and starts repeating $2\pi/3$ units to the right of this point. (Be sure to grasp this point well. The period tells how many units there are along the x-axis *between* such corresponding points.) The value of y is zero when $x = \pi/3$ and when $x = \pi$. It is also zero halfway between these values of x, or when $x = 2\pi/3$. The curve reaches the amplitude value of 2 midway between $x = \pi/3$ and $x = 2\pi/3$, or when $x = \pi/2$. Extending the curve to the left to $x = 0$, we note that, since the period is $2\pi/3$, the curve passes through $(0, 0)$. Therefore, we have the following table of important values.

x	0	$\frac{\pi}{6}$	$\frac{\pi}{3}$	$\frac{\pi}{2}$	$\frac{2\pi}{3}$	$\frac{5\pi}{6}$	π
y	0	-2	0	2	0	-2	0

From these values we sketch the graph shown in Figure 6-13.

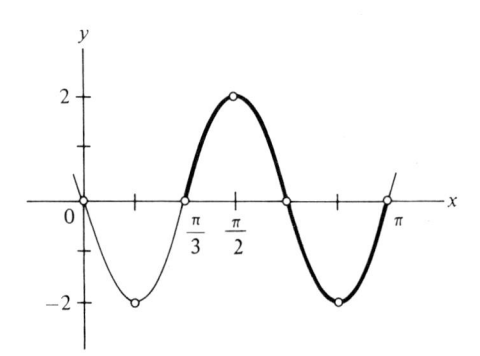

Figure 6-13

6-3 Graphs of y = a sin (bx + c) and y = a cos (bx + c)

Example C
Sketch the graph of the function $y = -\cos(2x + \pi/6)$.

First we determine that the amplitude is 1, the period is $2\pi/2 = \pi$, and the displacement is $(\pi/6) \div 2 = \pi/12$ to the left ($c > 0$). From these values we construct the following table, remembering that the curve starts repeating π units to the right of $-\pi/12$.

x	$-\dfrac{\pi}{12}$	$\dfrac{\pi}{6}$	$\dfrac{5\pi}{12}$	$\dfrac{2\pi}{3}$	$\dfrac{11\pi}{12}$
y	-1	0	1	0	-1

From this table we sketch the graph shown in Figure 6-14.

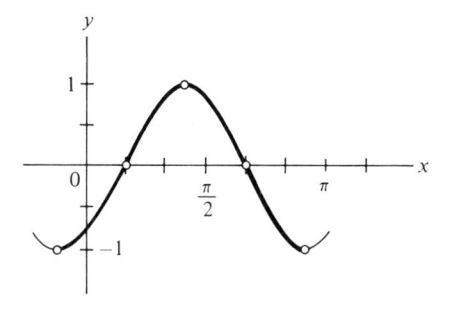

Figure 6-14

Example D
Sketch the graph of the function $y = 2\cos(x/2 - \pi/6)$.

From the values $a = 2$, $b = \tfrac{1}{2}$, and $c = -\pi/6$, we determine that the amplitude is 2, the period is $2\pi/\tfrac{1}{2} = 4\pi$, and the displacement is $(\pi/6) \div \tfrac{1}{2} = \pi/3$ to the right. From these values we construct the table of values.

x	$\dfrac{\pi}{3}$	$\dfrac{4\pi}{3}$	$\dfrac{7\pi}{3}$	$\dfrac{10\pi}{3}$	$\dfrac{13\pi}{3}$
y	2	0	-2	0	2

The graph is shown in Figure 6-15. We note that when the coefficient of x is less than 1, the period is greater than 2π.

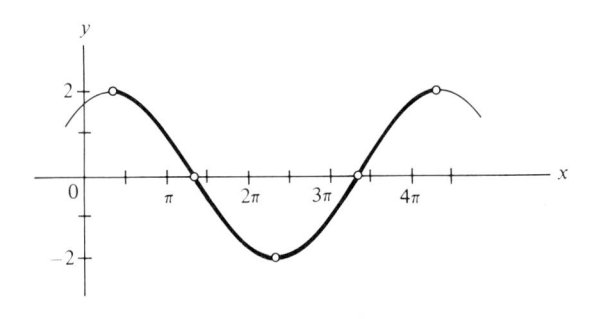

Figure 6-15

Example E
Sketch the graph of the function
$y = 0.7 \sin(\pi x + \pi/4)$.

From the values $a = 0.7$, $b = \pi$, and $c = \pi/4$, we determine that the amplitude is 0.7, the period is $2\pi/\pi = 2$, and the displacement is $(\pi/4) \div \pi = \frac{1}{4}$ to the left. From these values we construct the following table of values.

x	$-\frac{1}{4}$	$\frac{1}{4}$	$\frac{3}{4}$	$\frac{5}{4}$	$\frac{7}{4}$
y	0	0.7	0	-0.7	0

Since π is not used in the values of x, it is more convenient to use decimal number units for the graph (Figure 6-16).

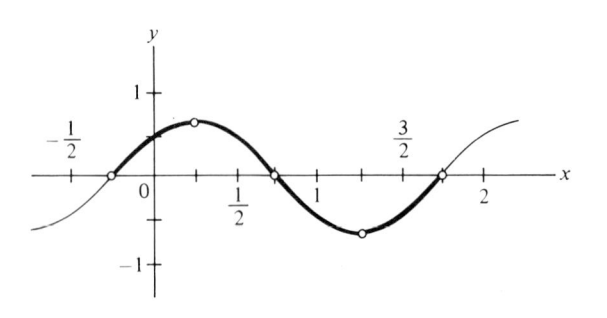

Figure 6-16

6-3 Graphs of y = a sin (bx + c) and y = a cos (bx + c)   195

The heavy portions of the graphs in Figures 6-13, 6-14, 6-15, and 6-16 are known as *cycles* of the curves.

Exercises 6-3

In the following exercises determine the amplitude, period, and displacement for each of the functions. Then sketch the graphs of the functions.

1. $y = \sin\left(x - \frac{\pi}{6}\right)$
2. $y = 3 \sin\left(x + \frac{\pi}{4}\right)$
3. $y = \cos\left(x + \frac{\pi}{6}\right)$
4. $y = 2 \cos\left(x - \frac{\pi}{8}\right)$
5. $y = 2 \sin\left(2x + \frac{\pi}{2}\right)$
6. $y = -\sin\left(3x - \frac{\pi}{2}\right)$
7. $y = -\cos(2x - \pi)$
8. $y = 4 \cos\left(3x + \frac{\pi}{3}\right)$
9. $y = \frac{1}{2} \sin\left(\frac{1}{2}x - \frac{\pi}{4}\right)$
10. $y = 2 \sin\left(\frac{1}{4}x + \frac{\pi}{2}\right)$
11. $y = 3 \cos\left(\frac{1}{3}x + \frac{\pi}{3}\right)$
12. $y = \frac{1}{3} \cos\left(\frac{1}{2}x - \frac{\pi}{8}\right)$
13. $y = \sin\left(\pi x + \frac{\pi}{8}\right)$
14. $y = -2 \sin(2\pi x - \pi)$
15. $y = \frac{3}{4} \cos\left(4\pi x - \frac{\pi}{5}\right)$
16. $y = 6 \cos\left(3\pi x + \frac{\pi}{2}\right)$
17. $y = -0.6 \sin(2\pi x - 1)$
18. $y = 1.8 \sin\left(\pi x + \frac{1}{3}\right)$
19. $y = 4 \cos(3\pi x + 2)$
20. $y = 3 \cos(6\pi x - 1)$
21. $y = \sin(\pi^2 x - \pi)$
22. $y = \pi \cos\left(\frac{1}{\pi}x + \frac{1}{3}\right)$
23. $y = 2\pi \cos(4x + 1)$
24. $y = -\sin(5x - 2)$

6-4 Graphs of $y = \tan x$, $y = \cot x$, $y = \sec x$, $y = \csc x$

In this section we briefly consider the graphs of the other trigonometric functions. We shall establish the basic form of each curve, and from these we shall be able to sketch other curves for these functions.

Considering the values and signs of the trigonometric functions as established in Chapter 3, we set up the following table for the function $y = \tan x$ (asterisks denote undefined values). The graph is shown in Figure 6-17.

x	0	$\frac{\pi}{6}$	$\frac{\pi}{4}$	$\frac{\pi}{3}$	$\frac{\pi}{2}$	$\frac{2\pi}{3}$	$\frac{3\pi}{4}$	$\frac{5\pi}{6}$	π	$\frac{7\pi}{6}$	$\frac{5\pi}{4}$	$\frac{4\pi}{3}$	$\frac{3\pi}{2}$	$\frac{5\pi}{3}$	$\frac{7\pi}{4}$	$\frac{11\pi}{6}$	2π
y	0	0.6	1	1.7	*	-1.7	-1	-0.6	0	0.6	1	1.7	*	-1.7	-1	-0.6	0

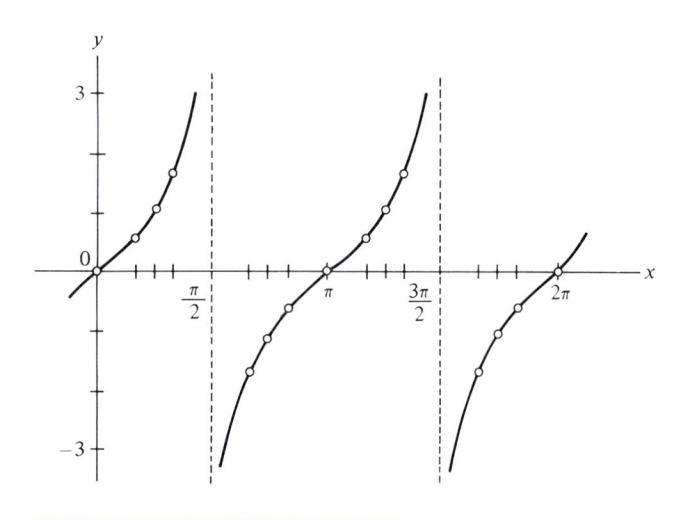

Figure 6-17

Since the curve is not defined for $x = \pi/2$, $x = 3\pi/2$, and so forth, we look at the table and note that the value of $\tan x$ becomes very large when x approaches the value $\pi/2$. We must keep in mind, however, that there is no point on the curve corresponding to $x = \pi/2$. We note that the period of the tangent curve is π. This differs from the period of the sine and cosine functions.

6-4 Graphs of y = tan x, y = cot x, y = sec x, y = csc x   197

From Figure 6-17 we can see that the domain of the tangent function is all real numbers, except $\pi/2$, $3\pi/2$, $5\pi/2$, and so forth. The range of the tangent function is all real numbers.

The graph of the tangent function can also be established in a manner similar to the alternative method we used for the graphs of the sine and cosine functions. This time, however, we have the x-coordinate maintain a value of 1 or -1 and observe the values of the y-coordinate. In this way, since the tangent of an angle equals y/x, the tangent of the angle is represented by the y-coordinate. See Figure 6-18.

As the angle increases through the first quadrant (Figure 6-18a), the y-coordinate of the point on the

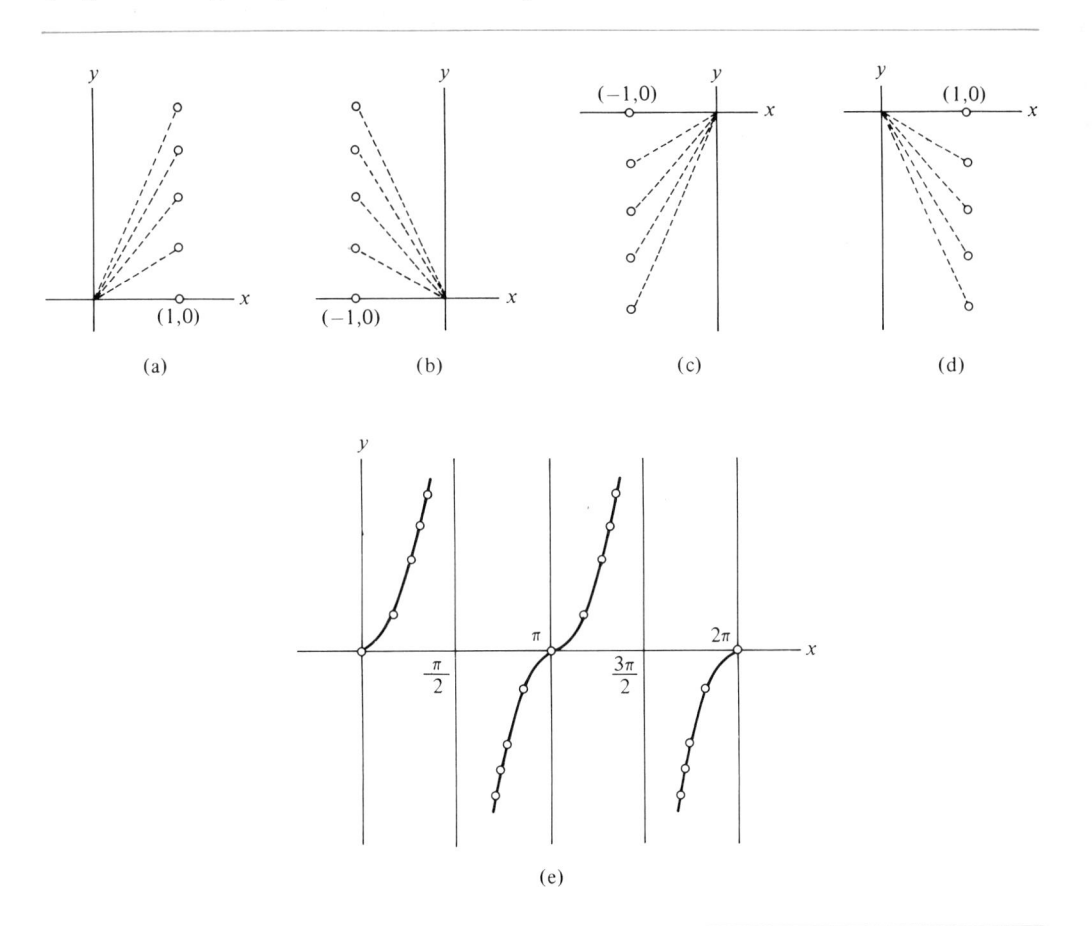

Figure 6-18

terminal side increases from zero until, when the angle is $\pi/2$, it is infinitely large. Here the terminal side of the angle is parallel to the line representing the y-coordinate. Since both the x- and y-coordinates are positive in the first quadrant, the tangent of the angle is positive.

As the angle increases through the second quadrant (Figure 6-18b), the y-coordinate decreases until it reaches a value of zero at π. Here, however, the x-coordinate is negative and the ratio of y to x is negative. Thus, the tangent of the angle increases through negative values to zero.

In the third quadrant both the y-coordinate and the x-coordinate are negative, so the tangent of the angle goes from zero to an infinitely large number. See Figure 6-18(c). In the fourth quadrant the ratio of y to x is negative and the tangent of the angle goes through negative values to zero. See

$y = \tan x$

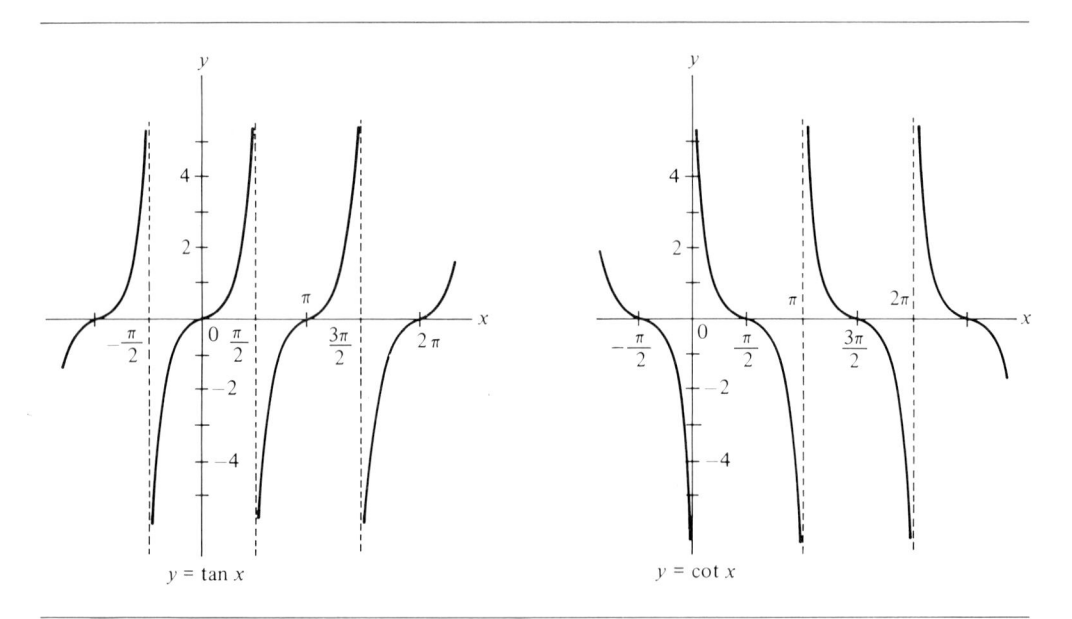

Figure 6-19

$y = \cot x$

Figure 6-20

6-4 Graphs of $y = \tan x$, $y = \cot x$, $y = \sec x$, $y = \csc x$ 199

Figure 6-18(d). As the angle continues to increase, the cycle is repeated. The values obtained are shown in Figure 6-18(e).

By following the same procedure, we can set up tables for the graphs of the other functions. We present in Figures 6-19 through 6-22 the graphs of $y = \tan x$, $y = \cot x$, $y = \sec x$, and $y = \csc x$ (the graph of $y = \tan x$ is shown again to illustrate it more completely). The dashed lines in these figures are called *asymptotes.* An asymptote is a line which a curve approaches as one of the variables approaches some particular value.

To sketch functions such as $y = a \sec x$, we first sketch $y = \sec x$ and then multiply each y-value by a. Here a is not an amplitude, since these functions are not limited in the values they take on, as are the sine and cosine functions.

$y = \sec x$

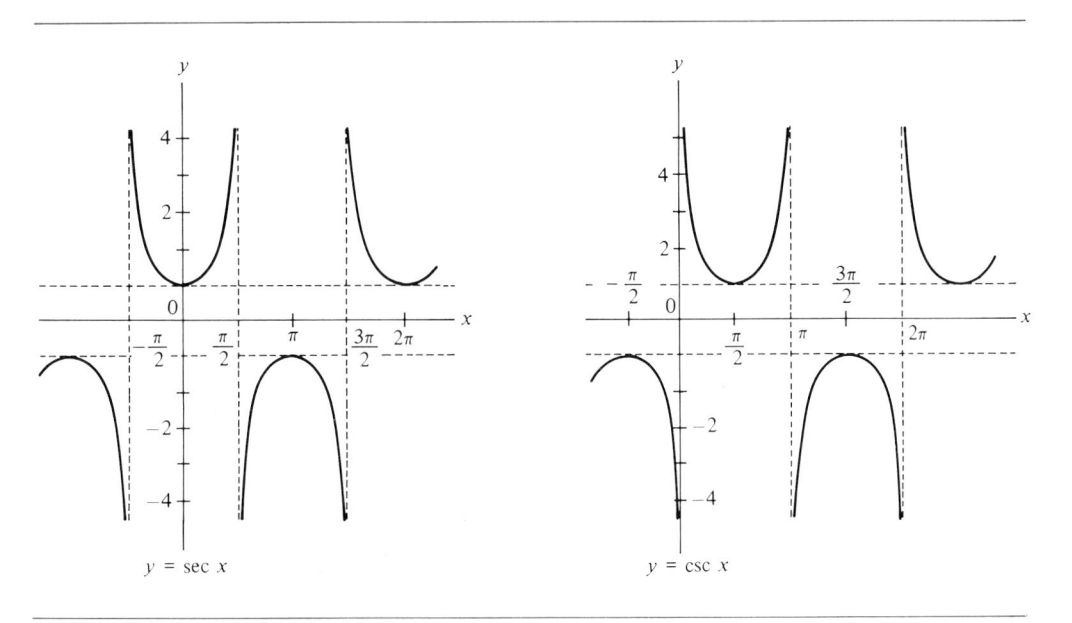

Figure 6-21

$y = \csc x$

Figure 6-22

Example A

Sketch the graph of $y = 2 \sec x$.

First we sketch in $y = \sec x$, shown as the light curves in Figure 6-23. Now we multiply the y-values of the secant function by 2 (approximately, of course). In this way we obtain the desired curve, shown as the heavy curve in Figure 6-23.

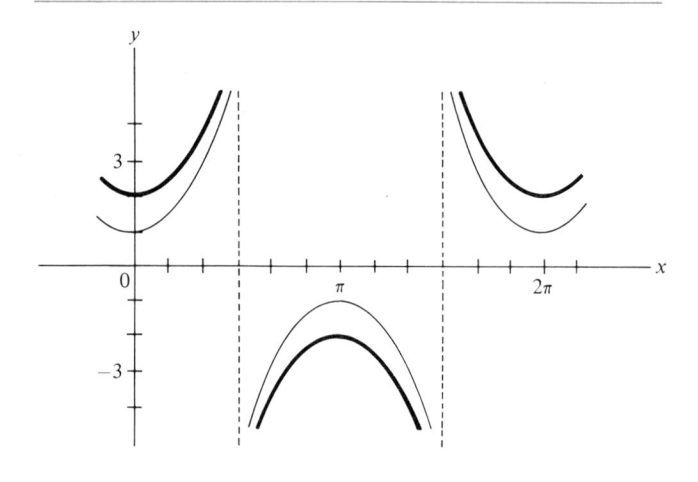

Figure 6-23

Example B

Sketch the graph of $y = -\frac{1}{2} \cot x$.

We sketch in $y = \cot x$, shown as the light curve in Figure 6-24. Now we multiply each y-value by $-\frac{1}{2}$. The effect of the negative sign is to invert the curve. The resulting curve is shown as the heavy curve in Figure 6-24.

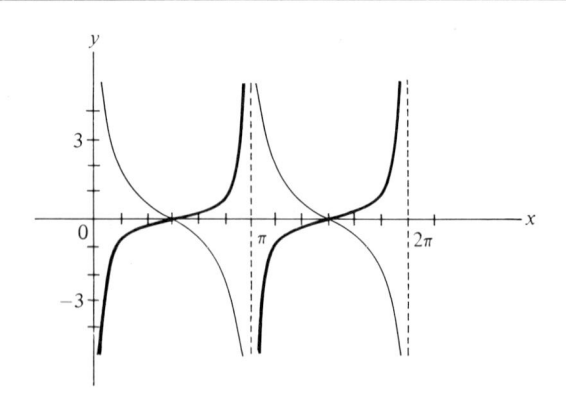

Figure 6-24

6-4 Graphs of y = tan x, y = cot x, y = sec x, y = csc x

By knowing the graphs of the sine, cosine, and tangent functions, it is possible to graph the other three functions. Remembering the definitions of the trigonometric functions (Equations 3-1), we see that sin x and csc x are reciprocals, cos x and sec x are reciprocals, and tan x and cot x are reciprocals. That is:

(6-1)
$$\csc x = \frac{1}{\sin x} \qquad \sec x = \frac{1}{\cos x} \qquad \cot x = \frac{1}{\tan x}$$

Thus, to sketch $y = \cot x$, $y = \sec x$, or $y = \csc x$, we sketch the corresponding reciprocal function and from this graph determine the necessary values.

Example C
Sketch the graph of $y = \csc x$.

We first sketch in the graph of $y = \sin x$ (light curve). Where sin x is 1, csc x will also be 1, since $\frac{1}{1} = 1$. Where sin x is zero, csc x is undefined, since $\frac{1}{0}$ is undefined. Where sin x is 0.5, csc x is 2, since $\frac{1}{0.5} = 2$. Thus, as sin x becomes larger, csc x becomes smaller; and as sin x becomes smaller, csc x becomes larger. The two functions always have the same sign. We sketch the graph of $y = \csc x$ from this information, as shown in Figure 6-25.

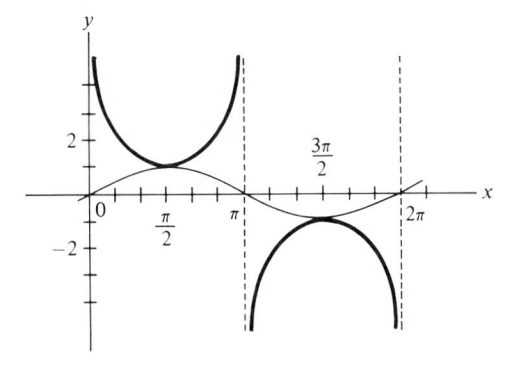

Figure 6-25

Exercises 6-4

In Exercises 1 through 4 fill in the following table for each function and then plot the curve from these points.

x	$-\frac{\pi}{2}$	$-\frac{\pi}{3}$	$-\frac{\pi}{4}$	$-\frac{\pi}{6}$	0	$\frac{\pi}{6}$	$\frac{\pi}{4}$	$\frac{\pi}{3}$	$\frac{\pi}{2}$	$\frac{2\pi}{3}$	$\frac{3\pi}{4}$	$\frac{5\pi}{6}$	π
y													

1. $y = \tan x$
2. $y = \cot x$
3. $y = \sec x$
4. $y = \csc x$

In Exercises 5 through 12 sketch the curves of the given functions by use of the basic curve forms (Figures 6-19, 6-20, 6-21, 6-22). See Examples A and B.

5. $y = 2 \tan x$
6. $y = 3 \cot x$
7. $y = \frac{1}{2} \sec x$
8. $y = \frac{3}{2} \csc x$
9. $y = -2 \cot x$
10. $y = -\tan x$
11. $y = -3 \csc x$
12. $y = -\frac{1}{2} \sec x$

In Exercises 13 through 20 plot the graphs by first making an appropriate table for $0 \le x \le \pi$.

13. $y = \tan 2x$
14. $y = 2 \cot 3x$
15. $y = \frac{1}{2} \sec 3x$
16. $y = \csc 2x$
17. $y = 2 \cot \left(2x + \frac{\pi}{6}\right)$
18. $y = \tan \left(3x - \frac{\pi}{2}\right)$
19. $y = \csc \left(3x - \frac{\pi}{3}\right)$
20. $y = 3 \sec \left(2x + \frac{\pi}{4}\right)$

In Exercises 21 through 24 sketch the given curves by first sketching the appropriate reciprocal function. See Example C.

21. $y = \sec x$
22. $y = \cot x$
23. $y = \csc 2x$
24. $y = \sec \pi x$

6-5 Graphing by Addition of Ordinates

There are a great many applications of the trigonometric functions and their graphs in the fields of optics, sound, and electricity. Many applications of the trigonometric functions involve the combination of two or more functions. In this section we discuss an important method by which trigonometric curves can be combined.

If we wish to find the curve of a function which itself is the sum of two other functions, we may find the resulting graph by first sketching the two individual functions and then adding the y-values graphically. This method, called *addition of ordinates*, is illustrated in the following examples.

Example A
Sketch the graph of $y = 2 \cos x + \sin 2x$.

On the same set of coordinate axes we sketch the curves $y = 2 \cos x$ and $y = \sin 2x$. These are shown as dashed curves in Figure 6-26. We then graphically add the y-values of these two curves for various values of x to obtain the points on the resulting curve shown as a heavy curve in Figure 6-26. We add the y-values for a sufficient number of x-values to obtain the proper representation. Some

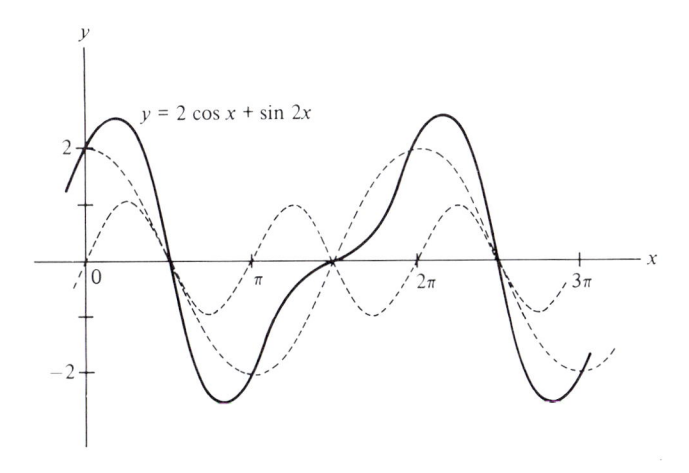

Figure 6-26

points are easily found. Where one curve crosses the x-axis, its y-value is zero, and therefore the resulting curve has its point on the other curve for this value of x. In this example, sin 2x is zero at x = 0, π/2, π, and so forth. We see that points on the resulting curve lie on the curve of 2 cos x. We should also add the values where each curve is at its amplitude values. In this case, sin 2x equals 1 at π/4, and the two y-values should be added together here to get a point on the resulting curve. At x = 5π/4, we must take care in adding the values, since sin 2x is positive and 2 cos x is negative. Reasonable care and accuracy are necessary to obtain a proper curve.

Example B
Sketch the graph of y = (x/2) − cos x.

The method of addition of ordinates is applicable regardless of the kinds of functions being added. Here we note that y = x/2 is a straight line that is to be combined with a trigonometric curve.

We could graph the functions y = x/2 and y = cos x and then subtract the ordinates of y = cos x from the ordinates of y = x/2. But it is easier, and far less confusing, to add values, so we shall sketch y = x/2 and y = −cos x and add the ordinates to obtain points on the resulting curve. These graphs are shown as dashed curves in Figure 6-27. The

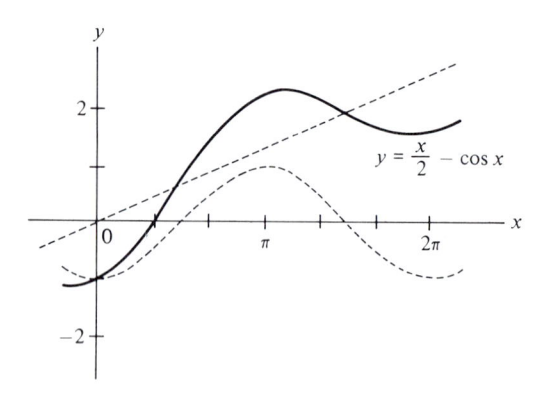

Figure 6-27

important points on the resulting curve are obtained by using the values of x corresponding to the points where the value of one function is zero and amplitude values of $y = -\cos x$.

Example C
Sketch the graph of $y = 2 \cos (2x + \frac{\pi}{2}) - 3 \sin (x - \frac{\pi}{4})$.

The graphs of $y = 2 \cos (2x + \frac{\pi}{2})$ and $y = -3 \sin (x - \frac{\pi}{4})$ are shown in Figure 6-28 as dashed curves. The graph of $y = 2 \cos (2x + \frac{\pi}{2}) - 3 \sin (x - \frac{\pi}{4})$ is shown as the solid curve in Figure 6-28.

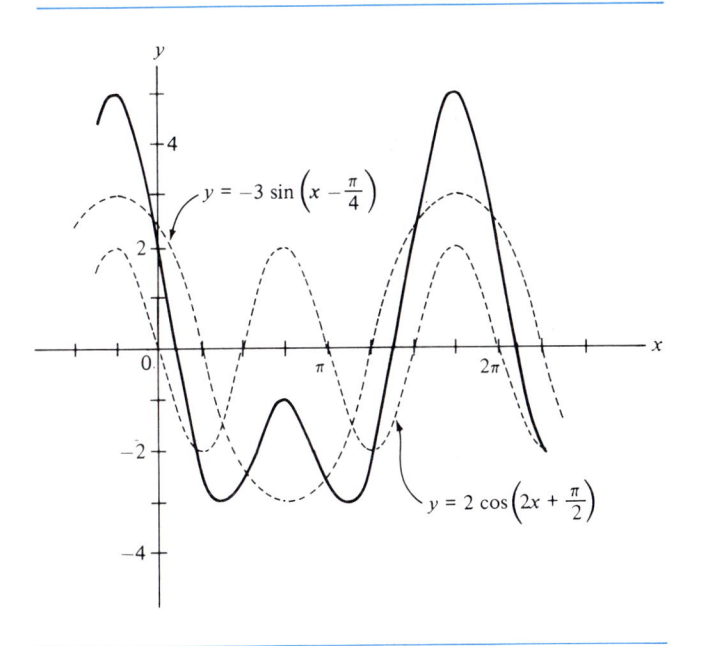

Figure 6-28

Example D
Sketch the graph of $y = \cos \pi x - 2 \sin 2x$.

The curves of $y = \cos \pi x$ and $y = -2 \sin 2x$ are shown as dashed curves in Figure 6-29. Points for the resulting curve are found primarily at the x-values where each of the curves has its zero or amplitude values. Again, special care should be taken where one curve is negative and the other is positive.

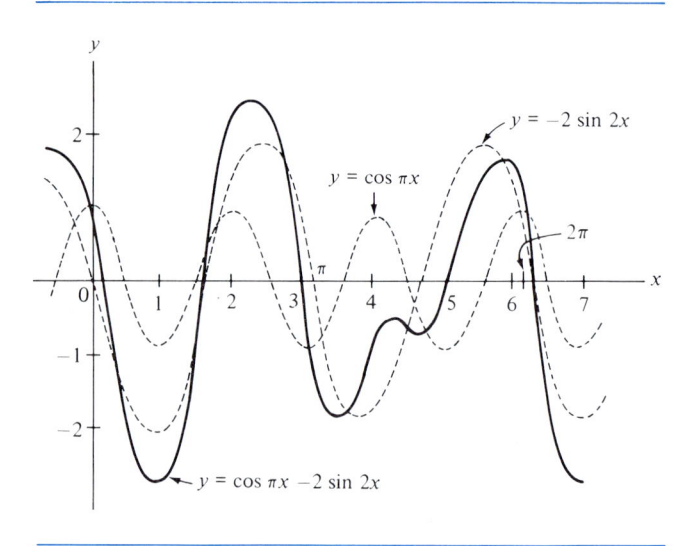

Figure 6-29

Exercises 6-5

In the following exercises use the method of addition of ordinates to sketch the given curves.

1. $y = \sin x + \cos x$
2. $y = \sin x + 2 \cos x$
3. $y = \sin 2x + 2 \cos x$
4. $y = \sin x + \cos 2x$
5. $y = 2 \sin x - \cos x$
6. $y = 2 \cos x - \sin x$
7. $y = \sin x - \sin 2x$
8. $y = \cos 3x - \sin x$
9. $y = x + \sin x$
10. $y = x + 2 \cos x$
11. $y = \frac{1}{10} x^2 - \sin 2x$
12. $y = \frac{1}{5} x^3 - \cos 2x$
13. $y = 2 \cos 2x + 3 \sin x$
14. $y = 2 \sin 3x + 3 \cos x$
15. $y = \frac{1}{2} \sin 4x + \cos 2x$
16. $y = \frac{1}{2} \cos 4x + \sin 2x$

6-5 Graphing by Addition of Ordinates   207

17. $y = 2 \sin \dfrac{x}{2} - \sin x$

18. $y = \cos x - 3 \cos \dfrac{x}{3}$

19. $y = 2 \cos \dfrac{x}{3} + \sin \dfrac{x}{4}$

20. $y = 3 \sin \dfrac{x}{3} + 2 \cos \dfrac{x}{2}$

21. $y = \sin 3x + \cos\left(x - \dfrac{\pi}{2}\right)$

22. $y = 3 \sin 2x + 2 \cos\left(x + \dfrac{\pi}{4}\right)$

23. $y = 2 \cos 4x - \cos\left(x - \dfrac{\pi}{4}\right)$

24. $y = 3 \sin 4x - \sin\left(x - \dfrac{\pi}{4}\right)$

25. $y = 3 \cos\left(2x + \dfrac{\pi}{2}\right) + 2 \sin 2x$

26. $y = 2 \sin\left(2x - \dfrac{\pi}{2}\right) + 3 \cos 2x$

27. $y = 2 \sin\left(2x - \dfrac{\pi}{6}\right) + \cos\left(2x + \dfrac{\pi}{3}\right)$

28. $y = 2 \cos\left(x + \dfrac{\pi}{3}\right) - 3 \sin\left(2x + \dfrac{\pi}{6}\right)$

29. $y = \sin \pi x - 2 \cos \pi x$

30. $y = 2 \sin 3\pi x - 3 \sin 2\pi x$

31. $y = 4 \cos \dfrac{3\pi}{2} x + 2 \sin x$

32. $y = \tfrac{1}{2} \sin 3\pi x + 2 \cos x$

33. The current i in a certain electric circuit is produced by two sources such that $i = 4 \sin 60\pi t + 2 \cos 120\pi t$. Sketch the graph representing the current i in amperes and the time t in seconds.

34. An object oscillating on a spring is subject to two forces such that its displacement y is given by $y = 0.4 \sin 4t + 0.3 \cos 4t$. Sketch y (in feet) versus t (in seconds).

Exercises for Chapter 6

In Exercises 1 through 32 sketch the curves of the given trigonometric functions.

1. $y = \frac{2}{3} \sin x$
2. $y = -4 \sin x$
3. $y = -2 \cos x$
4. $y = 2.3 \cos x$
5. $y = \sin 3x$
6. $y = \sin \frac{1}{2} x$
7. $y = \cos \frac{1}{3} x$
8. $y = \cos 4x$
9. $y = 3 \sin \frac{1}{2} x$
10. $y = 2 \sin 3x$
11. $y = 2 \cos 2x$
12. $y = 4 \cos 6x$
13. $y = \sin \pi x$
14. $y = 3 \sin 4\pi x$
15. $y = 5 \cos 2\pi x$
16. $y = -\cos 3\pi x$
17. $y = 2 \sin \left(3x - \frac{\pi}{2} \right)$
18. $y = -3 \sin \left(2x - \frac{\pi}{4} \right)$
19. $y = -\frac{3}{2} \sin \left(\frac{x}{2} + \frac{\pi}{2} \right)$
20. $y = \frac{4}{3} \sin \left(\frac{x}{4} + \frac{\pi}{4} \right)$
21. $y = -4 \cos (3x + \pi)$
22. $y = 2 \cos \left(2x + \frac{\pi}{2} \right)$
23. $y = 0.8 \cos \left(\frac{x}{6} - \frac{\pi}{2} \right)$
24. $y = -\frac{1}{2} \cos \left(\frac{x}{3} - \frac{\pi}{3} \right)$
25. $y = -\sin \left(\pi x + \frac{\pi}{6} \right)$
26. $y = 2 \sin (3\pi x - \pi)$
27. $y = 8 \cos \left(4\pi x - \frac{\pi}{2} \right)$
28. $y = 3 \cos (2\pi x + \pi)$
29. $y = 3 \tan x$
30. $y = -5 \cot x$
31. $y = -\frac{1}{3} \csc x$
32. $y = \frac{1}{4} \sec x$

Exercises for Chapter 6

In Exercises 33 through 40 sketch the given curves by the method of addition of ordinates.

33. $y = \sin 3x + 2 \cos 2x$
34. $y = \sin 2x + 3 \cos x$
35. $y = 2 \sin x - \cos 2x$
36. $y = \sin 3x - 2 \cos x$
37. $y = \frac{1}{2}x - \cos \frac{1}{3}x$
38. $y = \frac{1}{2} \sin 2x - x$
39. $y = \cos\left(x + \frac{\pi}{4}\right) - 2 \sin 2x$
40. $y = 2 \cos \pi x + \cos(2\pi x - \pi)$

In Exercises 41 through 44 sketch the indicated curves.

41. The voltage in a certain alternating current circuit is given by $y = 120 \sin 120\pi t$, where t represents time in seconds.
42. The angular displacement θ of a certain pendulum bob is given by $\theta = 0.1 \cos 2t$, where t represents time in seconds and θ is measured in radians.
43. An object is h meters high and d meters from an observer. The relationship giving d as a function of h is $d = h \cot \theta$, where θ is the angle of elevation of the top of the object. Sketch d as a function of θ for an object 100 m high.
44. In optics, two waves are said to interfere destructively if, when they pass through the same medium, the amplitude of the resulting wave is zero at all points in the medium. Sketch the curve of $y = \sin x + \cos(x + \frac{\pi}{2})$ and determine whether or not it would represent destructive interference of two waves.

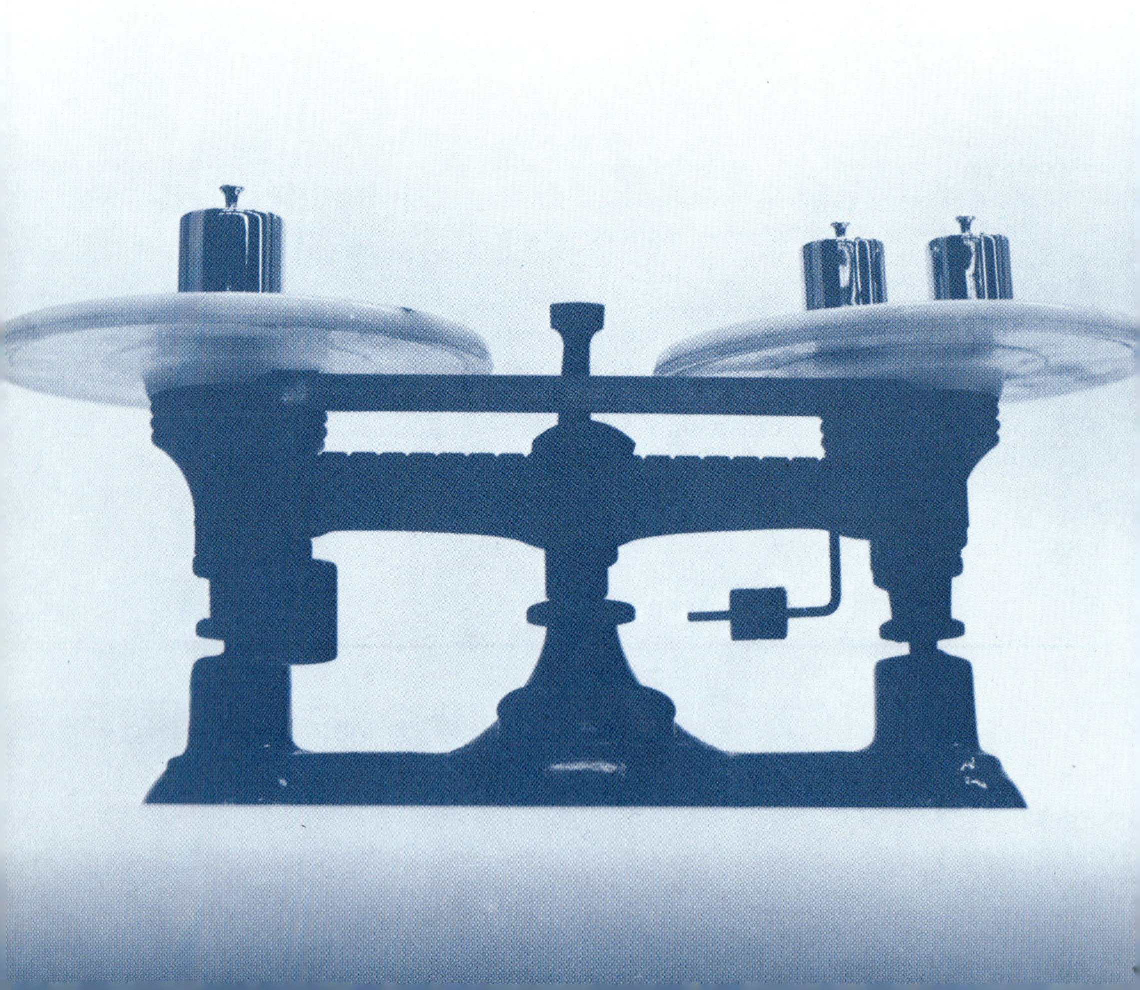

7

Basic Trigonometric Relationships

In trigonometric identities the two sides are equivalent, although this may not appear to be the case. We must maintain equality while we change one side into the form of the other side.

7-1 Fundamental Trigonometric Identities

The definitions of the trigonometric functions were introduced in Section 2-2 and were again summarized in Section 3-1. If we take a close look at these definitions, we find that there are many relationships among the various functions. For example, the definition of the sine of an angle is $\sin \theta = y/r$ and the definition of the cosecant of an angle is $\csc \theta = r/y$. But we know that $1/(r/y) = y/r$, which means that $\sin \theta = 1/\csc \theta$. In writing this down we made no reference to any particular angle, and since the definitions hold for *any* angle, this relation between $\sin \theta$ and $\csc \theta$ also holds for any angle. A relation such as this, which holds for any value of the variable, is called an *identity*. Of course, specific values where division by zero would be indicated are excluded.

Several important identities exist among the six trigonometric functions, and in this section we develop these identities. We also show how we can use the basic identities to verify other identities among the functions.

By the definitions, we have

$$\sin \theta \csc \theta = \frac{y}{r} \cdot \frac{r}{y} = 1 \quad \text{or} \quad \sin \theta = \frac{1}{\csc \theta} \quad \text{or} \quad \csc \theta = \frac{1}{\sin \theta}$$

$$\cos \theta \sec \theta = \frac{x}{r} \cdot \frac{r}{x} = 1 \quad \text{or} \quad \cos \theta = \frac{1}{\sec \theta} \quad \text{or} \quad \sec \theta = \frac{1}{\cos \theta}$$

$$\tan \theta \cot \theta = \frac{y}{x} \cdot \frac{x}{y} = 1 \quad \text{or} \quad \tan \theta = \frac{1}{\cot \theta} \quad \text{or} \quad \cot \theta = \frac{1}{\tan \theta}$$

$$\frac{\sin \theta}{\cos \theta} = \frac{y/r}{x/r} = \frac{y}{x} = \tan \theta$$

$$\frac{\cos \theta}{\sin \theta} = \frac{x/r}{y/r} = \frac{x}{y} = \cot \theta$$

Also, by the definitions and the Pythagorean theorem in the form of $x^2 + y^2 = r^2$, we arrive at the following identities.

By dividing the Pythagorean relation through by r^2, we have

$$\left(\frac{x}{r}\right)^2 + \left(\frac{y}{r}\right)^2 = 1$$

which leads us to $\cos^2 \theta + \sin^2 \theta = 1$.

By dividing the Pythagorean relation by x^2, we have

$$1 + \left(\frac{y}{x}\right)^2 = \left(\frac{r}{x}\right)^2$$

which leads us to $1 + \tan^2 \theta = \sec^2 \theta$.

By dividing the Pythagorean relation by y^2, we have

$$\left(\frac{x}{y}\right)^2 + 1 = \left(\frac{r}{y}\right)^2$$

which leads us to $\cot^2 \theta + 1 = \csc^2 \theta$.

Summarizing these results, we have the following important identities among the trigonometric functions:

(7-1) $\sin \theta = \dfrac{1}{\csc \theta}$	(7-2) $\cos \theta = \dfrac{1}{\sec \theta}$
(7-3) $\tan \theta = \dfrac{1}{\cot \theta}$	(7-4) $\tan \theta = \dfrac{\sin \theta}{\cos \theta}$
(7-5) $\cot \theta = \dfrac{\cos \theta}{\sin \theta}$	(7-6) $\sin^2 \theta + \cos^2 \theta = 1$
(7-7) $1 + \tan^2 \theta = \sec^2 \theta$	(7-8) $1 + \cot^2 \theta = \csc^2 \theta$

In using these identities, θ may stand for any angle or number or expression representing them.

7-1 Fundamental Trigonometric Identities

Example A

$$\sin(x+1) = \frac{1}{\csc(x+1)}$$

$$\tan 157° = \frac{\sin 157°}{\cos 157°}$$

$$1 + \tan^2\left(\frac{\pi}{6}\right) = \sec^2\left(\frac{\pi}{6}\right)$$

Example B

We shall verify three of the identities for particular values of θ.

From Table 3 in Appendix D, we find that $\cos 52° = 0.6157$ and $\sec 52° = 1.624$. Using Equation (7-2) and dividing, we find that

$$\cos 52° = \frac{1}{\sec 52°} = \frac{1}{1.624} = 0.6157$$

and this value checks.

Using Table 3, we find that $\sin 157° = 0.3907$ and $\cos 157° = -0.9205$. Using Equation (7-4) and dividing, we find that

$$\tan 157° = \frac{\sin 157°}{\cos 157°} = \frac{0.3907}{-0.9205} = -0.4245.$$

Checking with Table 3, we see that this value checks.
From Section 2-3, we recall that

$$\sin 45° = \frac{1}{\sqrt{2}} = \frac{\sqrt{2}}{2} \quad \text{and} \quad \cos 45° = \frac{\sqrt{2}}{2}$$

Using Equation (7-6), we have

$$\sin^2 45° + \cos^2 45° = \left(\frac{\sqrt{2}}{2}\right)^2 + \left(\frac{\sqrt{2}}{2}\right)^2 = \frac{1}{2} + \frac{1}{2} = 1.$$

We see that this identity checks for these values.

A great many identities exist among the trigonometric functions. A number of these are found in applications, where certain problems rely on a change of form for solution. This is especially true in calculus.

We are going to use the basic identities and a few additional ones given in later sections to prove the validity of these other identities. One of the reasons for doing this is for you to become familiar with the basic identities. The ability to prove such identities depends to a large extent on being very familiar with the basic identities, so that you can recognize them in somewhat different forms. If you do not learn these basic identities and learn them well, you will have difficulty in following the examples and doing the exercises. The more readily you recognize these forms, the more easily you will be able to prove such identities.

In proving identities, we should look for combinations which appear in, or are very similar to, those in the basic identities. Consider the following examples.

Example C
If a certain identity contained the expression $\sin^4 x - \cos^4 x$, we should note quickly that this may be factored into $(\sin^2 x - \cos^2 x)(\sin^2 x + \cos^2 x)$ and that $\sin^2 x + \cos^2 x$ may be replaced by 1. In this way, expression is simplified for use in proving the identity. Thus, $\sin^4 x - \cos^4 x = \sin^2 x - \cos^2 x$.

Example C also illustrates another important point. To prove identities, basic algebraic operations must be performed. Unless these operations are performed carefully and correctly, an improper algebraic step may make the proof appear impossible.

Example D
Either of the expressions $\tan^2 x$ or $\sec^2 x$ should bring to mind the possibility of using the squared relation of Equation (7-7).

$$\frac{\sec^2 y}{\cot y} - \tan^3 y = \frac{\sec^2 y}{1/\tan y} - \tan^3 y = \sec^2 y \tan y - \tan^3 y$$

$$= \tan y (\sec^2 y - \tan^2 y) = \tan y (1) = \tan y$$

Here we have used Equation (7-7) in the form $\sec^2 y - \tan^2 y = 1$.

Example E
The combination $1 - \sin x$ also suggests $1 - \sin^2 x$, since multiplying $(1 - \sin x)$ by $(1 + \sin x)$ gives $1 - \sin^2 x$, which can then be replaced by $\cos^2 x$.

$$\frac{1 - \sin x}{\sin x \cot x} = \frac{(1 - \sin x)(1 + \sin x)}{\sin x \cot x (1 + \sin x)}$$

$$= \frac{1 - \sin^2 x}{\sin x \frac{\cos x}{\sin x} (1 + \sin x)} = \frac{\cos^2 x}{\cos x (1 + \sin x)}$$

$$= \frac{\cos x}{1 + \sin x}$$

In problems involving proving identities, we are given both sides and must show by changing one side or the other that the two expressions are equal. Although this restriction is not entirely necessary, we shall restrict the method of proof to changing only one side into the same form as the other side. In this way, we know precisely what form we are to change to, and therefore, by looking ahead to what changes may be made, we are better able to make the proper changes.

There is no set procedure which can be stated for working with identities. The most important factors are to recognize the proper forms, see what effect any change may have before we actually perform it, and then perform it correctly. Normally it is more profitable to change the more complicated side of an identity to the same form as the less complicated side. If the two sides are of approximately the same complexity, a close look at each side usually suggests steps which will lead to the solution.

Example F
Prove the identity

$$\frac{\cos x \csc x}{\cot^2 x} = \tan x.$$

First, we note that left-hand side has several factors and the right-hand side has only one. Therefore, let us transform the left-hand side. Next, we note that we want $\tan x$ as the result. We know that $\cot x = 1/\tan x$. Thus,

$$\frac{\cos x \csc x}{\cot^2 x} = \frac{\cos x \csc x}{1/\tan^2 x} = \cos x \csc x \tan^2 x.$$

At this point, we have two factors of $\tan x$ on the left. Since we want only one, let us factor out one. Therefore,

$$\cos x \csc x \tan^2 x = \tan x \, (\cos x \csc x \tan x).$$

Now, replacing $\tan x$ within the parentheses by $\sin x / \cos x$, we have

$$\tan x \, (\cos x \csc x \tan x) = \frac{\tan x \, (\cos x \csc x \sin x)}{\cos x}.$$

Now we may cancel $\cos x$. Also, $\csc x \sin x = 1$ from Equation (7-1). Finally,

$$\frac{\tan x \, (\cos x \csc x \sin x)}{\cos x} = \tan x \left(\frac{\cos x}{\cos x}\right)(\csc x \sin x)$$

$$= \tan x (1)(1) = \tan x.$$

Since we have transformed the left-hand side into $\tan x$, we have proved the identity. Of course, it is not necessary to rewrite expressions as we did in this example. This was done here only to include the explanations between steps.

Example G
Prove the identity $\sec^2 x + \csc^2 x = \sec^2 x \csc^2 x$.

Here we note the presence of $\sec^2 x$ and $\csc^2 x$ on each side. This suggests the possible use of the square relationships. By replacing the $\sec^2 x$ on the right-hand side by $1 + \tan^2 x$, we can create $\csc^2 x$ plus another term. The left-hand side is the $\csc^2 x$ plus another term, so this procedure should help. Thus,

7-1 Fundamental Trigonometric Identities

$$\sec^2 x + \csc^2 x = \sec^2 x \csc^2 x$$
$$= (1 + \tan^2 x)(\csc^2 x)$$
$$= \csc^2 x + \tan^2 x \csc^2 x.$$

Now we note that $\tan x = \sin x/\cos x$ and $\csc x = 1/\sin x$. Thus,

$$\sec^2 x + \csc^2 x = \csc^2 x + \left(\frac{\sin^2 x}{\cos^2 x}\right)\left(\frac{1}{\sin^2 x}\right)$$
$$= \csc^2 x + \frac{1}{\cos^2 x}.$$

But $\sec x = 1/\cos x$, and therefore $\sec^2 x + \csc^2 x = \csc^2 x + \sec^2 x$, which therefore proves the identity. We could have used many other variations of this procedure, and they would have been perfectly valid.

Example H
Prove the identity

$$\frac{\csc x}{\tan x + \cot x} = \cos x.$$

Here we shall simplify the left-hand side until we have the expression which appears on the right-hand side.

$$\frac{\csc x}{\tan x + \cot x} = \frac{\csc x}{\tan x + \dfrac{1}{\tan x}} = \frac{\csc x}{\dfrac{\tan^2 x + 1}{\tan x}}$$

$$= \frac{\csc x \tan x}{\tan^2 x + 1} = \frac{\csc x \tan x}{\sec^2 x}$$

$$= \frac{\dfrac{1}{\sin x} \cdot \dfrac{\sin x}{\cos x}}{\dfrac{1}{\cos^2 x}} = \frac{1}{\sin x} \cdot \frac{\sin x}{\cos x} \cdot \frac{\cos^2 x}{1}$$

$$= \cos x.$$

Therefore, we have $\cos x = \cos x$, which proves the identity.

Exercises 7-1

In Exercises 1 through 4 verify the indicated basic identities for the given angles.

1. Verify Equation (7-3) for $\theta = 56°$.
2. Verify Equation (7-5) for $\theta = 80°$.
3. Verify Equation (7-6) for $\theta = \dfrac{2\pi}{3}$.
4. Verify Equation (7-8) for $\theta = \dfrac{7\pi}{6}$.

In Exercises 5 through 46 prove the given identities.

5. $\sin \theta \cot \theta = \cos \theta$
6. $\cos x \tan x = \sin x$
7. $\tan y \csc y = \sec y$
8. $\cot \theta \sec \theta = \csc \theta$
9. $\sin x \sec x = \tan x$
10. $\cos y \csc y = \cot y$
11. $\dfrac{\cot \theta}{\cos \theta} = \csc \theta$
12. $\dfrac{\tan y}{\sin y} = \sec y$
13. $\sin y \tan y + \cos y = \sec y$
14. $\cos \theta \cot \theta + \sin \theta = \csc \theta$
15. $\sec x \csc x - \cot x = \tan x$
16. $\csc x \sec x - \tan x = \cot x$
17. $\csc^2 \theta (1 - \cos^2 \theta) = 1$
18. $\cos^2 y (1 + \tan^2 y) = 1$
19. $\sin x (1 + \cot^2 x) = \csc x$
20. $\sec \theta (1 - \sin^2 \theta) = \cos \theta$
21. $\sin x (\csc x - \sin x) = \cos^2 x$
22. $\cos y (\sec y - \cos y) = \sin^2 y$
23. $\tan y (\cot y + \tan y) = \sec^2 y$
24. $\csc x (\csc x - \sin x) = \cot^2 x$
25. $\cot \theta \sec^2 \theta - \cot \theta = \tan \theta$
26. $\sin y + \sin y \cot^2 y = \csc y$
27. $\sin \theta \sec^2 \theta - \sin \theta \tan^2 \theta = \sin \theta$
28. $\cos \theta \csc^2 \theta - \cos \theta \cot^2 \theta = \cos \theta$
29. $\tan x + \cot x = \sec x \csc x$
30. $\tan x + \cot x = \tan x \csc^2 x$
31. $\cos^2 y - \sin^2 y = 1 - 2 \sin^2 y$
32. $\tan^2 y \sec^2 y - \tan^4 y = \tan^2 y$
33. $\dfrac{\sin x}{1 - \cos x} = \csc x + \cot x$

7-2 Functions of the Sum and Difference of Two Angles   221

34. $\dfrac{1 + \cos \theta}{\sin \theta} = \dfrac{\sin \theta}{1 - \cos \theta}$

35. $\dfrac{\sec \theta + \csc \theta}{1 + \tan \theta} = \csc \theta$

36. $\dfrac{\cot x + 1}{\cot x} = 1 + \tan x$

37. $\tan^2 y \cos^2 y + \cot^2 y \sin^2 y = 1$

38. $\cos^3 \theta \csc^3 \theta \tan^3 \theta = \csc^2 \theta - \cot^2 \theta$

39. $4 \sin x + \tan x = \dfrac{4 + \sec x}{\csc x}$

40. $\dfrac{1 + \tan y}{\sin y} - \sec y = \csc y$

41. $\sec \theta + \tan \theta + \cot \theta = \dfrac{1 + \sin \theta}{\cos \theta \sin \theta}$

42. $\sec y (\sec y - \cos y) + \dfrac{\cos y - \sin y}{\cos y} + \tan y = \sec^2 y$

43. $2 \sin^4 y - 3 \sin^2 y + 1 = \cos^2 y (1 - 2 \sin^2 y)$

44. $\dfrac{\sin^4 \theta - \cos^4 \theta}{1 - \cot^4 \theta} = \sin^4 \theta$

45. $\dfrac{(\sec x - 1)^2}{\tan^2 x} = \dfrac{\sec x - 1}{\sec x + 1}$

46. $\sec^4 x - \tan^4 x = \dfrac{1 + \sin^2 x}{\cos^2 x}$

7-2 Trigonometric Functions of the Sum and Difference of Two Angles

There are other important relations among the trigonometric functions. The most important and useful relations are those which involve twice an angle and half an angle. To obtain these relations, we first derive the expressions for the sine and cosine of the sum and difference of two angles. These expressions lead directly to the desired relations of double and half angles.

In Figure 7-1, the angle α is in standard position and the angle β has as its initial side the terminal side of α. Thus, the angle of interest, $\alpha + \beta$, is in standard position. From a point P on the terminal side of $\alpha + \beta$, perpendiculars are dropped to the x-axis and to the terminal side of α at points M and Q, respectively. Then perpendiculars are dropped from Q to the x-axis and to the line MP at points N and R, respectively. By this construction, $\angle RPQ$ is equal to $\angle \alpha$. (The $\angle RQO = \angle \alpha$ by alternate interior angles; $\angle RQO + \angle RQP = 90°$ by construction; $\angle RQP + \angle RPQ = 90°$ by the sum of the angles of a triangle being $180°$. Thus $\angle \alpha = \angle RPQ$.) By definition,

$$\sin(\alpha + \beta) = \frac{MP}{OP} = \frac{MR + RP}{OP} = \frac{NQ}{OP} + \frac{RP}{OP}$$

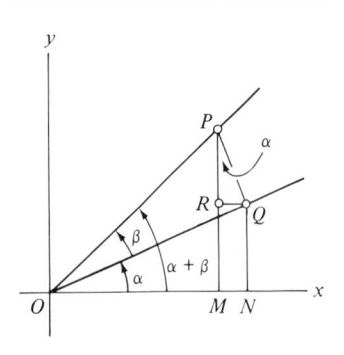

Figure 7-1

These last two fractions do not define any function of either α or β, and therefore we multiply the first fraction (numerator and denominator) by OQ and the second fraction by QP. When we do this and rearrange the fractions, we have functions of α and β:

$$\sin(\alpha + \beta) = \frac{NQ}{OP} + \frac{RP}{OP} = \frac{NQ}{OQ} \cdot \frac{OQ}{OP} + \frac{RP}{QP} \cdot \frac{QP}{OP}$$

$$= \sin \alpha \cos \beta + \cos \alpha \sin \beta.$$

Using the same figure, we can also obtain the expression for $\cos(\alpha + \beta)$. Thus, we have the relations:

(7-9) $\sin(\alpha + \beta) = \sin \alpha \cos \beta + \cos \alpha \sin \beta$

and

(7-10) $\cos(\alpha + \beta) = \cos \alpha \cos \beta - \sin \alpha \sin \beta$

To find an expression for $\tan(\alpha + \beta)$, we use the identity $\tan \theta = \sin \theta / \cos \theta$. Thus,

$$\tan(\alpha + \beta) = \frac{\sin(\alpha + \beta)}{\cos(\alpha + \beta)} = \frac{\sin \alpha \cos \beta + \cos \alpha \sin \beta}{\cos \alpha \cos \beta - \sin \alpha \sin \beta}.$$

7-2 Functions of the Sum and Difference of Two Angles

Now, dividing the numerator and the denominator by $\cos \alpha \cos \beta$, we have

$$\tan(\alpha + \beta) = \frac{\dfrac{\sin \alpha \cos \beta}{\cos \alpha \cos \beta} + \dfrac{\cos \alpha \sin \beta}{\cos \alpha \cos \beta}}{\dfrac{\cos \alpha \cos \beta}{\cos \alpha \cos \beta} - \dfrac{\sin \alpha \sin \beta}{\cos \alpha \cos \beta}} = \frac{\dfrac{\sin \alpha}{\cos \alpha} + \dfrac{\sin \beta}{\cos \beta}}{1 - \dfrac{\sin \alpha}{\cos \alpha} \cdot \dfrac{\sin \beta}{\cos \beta}}$$

Now, using the identities from Section 7-1, we have:

(7-11) $\quad \tan(\alpha + \beta) = \dfrac{\tan \alpha + \tan \beta}{1 - \tan \alpha \tan \beta}$

Example A
Find $\sin 75°$ from $\sin 75° = \sin(45° + 30°)$.

$\sin 75° = \sin(45° + 30°) = \sin 45° \cos 30° + \cos 45° \sin 30°$

$= \dfrac{\sqrt{2}}{2} \cdot \dfrac{\sqrt{3}}{2} + \dfrac{\sqrt{2}}{2} \cdot \dfrac{1}{2} = \dfrac{\sqrt{6}}{4} + \dfrac{\sqrt{2}}{4} = \dfrac{\sqrt{6} + \sqrt{2}}{4}$

$= 0.9659$

Example B
Verify that $\sin 90° = 1$ by finding $\sin(60° + 30°)$.

$\sin 90° = \sin(60° + 30°) = \sin 60° \cos 30° + \cos 60° \sin 30°$

$= \dfrac{\sqrt{3}}{2} \cdot \dfrac{\sqrt{3}}{2} + \dfrac{1}{2} \cdot \dfrac{1}{2} = \dfrac{3}{4} + \dfrac{1}{4} = 1$

It should be obvious from this example that $\sin(\alpha + \beta)$ is *not* equal to $\sin \alpha + \sin \beta$, something which many students assume before they are familiar with the formulas and ideas of this section. If we used such a formula, we would get $\sin 90° = \sqrt{3}/2 + \frac{1}{2} = 1.366$ for the combination $(60° + 30°)$. This is not possible, since the values of the sine never exceed 1 in value. Also, if we used the combination $(45° + 45°)$, we would get 1.414, a different value for the same function.

From Equations (7-9), (7-10), and (7-11), we can easily find expressions for sin $(\alpha - \beta)$, cos $(\alpha - \beta)$, and tan $(\alpha - \beta)$. This is done by finding the expressions for sin $[\alpha + (-\beta)]$, cos $[\alpha + (-\beta)]$, and tan $[\alpha + (-\beta)]$. Thus, we have

$$\sin(\alpha - \beta) = \sin[\alpha + (-\beta)] = \sin\alpha\cos(-\beta) + \cos\alpha\sin(-\beta).$$

Since cos $(-\beta) = \cos\beta$ and sin $(-\beta) = -\sin\beta$ (see Equations 3-6 of Section 3-2), we have:

$$(7\text{-}12) \quad \sin(\alpha - \beta) = \sin\alpha\cos\beta - \cos\alpha\sin\beta$$

In the same manner we find that:

$$(7\text{-}13) \quad \cos(\alpha - \beta) = \cos\alpha\cos\beta + \sin\alpha\sin\beta$$

and

$$(7\text{-}14) \quad \tan(\alpha - \beta) = \frac{\tan\alpha - \tan\beta}{1 + \tan\alpha\tan\beta}$$

Example C

Find cos 15° from cos (45° − 30°).

$$\cos 15° = \cos(45° - 30°) = \cos 45° \cos 30° + \sin 45° \sin 30°$$

$$= \frac{\sqrt{2}}{2} \cdot \frac{\sqrt{3}}{2} + \frac{\sqrt{2}}{2} \cdot \frac{1}{2} = \frac{\sqrt{6} + \sqrt{2}}{4} = 0.9659$$

We get the same results as in Example A, which should be the case since sin 75° = cos 15°. (See Section 2-4.)

By using Equations (7-9), (7-10), and (7-11), it is possible to find expressions for cot $(\alpha + \beta)$, sec $(\alpha + \beta)$, and csc $(\alpha + \beta)$. These expressions are generally less useful than those for the sine, cosine, and tangent. Therefore, we shall not derive them here, although an expression for cot $(\alpha + \beta)$ is found in the exercises at the end of this section. By using Equations

7-2 Functions of the Sum and Difference of Two Angles   225

(7-12), (7-13), and (7-14), we could also find similar expressions for the functions of $(\alpha - \beta)$.

Certain trigonometric identities can also be worked out by using the formulas derived in this section. The following examples illustrate the use of these formulas in identities.

Example D
Show that

$$\frac{\sin(\alpha - \beta)}{\sin \alpha \sin \beta} = \cot \beta - \cot \alpha.$$

By using Equation (7-12), we have

$$\frac{\sin(\alpha - \beta)}{\sin \alpha \sin \beta} = \frac{\sin \alpha \cos \beta - \cos \alpha \sin \beta}{\sin \alpha \sin \beta} = \frac{\sin \alpha \cos \beta}{\sin \alpha \sin \beta} - \frac{\cos \alpha \sin \beta}{\sin \alpha \sin \beta}$$

$$= \frac{\cos \beta}{\sin \beta} - \frac{\cos \alpha}{\sin \alpha} = \cot \beta - \cot \alpha.$$

Example E
Show that

$$\sin\left(\frac{\pi}{4} + x\right) \cos\left(\frac{\pi}{4} + x\right) = \frac{1}{2}(\cos^2 x - \sin^2 x).$$

The calculations are as follows:

$$\sin\left(\frac{\pi}{4} + x\right) \cos\left(\frac{\pi}{4} + x\right)$$

$$= \left(\sin\frac{\pi}{4} \cos x + \cos\frac{\pi}{4} \sin x\right)\left(\cos\frac{\pi}{4} \cos x - \sin\frac{\pi}{4} \sin x\right)$$

$$= \sin\frac{\pi}{4} \cos\frac{\pi}{4} \cos^2 x - \sin^2 \frac{\pi}{4} \sin x \cos x + \cos^2 \frac{\pi}{4} \sin x \cos x$$

$$\quad - \sin^2 x \sin\frac{\pi}{4} \cos\frac{\pi}{4}$$

$$= \frac{\sqrt{2}}{2} \frac{\sqrt{2}}{2} \cos^2 x - \left(\frac{\sqrt{2}}{2}\right)^2 \sin x \cos x + \left(\frac{\sqrt{2}}{2}\right)^2 \sin x \cos x$$

$$\quad - \frac{\sqrt{2}}{2} \frac{\sqrt{2}}{2} \sin^2 x$$

$$= \frac{1}{2} \cos^2 x - \frac{1}{2} \sin^2 x = \frac{1}{2}(\cos^2 x - \sin^2 x).$$

Example F

Show that $\sin(x+y)\cos y - \cos(x+y)\sin y = \sin x$.

If we let $x + y = z$, we note that the left-hand side of the above expression becomes $\sin z \cos y - \cos z \sin y$, which is the proper form for $\sin(z - y)$. By replacing z with $x + y$, we obtain $(x + y - y)$, which is $\sin x$. Therefore, the above expression has been shown to be true. We again see that proper recognition of a basic form leads to the solution.

Exercises 7-2

In Exercises 1 through 8 determine the values of the given functions as indicated.

1. Find $\sin 105°$ by using $105° = 60° + 45°$.
2. Find $\cos 75°$ by using $75° = 30° + 45°$.
3. Find $\tan 15°$ by using $15° = 60° - 45°$.
4. Find $\sin 15°$ by using $15° = 45° - 30°$.
5. Find $\cos 105°$ by using $105° = 135° - 30°$.
6. Find $\tan 210°$ by using $210° = 150° + 60°$.
7. Find $\sin 75°$ by using $75° = 120° - 45°$.
8. Find $\tan 165°$ by using $165° = 120° + 45°$.

In Exercises 9 through 12 evaluate the indicated functions with the following given information: $\sin \alpha = \frac{4}{5}$ (in the first quadrant) and $\cos \beta = -\frac{12}{13}$ (in the second quadrant).

9. $\sin(\alpha + \beta)$
10. $\cos(\beta - \alpha)$
11. $\cos(\alpha + \beta)$
12. $\tan(\alpha - \beta)$

In Exercises 13 through 20 reduce each of the given expressions to a single term. Expansion of any term is not necessary; proper recognition of the form of the expression leads to the proper result.

13. $\dfrac{\tan 4x - \tan 2x}{1 + \tan 4x \tan 2x}$

14. $\sin y \cos 2y + \cos y \sin 2y$

15. $\cos 4y \cos y + \sin 4y \sin y$

7-2 Functions of the Sum and Difference of Two Angles

16. $\dfrac{\tan 2x + \tan x}{1 - \tan 2x \tan x}$

17. $\sin 3x \cos y - \cos 3x \sin y$

18. $\cos 2x \cos 3y + \sin 2x \sin 3y$

19. $\cos(2x - y) \cos y - \sin(2x - y) \sin y$

20. $\sin(x + y) \cos y - \cos(x + y) \sin y$

In Exercises 21 through 32 prove the given identities.

21. $\sin(270° - x) = -\cos x$

22. $\sin(90° + x) = \cos x$

23. $\cos\left(\dfrac{\pi}{2} - x\right) = \sin x$

24. $\cos\left(\dfrac{3\pi}{2} + x\right) = \sin x$

25. $\tan(x + 180°) = \tan x$

26. $\tan(45° - x) = \dfrac{1 - \tan x}{1 + \tan x}$

27. $\sin\left(\dfrac{\pi}{4} + x\right) = \dfrac{\sin x + \cos x}{\sqrt{2}}$

28. $\cos\left(\dfrac{\pi}{3} + x\right) = \dfrac{\cos x - \sqrt{3} \sin x}{2}$

29. $\sin(x + y) \sin(x - y) = \sin^2 x - \sin^2 y$

30. $\cos(x + y) \cos(x - y) = \cos^2 x - \sin^2 y$

31. $\cos(x + y) + \cos(x - y) = 2 \cos x \cos y$

32. $\cos(x - y) + \sin(x + y) = (\cos x + \sin x)(\cos y + \sin y)$

In Exercises 33 through 36 additional trigonometric identities are shown. Derive in the indicated manner. Equations (7-16), (7-17), and (7-18) are known as the product formulas.

33. By dividing Equation (7-10) by Equation (7-9), show that

$$\cot(\alpha + \beta) = \dfrac{\cot \alpha \cot \beta - 1}{\cot \beta + \cot \alpha}. \qquad (7\text{-}15)$$

228 Basic Trigonometric Relationships

34. By adding Equations (7-9) and (7-12), derive the equation

(7-16) $$\sin\alpha\cos\beta = \frac{1}{2}[\sin(\alpha+\beta) + \sin(\alpha-\beta)].$$

35. By adding Equations (7-10) and (7-13), derive the equation

(7-17) $$\cos\alpha\cos\beta = \frac{1}{2}[\cos(\alpha+\beta) + \cos(\alpha-\beta)].$$

36. By subtracting Equation (7-10) from Equation (7-13), derive

(7-18) $$\sin\alpha\sin\beta = \frac{1}{2}[\cos(\alpha-\beta) - \cos(\alpha+\beta)].$$

In Exercises 37 through 40 additional trigonometric identities are shown. Derive them by letting $\alpha + \beta = x$ and $\alpha - \beta = y$, which leads to $\alpha = \frac{1}{2}(x+y)$ and $\beta = \frac{1}{2}(x-y)$. The resulting equations are known as the factor formulas.

37. Use Equation (7-16) and the substitutions above to derive the equation

(7-19) $$\sin x + \sin y = 2\sin\tfrac{1}{2}(x+y)\cos\tfrac{1}{2}(x-y).$$

38. Use Equations (7-9) and (7-12) and the substitutions above to derive the equation

(7-20) $$\sin x - \sin y = 2\sin\tfrac{1}{2}(x-y)\cos\tfrac{1}{2}(x+y).$$

39. Use Equation (7-17) and the substitutions above to derive the equation

(7-21) $$\cos x + \cos y = 2\cos\tfrac{1}{2}(x+y)\cos\tfrac{1}{2}(x-y).$$

40. Use Equation (7-18) and the substitutions above to derive the equation

(7-22) $$\cos x - \cos y = -2\sin\tfrac{1}{2}(x+y)\sin\tfrac{1}{2}(x-y).$$

7-3 Double-Angle Formulas

If we let $\beta = \alpha$ in Equations (7-9) and (7-10), we can derive the important double-angle formulas. Thus, by making this substitution in Equation (7-9), we have

$$\sin(\alpha + \alpha) = \sin(2\alpha) = \sin\alpha\cos\alpha + \cos\alpha\sin\alpha = 2\sin\alpha\cos\alpha.$$

Using the same substitution in Equation (7-10), we have

$$\cos(\alpha + \alpha) = \cos\alpha\cos\alpha - \sin\alpha\sin\alpha = \cos^2\alpha - \sin^2\alpha.$$

By using the basic identity (7-6), other forms of this last equation may be derived. Thus,

$$\cos 2\alpha = \cos^2\alpha - \sin^2\alpha$$
$$= \cos^2\alpha - (1 - \cos^2\alpha) = 2\cos^2\alpha - 1$$

or

$$\cos 2\alpha = (1 - \sin^2\alpha) - \sin^2\alpha = 1 - 2\sin^2\alpha.$$

Also,

$$\tan 2\alpha = \tan(\alpha + \alpha) = \frac{\tan\alpha + \tan\alpha}{1 - \tan\alpha\tan\alpha} = \frac{2\tan\alpha}{1 - \tan^2\alpha}.$$

Summarizing these formulas, we have:

(7-23) $\quad\sin 2\alpha = 2\sin\alpha\cos\alpha$

(7-24) $\quad\cos 2\alpha = \cos^2\alpha - \sin^2\alpha$

(7-25) $\quad\cos 2\alpha = 2\cos^2\alpha - 1$

(7-26) $\quad\cos 2\alpha = 1 - 2\sin^2\alpha$

(7-27) $\quad\tan 2\alpha = \dfrac{2\tan\alpha}{1 - \tan^2\alpha}$

Note carefully that these equations give expressions for the sine, cosine, and tangent of twice an angle in terms of functions of the angle. They can be used any time we have expressed one angle as twice another. These double-angle formulas are widely used in applications of trigonometry, especially in the calculus. They should be known and recognized quickly in any of the various forms.

Example A
If $\alpha = 30°$, we have $\cos 2(30°) = \cos 60° = \cos^2 30° - \sin^2 30°$. If $\alpha = 3x$, we have $\sin 2(3x) = \sin 6x = 2 \sin 3x \cos 3x$. If $2\alpha = x$, we may write $\alpha = x/2$, which means that

$$\tan 2\left(\frac{x}{2}\right) = \tan x = \frac{2 \tan \frac{x}{2}}{1 - \tan^2 \frac{x}{2}}.$$

Example B
Using the double-angle formulas, simplify the expression

$$\cos^2 2x - \sin^2 2x.$$

By using Equation (7-24) and letting $\alpha = 2x$, we have

$$\cos^2 2x - \sin^2 2x = \cos 2(2x) = \cos 4x.$$

Example C
Verify the values of $\sin 90°$, $\cos 90°$, and $\tan 90°$ by use of the functions of $45°$.

$$\sin 90° = \sin 2(45°) = 2 \sin 45° \cos 45° = 2 \left(\frac{\sqrt{2}}{2}\right)\left(\frac{\sqrt{2}}{2}\right) = 1$$

$$\cos 90° = \cos 2(45°) = \cos^2 45° - \sin^2 45° = \left(\frac{\sqrt{2}}{2}\right)^2 - \left(\frac{\sqrt{2}}{2}\right)^2 = 0$$

$$\tan 90° = \tan 2(45°) = \frac{2 \tan 45°}{1 - \tan^2 45°} = \frac{2(1)}{1 - 1^2} = \frac{2}{0} = \text{undefined}$$

7-3 Double-Angle Formulas

Example D
Given that $\cos \alpha = \frac{3}{5}$ (in the fourth quadrant), find $\tan 2\alpha$.

Knowing that $\cos \alpha = \frac{3}{5}$ for an angle in the fourth quadrant, we then determine that

$$\tan \alpha = -\frac{4}{3}.$$

(See Figure 7-2.) Thus,

$$\tan 2\alpha = \frac{2\left(-\frac{4}{3}\right)}{1 - \left(-\frac{4}{3}\right)^2} = \frac{24}{7}.$$

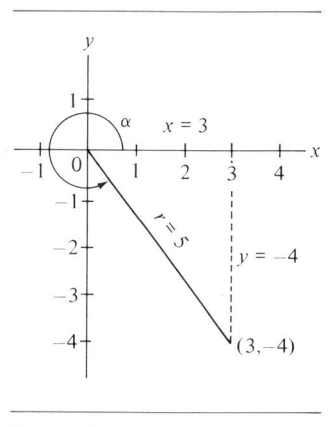

Figure 7-2

Example E
Prove the identity

$$\frac{2}{1 + \cos 2x} = \sec^2 x.$$

The solution is

$$\frac{2}{1 + \cos 2x} = \frac{2}{1 + (2\cos^2 x - 1)} = \frac{2}{2\cos^2 x} = \sec^2 x.$$

Example F
Show that $\dfrac{\sin 3x}{\sin x} + \dfrac{\cos 3x}{\cos x} = 4 \cos 2x$.

The first step is to combine the two fractions on the left so that we can see if any usable forms will emerge:

$$\frac{\sin 3x \cos x + \cos 3x \sin x}{\sin x \cos x} = 4 \cos 2x.$$

We now note that the numerator is of the form $\sin (A + x)$, where $A = 3x$. Also, the denominator is $\frac{1}{2} \sin 2x$. Making these substitutions, we have

$$\frac{\sin (3x + x)}{\frac{1}{2} \sin 2x} = \frac{2 \sin 4x}{\sin 2x} = 4 \cos 2x.$$

By expanding $\sin 4x$ into $2 \sin 2x \cos 2x$, we obtain

$$\frac{2(2 \sin 2x \cos 2x)}{\sin 2x} = 4 \cos 2x.$$

Therefore, the expression is shown to be valid.

Exercises 7-3

In Exercises 1 through 8 determine the values of the indicated functions in the given manner.

1. Find sin 60° by using the functions of 30°.
2. Find cos 60° by using the functions of 30°.
3. Find tan 120° by using the functions of 60°.
4. Find sin 120° by using the functions of 60°.
5. Find cos 240° by using the functions of 120°.
6. Find tan 240° by using the functions of 120°.
7. Find sin 300° by using the functions of 150°.
8. Find cos 300° by using the functions of 150°.

In Exercises 9 through 16 evaluate the indicated functions with the given information.

9. Find sin $2x$ if cos $x = \frac{4}{5}$ (in the first quadrant).
10. Find cos $2x$ if sin $x = -\frac{12}{13}$ (in the third quadrant).
11. Find tan $2x$ if tan $x = -\frac{3}{4}$ (in the second quadrant).
12. Find sin $2x$ if tan $x = \frac{12}{5}$ (in the third quadrant).
13. Find cos $2x$ if cos $x = \frac{3}{5}$ (in the fourth quadrant).
14. Find tan $2x$ if sin $x = \frac{5}{13}$ (in the second quadrant).
15. Find sin $4x$ if sin $x = \frac{3}{5}$ (in the first quadrant) [$4x = 2(2x)$].
16. Find cos $4x$ if tan $x = -\frac{4}{3}$ (in the fourth quadrant).

In Exercises 17 through 24 reduce the given expressions to a single term. Expansion of terms is not necessary; proper recognition of the expression leads to the proper result.

17. $4 \sin 4x \cos 4x$
18. $\dfrac{2 \tan 2x}{1 - \tan^2 2x}$
19. $4 \sin 3x \cos 3x$
20. $1 - 2 \sin^2 4x$

21. $\dfrac{\tan 3x}{1 - \tan^2 3x}$ 22. $4 \sin^2 x \cos^2 x$

23. $\sin^2 4x - \cos^2 4x$ 24. $2 - 4 \cos^2 2x$

In Exercises 25 through 32 prove the given identities.

25. $\cos^2 \alpha - \sin^2 \alpha = 2 \cos^2 \alpha - 1$

26. $\cos^2 \alpha - \sin^2 \alpha = 1 - 2 \sin^2 \alpha$

27. $\cot 2x = \dfrac{\cot x - \tan x}{2}$

28. $\cos^4 x - \sin^4 x = \cos 2x$

29. $(\sin x + \cos x)^2 = 1 + \sin 2x$

30. $2 \csc 2x \tan x = \sec^2 x$

31. $2 \sin x + \sin 2x = \dfrac{2 \sin^3 x}{1 - \cos x}$

32. $\dfrac{2 \tan x}{\sec^2 x - 2 \tan^2 x} = \tan 2x$

In Exercises 33 through 36 prove the given identities by letting $3x = 2x + x$.

33. $\dfrac{\sin 3x}{\sin x} - \dfrac{\cos 3x}{\cos x} = 2$

34. $\dfrac{\sin 3x}{\sin x} + \dfrac{\cos 3x}{\cos x} = 4 \cos 2x$

35. $\sin 3x = 3 \cos^2 x \sin x - \sin^3 x$

36. $\cos 3x = \cos^3 x - 3 \sin^2 x \cos x$

7-4 Half-Angle Formulas

If we let $\theta = \alpha/2$ in the identity $\cos 2\theta = 1 - 2 \sin^2 \theta$ and then solve for $\sin (\alpha/2)$, we obtain:

(7-28) $\sin \dfrac{\alpha}{2} = \pm \sqrt{\dfrac{1 - \cos \alpha}{2}}$

Also, with the same substitution in the identity $\cos 2\theta = 2\cos^2\theta - 1$, which is then solved for $\cos(\alpha/2)$, we have:

$$(7\text{-}29) \qquad \cos\frac{\alpha}{2} = \pm\sqrt{\frac{1+\cos\alpha}{2}}$$

To find an expression for $\tan(\alpha/2)$, we let

$$\tan\frac{\alpha}{2} = \frac{\sin(\alpha/2)}{\cos(\alpha/2)}$$

and substitute Equations (7-28) and (7-29) into the expression and simplify. We obtain:

$$(7\text{-}30) \qquad \tan\frac{\alpha}{2} = \pm\sqrt{\frac{1-\cos\alpha}{1+\cos\alpha}}$$

In each of Equations (7-28), (7-29), and (7-30), the sign to be used depends on the quadrant in which $\alpha/2$ lies.

We can use these half-angle formulas to find values of the functions of angles which are half of those of which the functions are known. The following examples illustrate how these identities are used in evaluations and identities.

Example A
We can find $\sin 15°$ by using the relation

$$\sin 15° = \sqrt{\frac{1-\cos 30°}{2}} = \sqrt{\frac{1-0.8660}{2}} = 0.2588.$$

Here the plus sign is used, since $15°$ is in the first quadrant.

7-4 Half-Angle Formulas

Example B
We can find cos 165° by using the relation

$$\cos 165° = -\sqrt{\frac{1 + \cos 330°}{2}} = -\sqrt{\frac{1 + 0.8660}{2}} = -0.9659.$$

Here the minus sign is used, since 165° is in the second quadrant, and the cosine of a second-quadrant angle is negative.

Example C
We can find tan 75° by using the relation

$$\tan 75° = \sqrt{\frac{1 - \cos 150°}{1 + \cos 150°}} = \sqrt{\frac{1 - (-0.8660)}{1 + (-0.8660)}} = 3.732.$$

Here we used the plus sign, since 75° is in the first quadrant.

Example D
Simplify the expression $\sqrt{18 - 18 \cos 4x}$.

First we factor the 18 from each term under the radical and note that $18 = 9(2)$ and 9 is a perfect square. This leads to

$$\sqrt{18 - 18 \cos 4x} = \sqrt{9(2)(1 - \cos 4x)} = 3\sqrt{2(1 - \cos 4x)}.$$

This last expression is very similar to that for $\sin(\alpha/2)$, except that no 2 appears in the denominator. Therefore, multiplying the numerator and the denominator under the radical by 2 leads to the solution:

$$3\sqrt{2(1 - \cos 4x)} = 3\sqrt{\frac{4(1 - \cos 4x)}{2}} = 6\sqrt{\frac{1 - \cos 4x}{2}}$$

$$= 6 \sin \frac{4x}{2} = 6 \sin 2x.$$

Example E
Prove the identity

$$\sec \frac{\alpha}{2} + \csc \frac{\alpha}{2} = \frac{2[\sin(\alpha/2) + \cos(\alpha/2)]}{\sin \alpha}.$$

Using Equation (7-23) the solution is as follows:

$$\frac{2[\sin(\alpha/2) + \cos(\alpha/2)]}{2 \sin(\alpha/2) \cos(\alpha/2)} = \frac{1}{\cos(\alpha/2)} + \frac{1}{\sin(\alpha/2)} = \sec \frac{\alpha}{2} + \csc \frac{\alpha}{2}.$$

Example F
We can find relations for the other functions of $\alpha/2$ by expressing these functions in terms of $\sin(\alpha/2)$ and $\cos(\alpha/2)$. For example:

$$\sec\frac{\alpha}{2} = \frac{1}{\cos(\alpha/2)} = \pm\frac{1}{\sqrt{(1+\cos\alpha)/2}} = \pm\sqrt{\frac{2}{1+\cos\alpha}}.$$

Example G
Show that $2\cos^2(x/2) - \cos x = 1$.

The first step is to substitute for $\cos(x/2)$, which will result in each term containing x on the left being in terms of x, and no $x/2$ terms will exist. This might allow us to combine terms. So we perform this operation, and get

$$2\left(\frac{1+\cos x}{2}\right) - \cos x = 1.$$

Combining terms, we can complete the proof:

$$1 + \cos x - \cos x = 1.$$

Exercises 7-4

In Exercises 1 through 8 use the half-angle formulas to evaluate the given functions.

1. $\cos 15°$
2. $\sin 22.5°$
3. $\sin 75°$
4. $\cos 112.5°$
5. $\tan 67.5°$
6. $\tan 105°$
7. $\sin 165°$
8. $\cos 157.5°$

In Exercises 9 through 16 use the half-angle formulas to simplify the given expressions.

9. $\sqrt{\dfrac{1-\cos 6x}{2}}$
10. $\sqrt{\dfrac{1+\cos 8x}{2}}$
11. $\sqrt{\dfrac{1-\cos 2\alpha}{1+\cos 2\alpha}}$
12. $\sqrt{\dfrac{1+\cos 4\beta}{1-\cos 4\beta}}$
13. $\sqrt{\dfrac{4+4\cos 8\beta}{2}}$
14. $\sqrt{\dfrac{18-18\cos 6\alpha}{2+2\cos 6\alpha}}$
15. $\sqrt{8+8\cos 4x}$
16. $\sqrt{2-2\cos 16x}$

7-4 Half-Angle Formulas

In Exercises 17 through 20 evaluate the indicated functions with the information given.

17. Find the value of $\sin(\alpha/2)$ if $\cos \alpha = \dfrac{12}{13}$ (in the first quadrant).

18. Find the value of $\cos(\alpha/2)$ if $\sin \alpha = -\dfrac{4}{5}$ (in the third quadrant).

19. Find the value of $\tan(\alpha/2)$ if $\sin \alpha = \dfrac{3}{5}$ (in the second quadrant).

20. Find the value of $\sin(\alpha/2)$ if $\tan \alpha = -\dfrac{5}{12}$ (in the fourth quadrant).

In Exercises 21 through 28 prove the given identities.

21. $2 \sin^2 \dfrac{x}{2} = 1 - \cos x$

22. $\sec^2 \dfrac{\alpha}{2} = \dfrac{2(1 - \cos \alpha)}{\sin^2 \alpha}$

23. $\tan \dfrac{\alpha}{2} = \dfrac{\sin \alpha}{1 + \cos \alpha}$

24. $\cos x = \dfrac{2 - \sec^2 \dfrac{x}{2}}{\sec^2 \dfrac{x}{2}}$

25. $\dfrac{1 - \cos \beta}{2 \sin \dfrac{\beta}{2}} = \sin \dfrac{\beta}{2}$

26. $2 \cos \dfrac{x}{2} = (1 + \cos x) \sec \dfrac{x}{2}$

27. $\tan^2 \dfrac{\alpha}{2} = \csc^2 \alpha - 2 \cot \alpha \csc \alpha + \cot^2 \alpha$

28. $\cot^2 \dfrac{\beta}{2} = (\csc \beta + \cot \beta)^2$

7-5 Trigonometric Equations

One of the most important uses of the trigonometric identities is in the solution of equations involving the trigonometric functions. When an equation is written in terms of more than one function, the identities provide a way of transforming many of them to equations or factors involving only one function of the same angle. If we can accomplish this we can employ algebraic methods from then on to complete the solution. No general methods exist for the solution of such equations, but the following examples illustrate methods which prove to be useful.

Example A
Solve the equation $2 \cos \theta - 1 = 0$ for all values of θ such that $0 \leq \theta < 2\pi$.

Solving the equation for $\cos \theta$, we obtain $\cos \theta = \frac{1}{2}$. The problem asks for all values of θ between 0 and 2π that satisfy the equation. We know that the cosine of angles in the first and fourth quadrants is positive. We know also that $\cos(\pi/3) = \frac{1}{2}$. Therefore, $\theta = \pi/3$ and $\theta = 5\pi/3$.

Example B
Solve the equation
$2 \cos^2 x - \sin x - 1 = 0 \ (0 \leq x < 2\pi)$.

By use of the identity $\sin^2 x + \cos^2 x = 1$, this equation may be put in terms of $\sin x$ only. Thus, we have $2(1 - \sin^2 x) - \sin x - 1 = 0$.

$$-2 \sin^2 x - \sin x + 1 = 0$$
$$2 \sin^2 x + \sin x - 1 = 0$$

or

$$(2 \sin x - 1)(\sin x + 1) = 0$$

Just as in solving algebraic equations, we can set each factor equal to zero to find valid solutions. Thus, $\sin x = \frac{1}{2}$ and $\sin x = -1$. For the range between 0 and 2π, the value $\sin x = \frac{1}{2}$ gives values of x as $\pi/6$ and $5\pi/6$, and $\sin x = -1$ gives the value $x = 3\pi/2$. Thus, the complete solution is $x = \pi/6$, $x = 5\pi/6$, and $x = 3\pi/2$.

7-5 Trigonometric Equations

Example C
Solve the equation $\sin 2x + \sin x = 0$ $(0 \leq x < 2\pi)$.

By using the double-angle formula for $\sin 2x$, we can write the equation in the form

$$2 \sin x \cos x + \sin x = 0 \quad \text{or} \quad \sin x (2 \cos x + 1) = 0.$$

The first factor gives $x = 0$ or $x = \pi$. The second factor, for which $\cos x = -\frac{1}{2}$, gives $x = 2\pi/3$ and $x = 4\pi/3$. Thus, the complete solution is $x = 0$, $x = 2\pi/3$, $x = \pi$, and $x = 4\pi/3$.

Example D
Solve the equation $\cos(x/2) = 1 + \cos x$ $(0 \leq x < 2\pi)$.

By using the half-angle formula for $\cos x/2$ and then squaring both sides of the resulting equation, this equation can be solved.

$$\pm \sqrt{\frac{1 + \cos x}{2}} = 1 + \cos x$$

$$\frac{1 + \cos x}{2} = 1 + 2 \cos x + \cos^2 x$$

Simplifying this last equation, we have

$$2 \cos^2 x + 3 \cos x + 1 = 0$$

$$(2 \cos x + 1)(\cos x + 1) = 0.$$

The values of the cosine which come from these factors are $\cos x = -\frac{1}{2}$ and $\cos x = -1$. Thus, the values of x which satisfy the last equation are $x = 2\pi/3$, $x = 4\pi/3$, and $x = \pi$. However, when we square both sides of an equation, roots may be introduced into a subsequent equation which are not roots of the original equation. These roots are called *extraneous roots*. Thus, we must check each solution in the original equation to see if it is valid. Hence,

$$\cos \frac{\pi}{3} \stackrel{?}{=} 1 + \cos \frac{2\pi}{3} \quad \text{or} \quad \frac{1}{2} \stackrel{?}{=} 1 + (-\frac{1}{2}) \quad \text{or} \quad \frac{1}{2} = \frac{1}{2}$$

$$\cos \frac{2\pi}{3} \stackrel{?}{=} 1 + \cos \frac{4\pi}{3} \quad \text{or} \quad -\frac{1}{2} \stackrel{?}{=} 1 + (-\frac{1}{2}) \quad \text{or} \quad -\frac{1}{2} \neq \frac{1}{2}$$

$$\cos \frac{\pi}{2} \stackrel{?}{=} 1 + \cos \pi \quad \text{or} \quad 0 \stackrel{?}{=} 1 - 1 \quad \text{or} \quad 0 = 0.$$

Thus, the apparent solution $x = 4\pi/3$ is not a solution of the original equation. The correct solutions are $x = 2\pi/3$ and $x = \pi$.

Example E
Solve the equation $\tan 3\theta - \cot 3\theta = 0$ $(0 \leq \theta < 2\pi)$. We replace $\cot 3\theta$ by $1/\tan 3\theta$ and solve the resulting equation.

$$\tan 3\theta - \frac{1}{\tan 3\theta} = 0 \quad \text{or} \quad \tan^2 3\theta = 1 \quad \text{or} \quad \tan 3\theta = \pm 1$$

Thus,

$$3\theta = \frac{\pi}{4}, \frac{3\pi}{4}, \frac{5\pi}{4}, \frac{7\pi}{4}, \frac{9\pi}{4}, \frac{11\pi}{4}, \frac{13\pi}{4}, \frac{15\pi}{4}, \frac{17\pi}{4}, \frac{19\pi}{4}, \frac{21\pi}{4}, \frac{23\pi}{4}.$$

Here we must include values of angles which when divided by 3 give angles between 0 and 2π. Thus, values of 3θ from 0 to 6π are necessary. The solutions are

$$\theta = \frac{\pi}{12}, \frac{\pi}{4}, \frac{5\pi}{12}, \frac{7\pi}{12}, \frac{3\pi}{4}, \frac{11\pi}{12}, \frac{13\pi}{12}, \frac{5\pi}{4}, \frac{17\pi}{12}, \frac{19\pi}{12}, \frac{7\pi}{4}, \frac{23\pi}{12}.$$

Note that these values satisfy the original equation. Since we multiplied through by $\tan 3\theta$ in the solution, any value of θ which leads to $\tan 3\theta = 0$ would not be valid, because this would indicate division by zero in the original equation.

Example F
Solve the equation $\cos 3x \cos x + \sin 3x \sin x = 1$ $(0 \leq x < 2\pi)$.

The left side of this equation is of the form $\cos (A - x)$, where $A = 3x$. Therefore,

$$\cos 3x \cos x + \sin 3x \sin x = \cos (3x - x) = \cos 2x.$$

The original equation becomes

$$\cos 2x = 1.$$

This equation is satisfied if $2x = 0$ and $2x = 2\pi$. The solutions are $x = 0$ and $x = \pi$. Only by recognizing the proper trigonometric form can we readily solve this equation.

7-5 Trigonometric Equations

Exercises 7-5

In the following exercises solve the given trigonometric equations for values of x so that $0 \leq x < 2\pi$.

1. $\sin x - 1 = 0$
2. $2 \sin x + 1 = 0$
3. $\tan x + 1 = 0$
4. $2 \cos x + 1 = 0$
5. $4 \sin^2 x - 3 = 0$
6. $3 \tan^2 x - 1 = 0$
7. $2 \sin^2 x - \sin x = 0$
8. $3 \cos x - 4 \cos^2 x = 0$
9. $\sin 4x - \cos 2x = 0$
10. $\sin 4x - \sin 2x = 0$
11. $\sin 2x \sin x + \cos x = 0$
12. $\cos 2x + \sin^2 x = 0$
13. $2 \sin x - \tan x = 0$
14. $\sin x - \sin \frac{x}{2} = 0$
15. $2 \cos^2 x - 2 \cos 2x - 1 = 0$
16. $2 \cos^2 2x + 1 = 3 \cos 2x$
17. $\sin^2 x - 2 \sin x - 1 = 0$
18. $\tan^2 x - 5 \tan x + 6 = 0$
19. $4 \tan x - \sec^2 x = 0$
20. $\tan^2 x - 2 \sec^2 x + 4 = 0$
21. $\sin 2x \cos x - \cos 2x \sin x = 0$
22. $\cos 3x \cos x - \sin 3x \sin x = 0$
23. $\sin 2x + \cos 2x = 0$
24. $2 \sin 4x + \csc 4x = 3$
25. $\tan x + 3 \cot x = 4$
26. $\sin x \sin \frac{x}{2} = 1 - \cos x$

7-6 Introduction to the Inverse Trigonometric Functions

When we studied logarithms, we found that we often wished to change a given expression from exponential to logarithmic form or from logarithmic to exponential form. Each of these forms has its advantages for particular purposes. We found that the exponential function $y = b^x$ can also be written in logarithmic form with x as a function of y, or $x = \log_b y$. We then represented both of these functions as y in terms of x, saying that the letter used for the dependent and independent variables did not matter when we wished to express a functional relationship. Since y is normally the dependent variable, we wrote the logarithmic function as $y = \log_b x$.

These two functions, the exponential function $y = b^x$ and the logarithmic function $y = \log_b x$, are called *inverse functions*. This means that if we solve for the independent variable in terms of the dependent variable in one, we will arrive at the functional relationship expressed by the other. It also means that for every value of x there is only one corresponding value of y.

Just as we are able to solve $y = b^x$ for the exponent by writing it in logarithmic form, there are times when it is necessary to solve for the independent variable (the *angle*) in trigonometric functions. Therefore, we define the *inverse sine of x* by the relation:

(7-31)

$y = \arcsin x$ (the notation $y = \sin^{-1} x$ is also used for $y = \arcsin x$).

Similar relations exist for the other inverse trigonometric relations. In Equation (7-31), x is the value of the sine of the angle y, and therefore the most meaningful way of reading it is "y is the angle whose sine is x."

7-6 Introduction to the Inverse Trigonometric Functions

Example A
The equation $y = \arccos x$ would be read as "y is the angle whose cosine is x." The equation $y = \arctan 2x$ would be read as "y is the angle whose tangent is $2x$."

It is important to emphasize that $y = \arcsin x$ and $x = \sin y$ express the same relationship between x and y. The advantage of having both forms is that a trigonometric relation may be expressed in terms of a function of an angle or in terms of the angle itself.

If we consider closely the equation $y = \arcsin x$ and possible values of x, we note that there are an unlimited number of possible values of y for a given value of x. Consider the following example.

Example B
For $y = \arcsin x$, if $x = \frac{1}{2}$, we have $y = \arcsin \frac{1}{2}$. This means that we are to find an angle whose sine is $\frac{1}{2}$. We know that $\sin(\pi/6) = \frac{1}{2}$. Therefore, $y = \pi/6$.

However, we also know that $\sin(5\pi/6) = \frac{1}{2}$. Therefore, $y = 5\pi/6$ is also a proper value. If we consider negative angles, such as $-7\pi/6$, or angles generated by additional rotations, such as $13\pi/6$, we conclude that there are an unlimited number of possible values for y.

To have a properly defined *function* in mathematics, there must be only one value of the dependent variable for a given value of the independent variable. A *relation*, on the other hand, may have more than one such value. Therefore, we see that $y = \arcsin x$ is not really a function, although it is properly a relation. It is necessary to restrict the values of y in order to define the *inverse trigonometric functions*, and this is done in the following section. It is the purpose of this section to introduce the necessary notation and develop an understanding of the basic concept. The following examples further illustrate the meaning of the notation.

Example C
If $y = \arccos 0$, y is the angle whose cosine is zero. The smallest positive angle for which this is true is $\pi/2$. Therefore, $y = \pi/2$ is an acceptable value.

If $y = \arctan 1$, an acceptable value for y is $\pi/4$. This is the same as saying $\tan \pi/4 = 1$.

Example D
Given that $y = \sec 2x$, solve for x.

We first express the inverse relation as $2x = \text{arcsec } y$. Then we solve for x by dividing through by 2. Thus, we have $x = \frac{1}{2} \text{arcsec } y$. Note that we first wrote the inverse relation by writing the expression for the angle, which in this case was $2x$. Just as $\sec 2x$ and $2 \sec x$ are different relations, so are $\text{arcsec } 2x$ and $2 \text{arcsec } x$.

Example E
Given that $4y = \text{arccot } 2x$, solve for x.

Writing this as the cotangent of $4y$ (since the given expression means "$4y$ is the angle whose cotangent is $2x$"), we have

$$2x = \cot 4y \quad \text{or} \quad x = \tfrac{1}{2} \cot 4y.$$

Example F
Given that $\pi - y = \text{arccsc } \tfrac{1}{3}x$, solve for x.

$$\tfrac{1}{3}x = \csc(\pi - y) \quad \text{or} \quad x = 3 \csc y$$

since $\csc(\pi - y) = \csc y$.

Exercises 7-6

In Exercises 1 through 6 write down the meaning of each of the given equations. See Example A.

1. $y = \arctan x$
2. $y = \text{arcsec } x$
3. $y = \text{arccot } 3x$
4. $y = \text{arccsc } 4x$
5. $y = 2 \arcsin x$
6. $y = 3 \arctan x$

In Exercises 7 through 16 find the smallest positive angle (in terms of π) for each of the given expressions.

7. $\arccos \frac{1}{2}$
8. $\arcsin 1$
9. $\operatorname{arcsec}(-\sqrt{2})$
10. $\arccos \frac{\sqrt{2}}{2}$
11. $\arctan(-1)$
12. $\operatorname{arccsc}(-1)$
13. $\arctan \sqrt{3}$
14. $\operatorname{arcsec} 2$
15. $\operatorname{arccot}(-\sqrt{3})$
16. $\arcsin\left(-\frac{\sqrt{3}}{2}\right)$

In Exercises 17 through 24 solve the given equations for x.

17. $y = \sin 3x$
18. $y = \cos(x - \pi)$
19. $y = \arctan\left(\frac{x}{4}\right)$
20. $y = 2 \arcsin\left(\frac{x}{6}\right)$
21. $y = 1 + \sec 3x$
22. $4y = 5 - \csc 8x$
23. $1 - y = \arccos(1 - x)$
24. $2y = \operatorname{arccot} 3x - 5$

In Exercises 25 through 28 determine the required quadrants.

25. In which quadrants is $\arcsin x$ if $0 < x < 1$?
26. In which quadrants is $\arctan x$ if $0 < x < 1$?
27. In which quadrants is $\arccos x$ if $-1 < x < 0$?
28. In which quadrants is $\arcsin x$ if $-1 < x < 0$?

7-7 The Inverse Trigonometric Functions

We noted in the preceding section that we could find many values of y if we assumed some value for x in the relation $y = \arcsin x$. As these relations were defined, any of the various possibilities would be considered correct. This, however, does not meet a basic requirement for a function, and it also leads to ambiguity. To define *inverse trigonometric functions* properly, so that this ambiguity does not exist, there must be only a single value of y for any given value of x. Therefore, the following values are defined for the given functions:

(7-32)

$$-\frac{\pi}{2} \leq \text{Arcsin } x \leq \frac{\pi}{2} \quad 0 \leq \text{Arccos } x \leq \pi \quad -\frac{\pi}{2} < \text{Arctan } x < \frac{\pi}{2}$$

$$0 < \text{Arccot } x < \pi \quad 0 \leq \text{Arcsec } x \leq \pi \quad -\frac{\pi}{2} \leq \text{Arccsc } x \leq \frac{\pi}{2}$$

(We should note that Arcsec $x \neq \pi/2$ and Arcsec $x \neq 0$ for these would require division by zero.)

This means that when we are looking for a value of y to correspond to a given value for x, we must use a value of y as defined in Equations (7-32). The capital letter designates the use of the inverse trigonometric *function*.

Example A

$$\text{Arcsin}\left(\frac{1}{2}\right) = \frac{\pi}{6}$$

This is the only value of the function which lies within the defined range. The value $5\pi/6$ is not correct, since it lies outside the defined range of values.

Example B

$$\text{Arccos}\left(-\frac{1}{2}\right) = \frac{2\pi}{3}$$

Other values such as $4\pi/3$ and $-2\pi/3$ are not correct, since they are not within the defined range of values for the function Arccos x.

7-7 The Inverse Trigonometric Functions

Example C

$$\text{Arctan}(-1) = -\frac{\pi}{4}$$

This is the only value within the defined range for the function Arctan x. We must remember that when x is negative for Arcsin x and Arctan x, the value of y is a fourth-quadrant angle, expressed as a *negative angle*. This is a direct result of the definition.

Example D

$$\text{Arcsin}\left(-\frac{\sqrt{3}}{2}\right) = -\frac{\pi}{3} \qquad \text{Arccos}(-1) = \pi$$

$$\text{Arctan } 0 = 0 \qquad \text{Arcsin}(-0.1564) = -\frac{\pi}{20}$$

$$\text{Arccos}(-0.8090) = \frac{4\pi}{5} \qquad \text{Arctan}(\sqrt{3}) = \frac{\pi}{3}$$

One might logically ask why these values are chosen when there are so many different possibilities. The values are so chosen that if x is positive, the resulting answer gives an angle in the first quadrant. We must, however, account for the possibility that x might be negative. We could not choose second-quadrant angles for Arcsin x. Since the sine of second-quadrant angles is also positive, this would lead to ambiguity. The sine is negative for fourth-quadrant angles, and to have a continuous range of values of x we express the fourth-quadrant angles in the form of negative angles. This range is also chosen for Arctan x, for similar reasons. However, Arccos x cannot be chosen in this way since the cosine of fourth-quadrant angles is also positive. Thus, again to keep a continuous range of values for Arccos x, the second-quadrant angles are chosen for negative values of x.

As for the values for the other functions, we chose values such that if x is positive, the result is also an angle in the first quadrant. As for negative values of

x, it rarely makes any difference, since either positive values of x arise or we can use one of the other functions. Our definitions, however, are those which are generally used.

The graphs of the inverse trigonometric relations can be used to show that many values of y correspond to a given value of x. We can also show the choice of the ranges used in defining the inverse trigonometric functions and that it is a specific section of the curve.

Since $y = \arcsin x$ and $x = \sin y$ are equivalent equations, we can obtain the graph of the inverse sine by sketching the sine curve *along the y-axis*. In Figures (7-3), (7-4), and (7-5), the graphs of the inverse trigonometric relations are shown, with the darker portions indicating the graphs of the inverse trigonometric functions. The graphs of the other inverse relations are found in the same way.

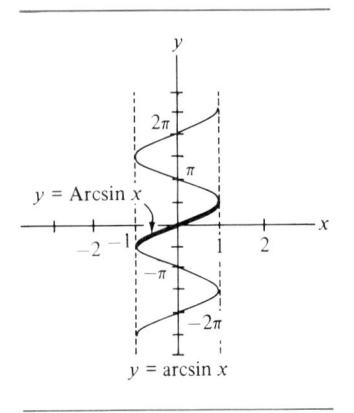

Figure 7-3

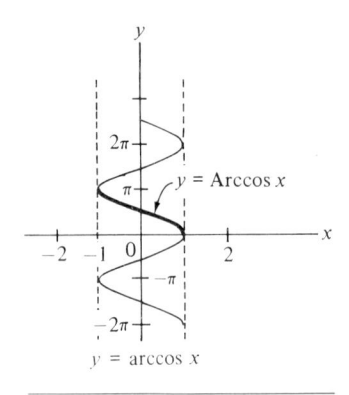

Figure 7-4

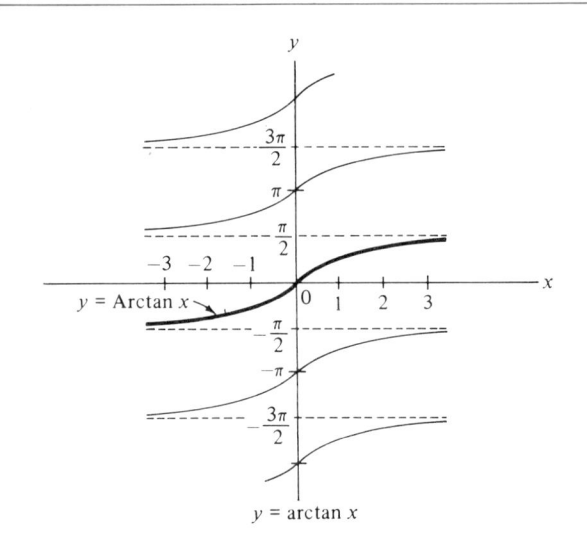

Figure 7-5

7-7 The Inverse Trigonometric Functions

If we know the value of x for one of the inverse functions, we can find the trigonometric functions of the angle. If general relations are desired, a representative triangle is very useful. The following examples illustrate these methods.

Example E
Find cos (Arcsin 0.5). [*Remember*: The inverse functions are *angles*.]

We know Arcsin 0.5 is a first-quadrant angle, since 0.5 is positive. Thus, we find Arcsin $0.5 = \pi/6$. The problem now becomes one of finding $\cos(\pi/6)$. This is, of course, $\sqrt{3}/2$ or 0.8660.

Example F

$$\sin(\text{Arccot } 1) = \sin \frac{\pi}{4} = \frac{\sqrt{2}}{2} = 0.7071$$

$$\tan[\text{Arccos}(-1)] = \tan \pi = 0$$

Example G
Find sin (Arctan x).

We know that Arctan x is another way of stating "the angle whose tangent is x." Thus, let us draw a right triangle (as in Figure 7-6) and label one of the acute angles θ, the side opposite θ as x, and the side adjacent to θ as 1. In this way we see that, by definition, $\tan \theta = x/1$, or $\theta = \text{Arctan } x$, which means θ is the desired angle. By the Pythagorean theorem, the hypotenuse of this triangle is $\sqrt{x^2 + 1}$. Now we find that $\sin \theta$, which is the same as $\sin(\text{Arctan } x)$, is $x/\sqrt{x^2 + 1}$ from the definition of the sine. Thus $\sin(\text{Arctan } x) = x/\sqrt{x^2 + 1}$.

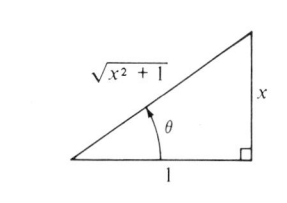

Figure 7-6

Example H
Find cos (2 Arcsin x).

From Figure 7-7, we see that $\theta = \text{Arcsin } x$. From the double-angle formulas, we have $\cos 2\theta = 1 - 2\sin^2\theta$. Thus, since $\sin \theta = x$, we have

$$\cos(2 \text{ Arcsin } x) = 1 - 2x^2$$

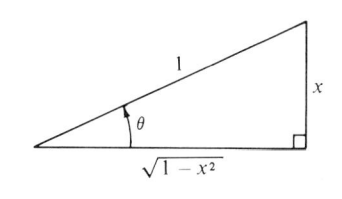

Figure 7-7

Example I
Find $\sin(\frac{1}{2} \text{Arctan } \frac{3}{4})$.

We let $\theta = \text{Arctan } \frac{3}{4}$. We know θ is in the first quadrant. Therefore, $\frac{1}{2}\theta$ is also in the first quadrant and $\sin \frac{1}{2}\theta$ is positive. Using the half-angle formula $\sin \frac{1}{2}\theta = \pm\sqrt{(1-\cos\theta)/2}$ we have

$$\sin\left(\frac{1}{2}\text{Arctan }\frac{3}{4}\right) = \sqrt{\frac{1-\left(\frac{4}{5}\right)}{2}} = \frac{1}{\sqrt{10}}$$

Exercises 7-7

In Exercises 1 through 28 evaluate the given expressions.

1. $\text{Arccos } \frac{1}{2}$
2. Arcsin 1
3. Arcsin 0
4. Arccos 0
5. $\text{Arctan}(-\sqrt{3})$
6. $\text{Arcsin}\left(-\frac{1}{2}\right)$
7. Arcsec 2
8. $\text{Arccot }\sqrt{3}$
9. $\text{Arctan }\frac{\sqrt{3}}{3}$
10. Arctan 1
11. $\text{Arcsin}\left(-\frac{\sqrt{2}}{2}\right)$
12. $\text{Arccos}\left(-\frac{\sqrt{3}}{2}\right)$
13. $\text{Arccsc }\sqrt{2}$
14. Arccot 1
15. $\text{Arctan}(-3.732)$
16. $\text{Arccos}(-0.5878)$
17. $\sin(\text{Arctan }\sqrt{3})$
18. $\tan\left(\text{Arcsin }\frac{\sqrt{2}}{2}\right)$
19. $\cos[\text{Arctan}(-1)]$
20. $\sec\left[\text{Arccos}\left(-\frac{1}{2}\right)\right]$
21. $\cos(\text{Arctan } 0.6088)$
22. $\sin(\text{Arccos } 0.3118)$

23. tan [Arccos (−0.6561)]
24. cot [Arcsin (−0.3827)]
25. cos (2 Arcsin 1)
26. sin (2 Arctan 2)
27. $\tan \left[\frac{1}{2} \text{Arccos} \left(-\frac{1}{2} \right) \right]$
28. $\sec \left[\frac{1}{2} \text{Arcsin} \left(-\frac{\sqrt{3}}{2} \right) \right]$

In Exercises 29 through 36 find an algebraic expression for each of the expressions given.

29. tan (Arcsin x) 30. sin (Arccos x)
31. cos (Arcsec x) 32. cot (Arccot x)
33. sec (Arccsc 3x) 34. tan (Arcsin 2x)
35. sin (2 Arcsin x) 36. cos (2 Arctan x)

In Exercises 37 through 44 find the exact value of each of the expressions given.

37. $\cos \left(\frac{1}{2} \text{Arcsin} \frac{3}{5} \right)$ 38. $\tan \left[\frac{1}{2} \text{Arccos} \left(-\frac{8}{17} \right) \right]$
39. $\sin \left(2 \text{ Arctan } \frac{2}{3} \right)$ 40. $\tan \left(2 \text{ Arccos } \frac{3}{4} \right)$
40. $\tan \left(2 \text{ Arccos } \frac{3}{4} \right)$
41. $\cos \left(\text{Arctan } \frac{4}{3} - \text{Arcsin } \frac{2}{3} \right)$
42. $\sin \left[\text{Arctan } \frac{2}{5} - \text{Arccos} \left(-\frac{5}{13} \right) \right]$
43. $\tan \left[\text{Arcsin} \left(-\frac{5}{6} \right) - \text{Arccos} \left(-\frac{5}{6} \right) \right]$
44. $\sin \left[\text{Arctan} \left(-\frac{3}{5} \right) + \text{Arcsin } \frac{8}{17} \right]$

Exercises for Chapter 7

In Exercises 1 through 12 determine the values of the indicated functions in the given manner.

1. Find sin 120° by using 120° = 90° + 30°.
2. Find cos 30° by using 30° = 90° − 60°.
3. Find tan 105° by using 105° = 45° + 60°.
4. Find sin 105° by using 105° = 150° − 45°.
5. Find cos 180° by using 180° = 2(90°).
6. Find tan 120° by using 120° = 2(60°).
7. Find sin 360° by using 360° = 2(180°).
8. Find cos 120° by using 120° = 2(60°).
9. Find tan 22.5° by using $22.5° = \frac{1}{2}(45°)$.
10. Find sin 45° by using $45° = \frac{1}{2}(90°)$.
11. Find cos 45° by using $45° = \frac{1}{2}(90°)$.
12. Find tan 60° by using $60° = \frac{1}{2}(120°)$.

In Exercises 13 through 20 reduce each of the given expressions to a single term. Expansion of terms is not necessary; proper recognition of the form of the expression leads to the proper result.

13. $\sin 2x \cos 3x + \cos 2x \sin 3x$
14. $\cos 7x \cos 3x + \sin 7x \sin 3x$
15. $8 \sin 6x \cos 6x$
16. $\cos^2 2x - \sin^2 2x$
17. $\sqrt{2 + 2 \cos 2x}$
18. $\sqrt{32 - 32 \cos 4x}$
19. $\dfrac{2 \tan 3x - 2 \tan x}{1 + \tan 3x \tan x}$
20. $\dfrac{\tan 2x}{2 - 2 \tan^2 2x}$

Exercises for Chapter 7

In Exercises 21 through 28 evaluate the given expressions.

21. Arcsin (-1) 22. Arcsec $\sqrt{2}$
23. Arccos 0.9659 24. Arctan (-0.6249)
25. $\tan\left[\text{Arcsin}\left(-\frac{1}{2}\right)\right]$ 26. $\cos\left[\text{Arctan}\left(-\sqrt{3}\right)\right]$
27. Arcsin $(\tan \pi)$ 28. $\text{Arccos}\left[\tan\left(-\frac{\pi}{4}\right)\right]$

In Exercises 29 through 48 prove the given identities.

29. $\dfrac{1}{\sin\theta} - \sin\theta = \cot\theta\,\cos\theta$

30. $\sin\theta\,\sec\theta\,\csc\theta\,\cos\theta = 1$
31. $\cos\theta\,\cot\theta + \sin\theta = \csc\theta$

32. $\dfrac{\sin x\,\cot x + \cos x}{\cot x} = 2\sin x$

33. $\dfrac{\sec^4 x - 1}{\tan^2 x} = 2 + \tan^2 x$

34. $\cos^2 y - \sin^2 y = \dfrac{1 - \tan^2 y}{1 + \tan^2 y}$

35. $\dfrac{1 + \tan^2 \frac{x}{2}}{1 - \tan^2 \frac{x}{2}} = \sec x$

36. $\csc 2x + \cot 2x = \cot x$
37. $2\csc 2x\,\cot x = 1 + \cot^2 x$
38. $\cos^8 x - \sin^8 x = (\cos^4 x + \sin^4 x)\cos 2x$

39. $\sin\dfrac{\theta}{2}\cos\dfrac{\theta}{2} = \dfrac{\sin\theta}{2}$

40. $\sin\dfrac{x}{2} = \dfrac{\sec x - 1}{2\sec x\,\sin(x/2)}$

41. $\sec x + \tan x = \dfrac{\cos x}{1 - \sin x}$

42. $\dfrac{\cos\theta - \sin\theta}{\cos\theta + \sin\theta} = \dfrac{\cot\theta - 1}{\cot\theta + 1}$

43. $\cos(x-y)\cos y - \sin(x-y)\sin y = \cos x$

44. $\sin 3y \cos 2y - \cos 3y \sin 2y = \sin y$

45. $\sin 4x(\cos^2 2x - \sin^2 2x) = \dfrac{\sin 8x}{2}$

46. $\csc 2x + \cot 2x = \cot x$

47. $\dfrac{\sin x}{\csc x - \cot x} = 1 + \cos x$

48. $\cos x - \sin \dfrac{x}{2} = \left(1 - 2\sin \dfrac{x}{2}\right)\left(1 + \sin \dfrac{x}{2}\right)$

In Exercises 49 through 52 solve for x.

49. $y = 2\cos 2x$
50. $y - 2 = 2\tan(x - \pi/2)$
51. $y = (\pi/4) - 3\arcsin 5x$
52. $2y = \operatorname{arcsec} 4x - 2$

In Exercises 53 through 60 solve the given equations for x such that $0 \leq x < 2\pi$.

53. $4\cos^2 x - 3 = 0$
54. $\cos 2x = \sin x$
55. $\sin^2 x - \cos^2 x + 1 = 0$
56. $\cos 3x \cos x + \sin 3x \sin x = 0$
57. $\sin^2(x/2) - \cos x + 1 = 0$
58. $\sin x + \cos x = 1$
59. $\tan 2x = \cot x$
60. $\cot x \tan 2x = 3$

In Exercises 61 through 64 find an algebraic expression for each of the expressions.

61. $\sin(2\operatorname{Arccos} x)$
62. $\cos(\pi - \operatorname{Arctan} x)$
63. $\tan(2\operatorname{Arcsin} x)$
64. $\sin(2\operatorname{Arccos} x)$

In Exercises 65 through 68 find the exact value of each of the expressions given.

65. $\tan\left[\frac{1}{2} \text{Arcsin}\left(-\frac{15}{17}\right)\right]$

66. $\cos\left(\text{Arcsin } \frac{12}{13} + \text{Arctan } 2\right)$

67. $\cos\left[2 \text{ Arctan}\left(-\frac{4}{5}\right)\right]$

68. $\tan\left[\text{Arccos}\left(-\frac{5}{13}\right) - \text{Arcsin } \frac{4}{5}\right]$

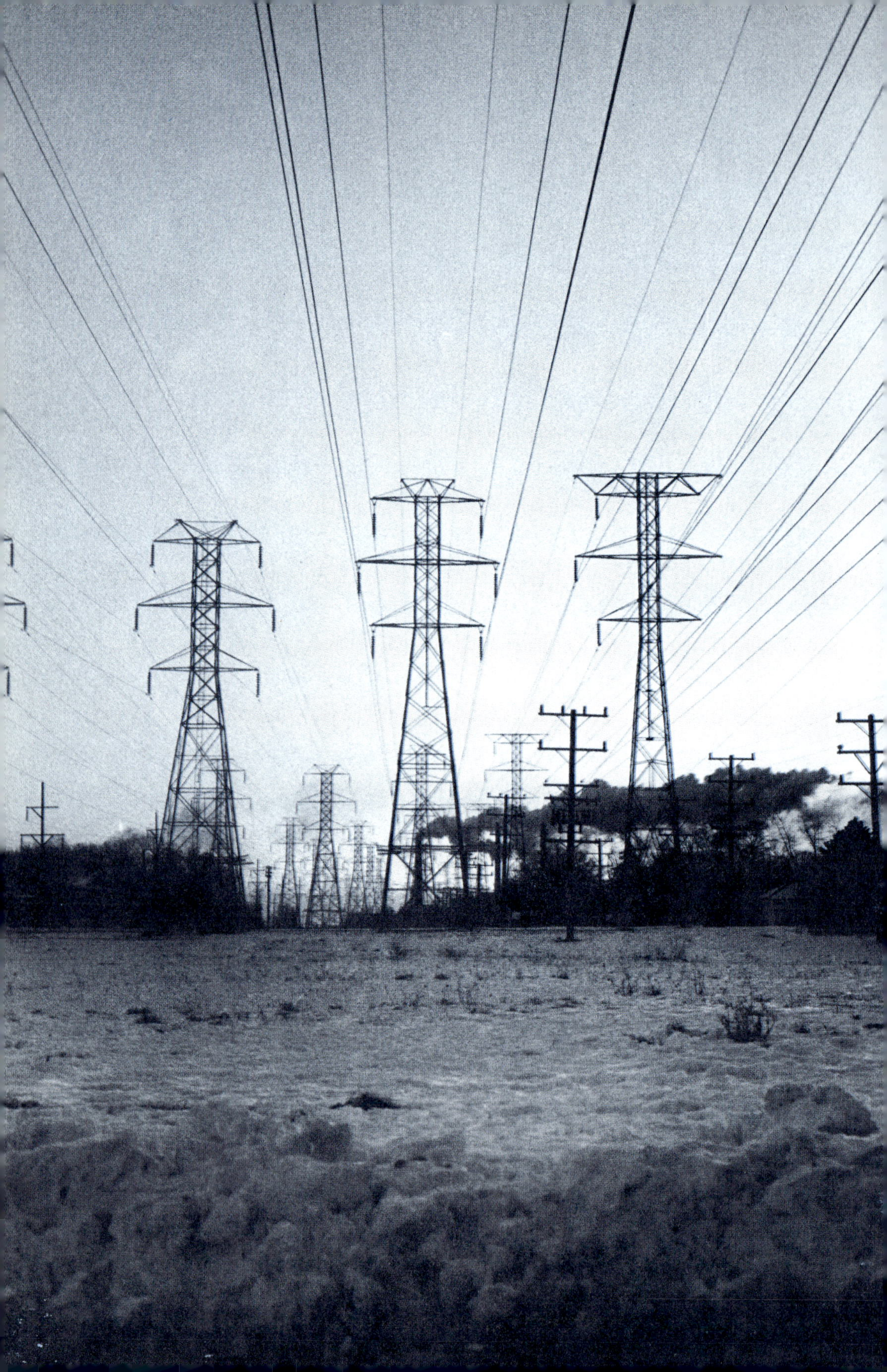

8

Complex Numbers

Complex numbers are of great importance in the theory used in developing many applications in electricity and electronics.

8-1 Imaginary and Complex Numbers

To this point, when considering square roots we have concerned ourselves only with the square roots of positive numbers. We have purposely avoided any extended discussion of square roots of negative numbers until now. In this chapter we discuss the properties of these numbers and show some of the ways in which they may be applied.

When we define radicals we are able to define square roots of positive numbers easily, since any positive or negative real number squared equals a positive real number. (See Appendix A for a discussion of exponents and radicals.) For this reason we can see that it is impossible to square any real number and have the product equal a negative number. We must define a new number system if we wish to include square roots of negative numbers.

If the radicand in a square root is negative, we can express the indicated root as the product of $\sqrt{-1}$ and the square root of a positive real number. The symbol $\sqrt{-1}$ is defined as the *imaginary* unit and is denoted by the symbol i. In keeping with this definition of i, we have:

(8-1) $$i^2 = -1$$

Example A

$$\sqrt{-9} = \sqrt{(9)(-1)} = \sqrt{9}\sqrt{-1} = 3i$$
$$\sqrt{-16} = \sqrt{16}\sqrt{-1} = 4i$$

Example B

$$(\sqrt{-4})^2 = (\sqrt{4}i)^2 = 4i^2 = -4$$

Numbers which contain the imaginary unit are called *imaginary numbers*. We note from Example B that those imaginary numbers which are

259

simply multiples of i do not follow the equation $\sqrt[n]{a}\sqrt[n]{b} = \sqrt[n]{ab}$. If the numbers in Example B did follow this equation, we would have $(\sqrt{-4})^2 = \sqrt{(-4)(-4)} = \sqrt{16} = 4$. But $4 \neq -4$. This is one reason why imaginary numbers are given special consideration.

From Example B we see that when we are dealing with the square roots of negative numbers, **each should be expressed in terms of i before proceeding**. To do this, for any positive real number a we write:

(8-2) $\qquad \sqrt{-a} = \sqrt{a}\, i \quad (a > 0)$

Example C

$$\sqrt{-6} = \sqrt{(6)(-1)} = \sqrt{6}\sqrt{-1} = \sqrt{6}\, i$$
$$\sqrt{-18} = \sqrt{(18)(-1)} = \sqrt{(9)(2)}\sqrt{-1} = 3\sqrt{2}\, i$$

In working with imaginary numbers, we often need to be able to raise these numbers to some power. Therefore, using the definitions of exponents and of i, we have the following results:

$$i = i \qquad\qquad i^4 = i^2 i^2 = (-1)(-1) = 1$$
$$i^2 = -1 \qquad\quad i^5 = i^4 i = i$$
$$i^3 = i^2 i = -i \qquad i^6 = i^4 i^2 = (1)(-1) = -1$$

The powers of i go through the cycle of i, -1, $-i$, 1, i, -1, $-i$, 1, and so forth. Remembering this fact, it is possible to raise i to any integral power almost on sight.

Example D

$$i^{10} = i^8 i^2 = (1)(-1) = -1$$
$$i^{45} = i^{44} i = (1)(i) = i$$
$$i^{531} = i^{528} i^3 = (1)(-i) = -i$$

Using real numbers and the imaginary unit i, we define a new kind of number. A *complex number* is any number which can be written in the form $a + bi$, where a and b are real numbers. If $b \neq 0$, the number is an imaginary number; if $a = 0$, we have a number of the form bi, which is a pure

8-1 Imaginary and Complex Numbers

imaginary number. If $b = 0$, then $a + bi$ is a real number. Thus, $3 + 0i$ is real, $0 + 3i$ is pure imaginary, and $2 - 3i$ is imaginary. The form $a + bi$ is known as the *rectangular form* of a complex number, where a is known as the *real part* and b is known as the *imaginary part.* We can see that complex numbers include all the real numbers and all the imaginary numbers.

For complex numbers written in terms of i to follow all the operations defined in algebra, we define equality of two complex numbers in a special way. Complex numbers are not positive or negative in the ordinary sense of these terms, but the real and imaginary parts of complex numbers *are* positive or negative. We define two complex numbers to be equal if the real parts are equal and the imaginary parts are equal. That is, two imaginary numbers, $a + bi$ and $x + yi$, are equal if $a = x$ and $b = y$.

Example E

$a + bi = 3 + 4i$ if $a = 3$ and $b = 4$

$x + yi = 5 - 3i$ if $x = 5$ and $y = -3$

Example F

Determine the values of x and y which satisfy the equation

$4 - 6i - x = i + yi.$

One way to solve this is to rearrange the terms so that all the known terms are on the right and all the terms containing the unknowns x and y are on the left. This leads to $-x - yi = -4 + 7i.$ From the definition of equality of complex numbers, $-x = -4$ and $-y = 7$, or $x = 4$ and $y = -7$.

Example G
Determine the values of x and y which satisfy the equation

$$x + 3(xi + y) = 5 - i - yi.$$

Rearranging the terms so that the known terms are on the right and terms containing x and y are on the left, we have

$$x + 3y + 3xi + yi = 5 - i.$$

Next, factoring i from the two terms on the left will put the expression on the left into proper form. This leads to

$$(x + 3y) + (3x + y)i = 5 - i.$$

Using the definition of equality, we have

$$x + 3y = 5 \quad \text{and} \quad 3x + y = -1.$$

We now solve this system of equations. The solution is $x = -1$ and $y = 2$. Actually, the solution can be obtained at any point by writing each side of the equation in the form $a + bi$ and then equating first the real parts and then the imaginary parts.

The *conjugate* of the complex number $a + bi$ is the complex number $a - bi$. We see that the sign of the imaginary part of a complex number is changed to obtain its conjugate.

Example H
The complex number $3 - 2i$ is the conjugate of $3 + 2i$. We may also say that $3 + 2i$ is the conjugate of $3 - 2i$. Thus, each is the conjugate of the other.

Exercises 8-1

In Exercises 1 through 8 express each number in terms of i.

1. $\sqrt{-81}$
2. $\sqrt{-121}$
3. $-\sqrt{-4}$
4. $-\sqrt{-0.01}$

5. $\sqrt{-8}$
6. $\sqrt{-48}$
7. $\sqrt{-\frac{7}{4}}$
8. $\sqrt{-\frac{5}{3}}$

In Exercises 9 through 16 simplify the given expressions.

9. i^7
10. i^{49}
11. $-i^{22}$
12. i^{408}
13. $i^2 - i^6$
14. $2i^5 - i^7$
15. $i^{15} - i^{13}$
16. $3i^{48} + i^{200}$

In Exercises 17 through 24 perform the indicated operations and simplify each complex number to its rectangular form $a + bi$.

17. $2 + \sqrt{-9}$
18. $-6 + \sqrt{-64}$
19. $2i^2 + 3i$
20. $i^3 - 6$
21. $\sqrt{18} - \sqrt{-8}$
22. $\sqrt{-27} + \sqrt{12}$
23. $(\sqrt{-2})^2 + i^4$
24. $(2\sqrt{2})^2 - (\sqrt{-1})^2$

In Exercises 25 through 28 find the conjugate of each complex number.

25. $6 - 7i$
26. $-3 + 2i$
27. $2i$
28. 6

In Exercises 29 through 36 find the values of x and y which satisfy the given equations.

29. $7x - 2yi = 14 + 4i$
30. $2x + 3yi = -6 + 12i$
31. $6i - 7 = 3 - x - yi$
32. $9 - i = xi + 1 - y$
33. $x - y = 1 - xi - yi - i$
34. $2x - 2i = 4 - 2xi - yi$
35. $x + 2 + 7i = yi - 2xi$
36. $2x + 6xi + 3 = yi - y + 7i$

In Exercises 37 and 38 answer the given questions.

37. What condition must be satisfied if a complex number and its conjugate are to be equal?
38. What type of number is a complex number if it is equal to the negative of its conjugate?

8-2 Basic Operations with Complex Numbers

The basic operations of addition, subtraction, multiplication, and division are defined in the same way for complex numbers in rectangular form as they are for real numbers. These operations are performed without regard for the fact that i has a special meaning. **We must be careful to express all complex numbers in terms of i before performing these operations.** But once this is done, we may proceed as with real numbers. We have the following definitions for these operations on complex numbers.

(8-3)

Addition and subtraction: $\quad (a + bi) + (c + di) = (a + c) + (b + d)i$

(8-4)

Multiplication: $\quad (a + bi)(c + di) = (ac - bd) + (ad + bc)i$

(8-5)

Division: $\quad \dfrac{a + bi}{c + di} = \dfrac{(a + bi)(c - di)}{(c + di)(c - di)} = \dfrac{(ac + bd) + (bc - ad)i}{c^2 + d^2}$

We note that in dividing two complex numbers we need merely to multiply numerator and denominator by the conjugate of the denominator. We use this procedure so that we can express any answer in the form of a complex number.

If we recall Example B of Section 8-1, we see the reason for expressing all complex numbers in terms of i before proceeding with any indicated operations.

8-2 Basic Operations with Complex Numbers

Example A

$(3 - 2i) + (-5 + 7i) = (3 - 5) + (-2 + 7)i$
$\qquad\qquad\qquad\qquad\; = -2 + 5i$

Example B

$(7 + 9i) - (6 - 4i) = 7 + 9i - 6 + 4i = 1 + 13i$

Example C

$(6 - \sqrt{-4})(\sqrt{-9}) = (6 - 2i)(3i) = 18i - 6i^2$
$\qquad\qquad\qquad\qquad\;\; = 18i - 6(-1) = 6 + 18i$

Example D

$(-9 - 6i)(2 + i) = -18 - 9i - 12i - 6i^2$
$\qquad\qquad\qquad\;\; = -18 - 21i - 6(-1)$
$\qquad\qquad\qquad\;\; = -12 - 21i$

Example E

$$\frac{7 - 2i}{3 + 4i} = \frac{7 - 2i}{3 + 4i} \cdot \frac{3 - 4i}{3 - 4i} = \frac{21 - 28i - 6i + 8i^2}{9 - 16i^2}$$

$$= \frac{21 - 34i + 8(-1)}{9 - 16(-1)} = \frac{13 - 34i}{25}$$

This could be written in the form $a + bi$ as $\frac{13}{25} - \frac{34}{25}i$, but it is generally left as a single fraction.

Example F

$$\frac{6 + i}{2i} = \frac{6 + i}{2i} \cdot \frac{-2i}{-2i} = \frac{-12i - 2i^2}{4} = \frac{2 - 12i}{4}$$

$$= \frac{1 - 6i}{2}$$

Example G

$$\frac{i^3 + 2i}{1 - i^5} = \frac{-i + 2i}{1 - i} = \frac{i}{1 - i} \cdot \frac{1 + i}{1 + i} = \frac{-1 + i}{2}$$

Exercises 8-2

In Exercises 1 through 40 perform the indicated operations, expressing all answers in the form $a + bi$.

1. $(3 - 7i) + (2 - i)$
2. $(-4 - i) + (-7 - 4i)$
3. $(7i - 6) - (3 + i)$
4. $(2 - 3i) - (2 + 3i)$
5. $(4 + \sqrt{-16}) + (3 - \sqrt{-81})$
6. $(-1 + 3\sqrt{-4}) + (8 - 4\sqrt{-49})$
7. $(5 - \sqrt{-9}) - (\sqrt{-4} + 5)$
8. $(\sqrt{-25} - 1) - \sqrt{-9}$
9. $i - (i - 7) - 8$
10. $(7 - i) - (4 - 4i) + (6 - i)$
11. $(7 - i)(7i)$
12. $(-2i)(i - 5)$
13. $\sqrt{-16}(2\sqrt{-1} - 5)$
14. $(\sqrt{-4} - 1)(\sqrt{-9})$
15. $(4 - i)(5 + 2i)$
16. $(3 - 5i)(6 + 7i)$
17. $(2\sqrt{-9} - 3)(3\sqrt{-4} + 2)$
18. $(5\sqrt{-64} - 5)(7 + \sqrt{-16})$
19. $\sqrt{-18}\sqrt{-4}\sqrt{-9}$
20. $(\sqrt{-36})^4$
21. $\sqrt{-108} - \sqrt{-27}$
22. $2\sqrt{-54} + \sqrt{-24}$
23. $7i^3 - 7\sqrt{-9}$
24. $i^2\sqrt{-7} - \sqrt{-28} + 8$
25. $(3 - 7i)^2$
26. $(4i + 5)^2$
27. $(1 - i)^3$
28. $(1 + i)(1 - i)^2$
29. $\dfrac{6i}{2 - 5i}$
30. $\dfrac{4}{3 + 7i}$
31. $\dfrac{2}{6 - \sqrt{-1}}$
32. $\dfrac{\sqrt{-4}}{2 + \sqrt{-9}}$
33. $\dfrac{1 - i}{1 + i}$
34. $\dfrac{9 - 8i}{i - 1}$
35. $\dfrac{\sqrt{-2} - 5}{\sqrt{-2} + 3}$
36. $\dfrac{1 - \sqrt{-4}}{2 + 9i}$
37. $\dfrac{i^2 - i}{2i - i^8}$

8-3 Graphical Representation of Complex Numbers

38. $\dfrac{i^5 - i^3}{3 + i}$ 39. $\dfrac{5}{i^6 - i}$ 40. $\dfrac{i^9 - 1}{4 - i^3}$

In Exercises 41 through 44 demonstrate the indicated properties.

41. Show that the sum of a complex number and its conjugate is a real number.
42. Show that the product of a complex number and its conjugate is a real number.
43. Show that the difference between a complex number and its conjugate is an imaginary number.
44. Show that the quotient of a complex number and its conjugate is a complex number.

8-3 Graphical Representation of Complex Numbers

We know that we can represent real numbers as points on a line. Because complex numbers include all real numbers as well as imaginary numbers, it is necessary to represent them graphically in a different way. Since there are two numbers associated with each complex number (the real part and the imaginary part), we find that we can represent complex numbers by representing the real parts by the x-values of the rectangular coordinate system and the imaginary parts by the y-values. In this way each complex number is represented as a point in the plane, the point being designated as $a + bi$. When the rectangular coordinate system is used in this manner it is called the *complex plane*.

Example A

In Figure 8-1, the point A represents the complex number $3 - 2i$. Point B represents $-1 + i$. Point C represents $-2 - 3i$. We note that these are equivalent to the points $(3, -2)$, $(-1, 1)$, and $(-2, -3)$. However, we must keep in mind that the meaning is different. Complex numbers were not included when we first graphed functions.

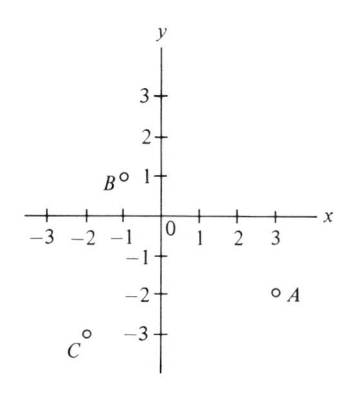

Figure 8-1

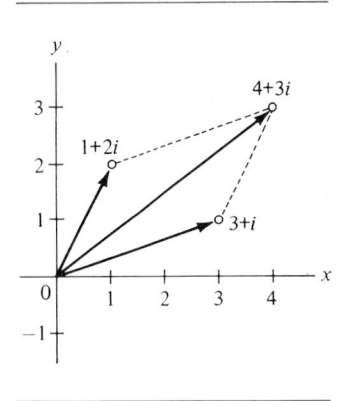

Figure 8-2

Let us represent two complex numbers and their sum in the complex plane. Consider, for example, the two complex numbers $1 + 2i$ and $3 + i$. By algebraic addition the sum is $4 + 3i$. When we draw lines from the origin to these points (see Figure 8-2), we note that if we think of the complex numbers as being vectors, their sum is the vector sum. Because complex numbers can be used to represent vectors, these numbers are particularly important. Any complex number can be thought of as representing a vector from the origin to its point in the complex plane. To add two complex numbers graphically, we find the point corresponding to one of them and draw a line from the origin to this point. We repeat this process for the second point. Next we complete a parallelogram with the lines drawn as adjacent sides. The resulting fourth vertex is the point representing the sum of the two complex numbers. Note that this is equivalent to adding vectors by graphical means.

Example B

Add the complex numbers $5 - 2i$ and $-2 - i$ graphically.

The solution is indicated in Figure 8-3. We can see that the fourth vertex of the parallelogram is $3 - 3i$, which is, of course, the algebraic sum.

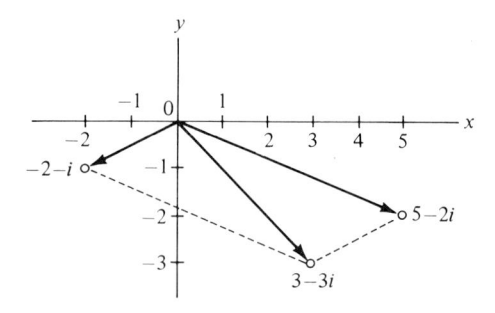

Figure 8-3

Example C

Subtract $4 - 2i$ from $2 - 3i$ graphically.

Subtracting $4 - 2i$ is equivalent to adding $-4 + 2i$. Thus, we complete the solution by adding $-4 + 2i$ and $2 - 3i$ (see Figure 8-4). The result is $-2 - i$.

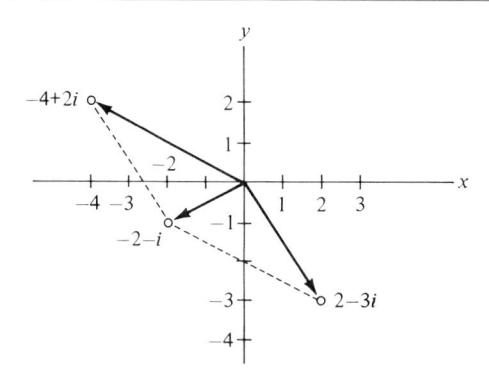

Figure 8-4

Example D

Show graphically that the sum of a complex number and its conjugate is a real number.

If we choose the complex number $a + bi$, we know that its conjugate is $a - bi$. The y-coordinate for the conjugate is as far below the x-axis as the y-coordinate of $a + bi$ is above it. Thus, the sum of the imaginary parts must be zero and the sum of the two numbers must therefore lie on the x-axis, as shown in Figure 8-5. Since any point on the x-axis is real, we have shown that the sum of $a + bi$ and $a - bi$ is real.

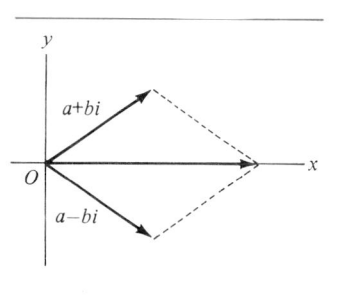

Figure 8-5

Exercises 8-3

In Exercises 1 through 12 perform the indicated operations graphically; check them algebraically.

1. $(5 - i) + (3 + 2i)$
2. $(3 - 2i) + (-1 - i)$
3. $(2 - 4i) + (-2 + i)$
4. $(-1 - 2i) + (6 - i)$
5. $(3 - 2i) - (4 - 6i)$
6. $(2 - i) - i$
7. $(1 + 4i) - (3 + i)$
8. $(-i - 2) - (-1 - 3i)$
9. $(4 - i) + (3 + 2i)$
10. $(5 + 2i) - (-4 - 2i)$
11. $(i - 6) - i + (i - 7)$
12. $i - (1 - i) + (3 + 2i)$

In Exercises 13 through 16 on the same coordinate system plot the given number, its negative, and its conjugate.

13. $3 + 2i$
14. $-2 + 4i$
15. $-3 - 5i$
16. $5 - i$

8-4 Polar Form of a Complex Number

We have just seen the relationship between complex numbers and vectors. Since one can be used to represent the other, we shall use this fact to write complex numbers in another way. The new form has certain advantages when basic operations are performed on complex numbers.

By drawing a vector from the origin to the point in the complex plane which represents the number $x + yi$, we see the relation between vectors and complex numbers. Further observation indicates that an angle in standard position has been formed. Also, the point $x + yi$ is r units from the origin. In fact, we can find any point in the complex plane by knowing this angle θ and the value of r. We have already developed the relations between x, y, r, and θ in Equations (4-1) to (4-3). Let us rewrite these equations in a slightly different form. By referring to Equations (4-1) through (4-3) and to Figure 8-6, we see that:

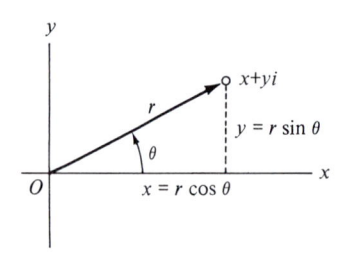

Figure 8-6

8-4 Polar Form of a Complex Number

(8-6) $\quad x = r \cos \theta \quad y = r \sin \theta$

and

(8-7) $\quad r^2 = x^2 + y^2 \quad \tan \theta = \dfrac{y}{x}$

Substituting Equations (8-6) into the rectangular form $x + yi$ of a complex number, we have

$$x + yi = r \cos \theta + i(r \sin \theta)$$

or

(8-8) $\quad x + yi = r(\cos \theta + i \sin \theta)$

The right side of Equation (8-8) is called the *polar form* of a complex number. Sometimes it is referred to as the *trigonometric form*. Other notations used to represent the polar form are $r \angle \theta$ and $r \text{cis} \theta$. The length r is called the *absolute value* or the *modulus,* and the angle θ is called the *argument* of the complex number. Therefore, Equation (8-8), along with Equations (8-7), defines the polar form of a complex number.

Example A
Represent the complex number $3 + 4i$ graphically and give its polar form.

From the rectangular form $3 + 4i$ we see that $x = 3$ and $y = 4$. Using Equations (8-7), we have $r = \sqrt{3^2 + 4^2} = 5$ and $\tan \theta = \frac{4}{3} = 1.333$, which means that $\theta = 53.1°$. Thus the polar form is $5(\cos 53.1° + i \sin 53.1°)$. The graphical representation is shown in Figure 8-7.

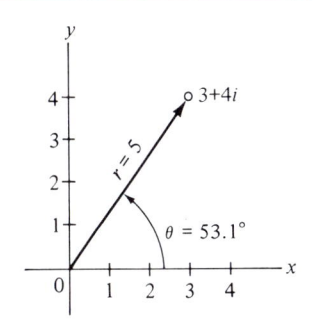

Figure 8-7

Example B

Represent the complex number $2 - 3i$ graphically and give its polar form.

From Equations (8-7) we have $r = \sqrt{2^2 + (-3)^2} = \sqrt{13} = 3.61$ and $\tan \theta = -\frac{3}{2} = -1.500$. In Figure 8-8 we see that θ is a fourth-quadrant angle. Therefore, since $\tan 56.3° = 1.500$, we have $\theta = 303.7°$. The polar form is

$$3.61(\cos 303.7° + i \sin 303.7°).$$

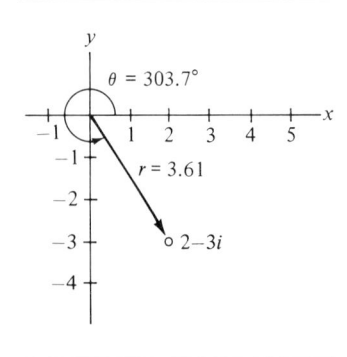

Figure 8-8

Example C

Express the complex number $3(\cos 120° + i \sin 120°)$ in rectangular form.

From the given polar form, we know that $r = 3$ and $\theta = 120°$. Using Equations (8-6), we have

$$x = 3 \cos 120° = 3(-0.500) = -1.50$$
$$y = 3 \sin 120° = 3(0.866) = 2.60.$$

Therefore, the rectangular form is $-1.50 + 2.60i$ (see Figure 8-9).

Example D

Represent the numbers 5, -5, $7i$, and $-7i$ in polar form.

Since any positive real number lies on the positive x-axis in the complex plane, real numbers are expressed in polar form by

$$a = a(\cos 0° + i \sin 0°).$$

Negative real numbers, being on the negative x-axis, are written as

$$a = |a|(\cos 180° + i \sin 180°).$$

Thus, $5 = 5(\cos 0° + i \sin 0°)$ and $-5 = 5(\cos 180° + i \sin 180°)$.

Positive pure imaginary numbers lie on the positive y-axis and are expressed in polar form by

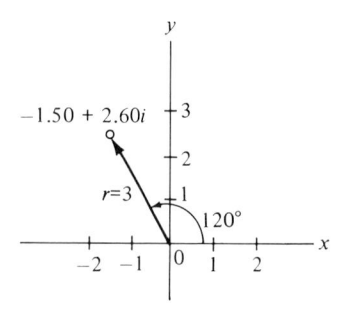

$$bi = b(\cos 90° + i \sin 90°).$$

Figure 8-9

8-4 Polar Form of a Complex Number

Similarly, negative pure imaginary numbers, being on the negative y-axis, are written as

$$bi = |b|(\cos 270° + i \sin 270°).$$

This means that $7i = 7(\cos 90° + i \sin 90°)$ and $-7i = 7(\cos 270° + i \sin 270°)$. The graphical representations of the *complex numbers* 5 and $7i$ are in Figure 8-10.

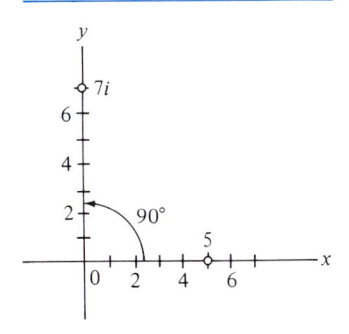

Figure 8-10

Exercises 8-4

In Exercises 1 through 12 represent each of the complex numbers graphically and give the polar form of each number.

1. $8 + 6i$
2. $3 + 5i$
3. $3 - 4i$
4. $-5 + 12i$
5. $-2 + 3i$
6. $7 - 5i$
7. $-5 - 2i$
8. $-4 - 4i$
9. $1 + \sqrt{3}\,i$
10. $\sqrt{2} - \sqrt{2}\,i$
11. -3
12. $-2i$

In Exercises 13 through 24 represent each of the complex numbers graphically and give the rectangular form of each number.

13. $5(\cos 54° + i \sin 54°)$
14. $3(\cos 232° + i \sin 232°)$
15. $1.6(\cos 150° + i \sin 150°)$
16. $2.5(\cos 315° + i \sin 315°)$
17. $10(\cos 345° + i \sin 345°)$
18. $2(\cos 155° + i \sin 155°)$
19. $6(\cos 180° + i \sin 180°)$
20. $7(\cos 270° + i \sin 270°)$
21. $4(\cos 200° + i \sin 200°)$
22. $1.5(\cos 62° + i \sin 62°)$
23. $\cos 240° + i \sin 240°$
24. $\cos 99° + i \sin 99°$

8-5 Exponential Form of a Complex Number

Another important form of a complex number is known as the *exponential form,* which is written $re^{i\theta}$. In this expression r and θ have the same meaning as given in the last section and e is the irrational number equal to about 2.718 which we used in Section 5-6. We now define $re^{i\theta}$:

(8-9) $\qquad re^{i\theta} = r(\cos\theta + i\sin\theta)$

When θ is expressed in radians, the expression $i\theta$ is an actual exponent, and it can be shown to obey all the laws of exponents. For this reason and because it is more meaningful in applications, we shall always express θ in radians when using the exponential form. The following examples show how complex numbers can be changed to and from exponential form.

Example A
Express the number $3 + 4i$ in exponential form.

From Example A of Section 8-4, we know that this complex number may be written in polar form as $5(\cos 53.1° + i \sin 53.1°)$. Therefore, we know that $r = 5$. We now express $53.1°$ in terms of radians as

$$\frac{53.1\pi}{180} = 0.927 \text{ rad.}$$

Thus, the exponential form is $5e^{0.927i}$. This means that

$$3 + 4i = 5(\cos 53.1° + i \sin 53.1°) = 5e^{0.927i}.$$

Example B
Express the number $3 - 7i$ in exponential form.

From the rectangular form of the number, we have $x = 3$ and $y = -7$. Therefore, $r = \sqrt{3^2 + (-7)^2} = \sqrt{58} = 7.62$. Also, $\tan\theta = -\frac{7}{3} = -2.333$. Since x is positive and y is negative, θ is a fourth-quadrant angle. From the tables we see that $2.333 = \tan 66.8°$. Therefore,

8-5 Exponential Form of a Complex Number

$\theta = 360° - 66.8° = 293.2°$. Converting 293.2° to radians we have 293.2° = 5.12 rad. Therefore, the exponential form is $7.62e^{5.12i}$. This means that

$$3 - 7i = 7.62e^{5.12i}.$$

Example C

Express the complex number $2e^{4.80i}$ in polar and rectangular forms.

We first express 4.80 rad in degrees: 4.80 rad = 275°. From the exponential form we know that $r = 2$. This leads to the polar form of $2(\cos 275° + i \sin 275°)$. By finding cos 275° and sin 275°, we find the rectangular form to be $2(0.0872 - 0.996i) = 0.174 - 1.99i$. This means that

$$2e^{4.80i} = 2(\cos 275° + i \sin 275°)$$
$$= 0.174 - 1.99i.$$

Example D

Express the complex number $3.40e^{2.46i}$ in polar and rectangular forms.

We first express 2.46 rad as 141°. From the exponential form we know that $r = 3.40$. Therefore, the polar form is $3.40(\cos 141° + i \sin 141°)$. Next we find that cos 141° = −0.777 and sin 141° = 0.629. The rectangular form is $3.40(-0.777 + 0.629i) = -2.64 + 2.14i$. This means that

$$3.40e^{2.46i} = 3.40(\cos 141° + i \sin 141°)$$
$$= -2.64 + 2.14i.$$

At this point we shall summarize the three important forms of a complex number:

Rectangular: $x + yi$
Polar: $r(\cos \theta + i \sin \theta)$
Exponential: $re^{i\theta}$

It follows that:

(8-10) $x + yi = r(\cos \theta + i \sin \theta) = re^{i\theta}$

where $r^2 = x^2 + y^2$ $\tan \theta = \dfrac{y}{x}$.

Exercises 8-5

In Exercises 1 through 12 express the given complex numbers in exponential form.

1. $3(\cos 60° + i \sin 60°)$
2. $5(\cos 135° + i \sin 135°)$
3. $4.5(\cos 282° + i \sin 282°)$
4. $2.1(\cos 228° + i \sin 228°)$
5. $3 - 4i$
6. $-1 - 5i$
7. $-3 + 2i$
8. $6 + i$
9. $5 + 2i$
10. $4 - i$
11. $-6 - 5i$
12. $-8 + 5i$

In Exercises 13 through 20 express the given complex numbers in polar and rectangular forms.

13. $3e^{0.5i}$
14. $2e^{i}$
15. $4e^{1.85i}$
16. $2.5e^{3.84i}$
17. $3.2e^{5.41i}$
18. $0.8e^{3.00i}$
19. $0.1e^{2.39i}$
20. $8.2e^{3.49i}$

In Exercises 21 and 22 perform the indicated operations.

21. The electric current in a certain alternating-current circuit is $0.500 + 0.220i$ amp. Write this current in exponential form.

22. The voltage in a certain alternating-current circuit is $125e^{1.31i}$. Write this voltage in rectangular form.

8-6 Products, Quotients, Powers, and Roots of Complex Numbers

We may find the product of two complex numbers by using the exponential form and the laws of exponents. Multiplying $r_1 e^{i\theta_1}$ by $r_2 e^{i\theta_2}$, we have:

(8-11)

$$r_1 e^{i\theta_1} \cdot r_2 e^{i\theta_2} = r_1 r_2 e^{i\theta_1 + i\theta_2} = r_1 r_2 e^{i(\theta_1 + \theta_2)}$$

We use Equation (8-11) to express the product of two complex numbers in polar form:

$$r_1 e^{i\theta_1} \cdot r_2 e^{i\theta_2} = r_1(\cos\theta_1 + i\sin\theta_1) \cdot r_2(\cos\theta_2 + i\sin\theta_2)$$

and

$$r_1 r_2 e^{i(\theta_1 + \theta_2)} = r_1 r_2 [\cos(\theta_1 + \theta_2) + i\sin(\theta_1 + \theta_2)].$$

Therefore, the polar expressions are equal:

(8-12)

$$r_1(\cos\theta_1 + i\sin\theta_1) \cdot r_2(\cos\theta_2 + i\sin\theta_2) = r_1 r_2 [\cos(\theta_1 + \theta_2) + i\sin(\theta_1 + \theta_2)]$$

Example A
Multiply the complex numbers $(2 + 3i)$ and $(1 - i)$ by using the polar form of each.

$r_1 = \sqrt{4 + 9} = 3.61$ $\tan\theta_1 = 1.500$ $\theta_1 = 56.3°$
$r_2 = \sqrt{1 + 1} = 1.41$ $\tan\theta_2 = -1.000$ $\theta_2 = 315°$

$3.61(\cos 56.3° + i\sin 56.3°)(1.41)(\cos 315° + i\sin 315°)$
        $= (3.61)(1.41)(\cos 371.3° + i\sin 371.3°)$
        $= 5.09(\cos 11.3° + i\sin 11.3°)$

Example B
When we use the exponential form to multiply the two complex numbers in Example A, we have

$$r_1 = 3.61 \quad \theta_1 = 56.3° = 0.983 \text{ rad}$$
$$r_2 = 1.41 \quad \theta_2 = 315° = 5.50 \text{ rad}$$
$$3.61e^{0.983i}1.41e^{5.50i} = 5.09e^{6.48i} = 5.09e^{0.20i}$$

If we wish to *divide* one complex number in exponential form by another, we arrive at the following result:

(8-13) $\quad r_1 e^{i\theta_1} \div r_2 e^{i\theta_2} = \dfrac{r_1}{r_2} e^{i(\theta_1 - \theta_2)}$

Therefore, the result of dividing one number in polar form by another is given by:

(8-14) $\quad \dfrac{r_1(\cos \theta_1 + i \sin \theta_1)}{r_2(\cos \theta_2 + i \sin \theta_2)} = \dfrac{r_1}{r_2}[\cos(\theta_1 - \theta_2) + i \sin(\theta_1 - \theta_2)]$

Example C
Divide the first complex number of Example A by the second.
  Using polar form, we have the following:

$$\dfrac{3.61(\cos 56.3° + i \sin 56.3°)}{1.41(\cos 315° + i \sin 315°)}$$
$$= 2.56[\cos(-258.7°) + i \sin(-258.7°)]$$
$$= 2.56(\cos 101.3° + i \sin 101.3°)$$

Example D
Repeating Example C and using exponential forms, we obtain

$$\dfrac{3.61e^{0.983i}}{1.41e^{5.50i}} = 2.56e^{-4.52i} = 2.56e^{1.76i}.$$

8-6 Products, Quotients, Powers, and Roots of Complex Numbers

To raise a complex number to a power, we simply multiply one complex number by itself the required number of times. For example, in Equation (8-11), if the two numbers being multiplied are equal, we have (letting $r_1 = r_2 = r$ and $\theta_1 = \theta_2 = \theta$):

(8-15) $$(re^{i\theta})^2 = r^2 e^{i2\theta}$$

Multiplying the expression in Equation (8-15) by $re^{i\theta}$ gives $r^3 e^{i3\theta}$. This leads to the general expression for raising a complex number to the nth power:

(8-16) $$(re^{i\theta})^n = r^n e^{in\theta}$$

Extending this to polar form, we have:

(8-17) $$[r(\cos\theta + i\sin\theta)]^n = r^n(\cos n\theta + i\sin n\theta)$$

Equation (8-17) is known as *DeMoivre's theorem*. It is valid for all real values of n and may also be used for finding the roots of complex numbers if n is a fractional exponent.

Example E
Using DeMoivre's theorem, find $(2 + 3i)^3$.

From Example A of this section, we know that $r = 3.61$ and $\theta = 56.3°$. Thus, we have

$$[3.61(\cos 56.3° + i\sin 56.3°)]^3$$
$$= 47.0(\cos 168.9° + i\sin 168.9°).$$

From Example B we know that $\theta = 0.983$ rad. Thus, in exponential form,

$$(3.61e^{0.983i})^3 = 47.0e^{2.95i}.$$

Example F
Find $\sqrt[3]{-1}$.

Since we know that -1 is a real number, we can find its cube root by means of the definition: $(-1)^3 = -1$. We shall check this by DeMoivre's theorem. Writing -1 in polar form, we have $-1 = 1(\cos 180° + i \sin 180°)$. Applying DeMoivre's theorem, with $n = \frac{1}{3}$, we obtain

$$(-1)^{1/3} = 1^{1/3}(\cos \tfrac{1}{3} \cdot 180° + i \sin \tfrac{1}{3} \cdot 180°)$$
$$= \cos 60° + i \sin 60° = 0.500 + 0.866i.$$

Note that we did not get -1 as an answer. If we check the answer which was obtained, in the form $\frac{1}{2}(1 + \sqrt{3}i)$, by actually cubing it, we obtain -1. Thus, it is a correct answer.

Note also that it is possible to take one-third of any angle up to 1080° and still have an angle less than 360°. Since 180° and 540° have the same terminal side, let us try writing -1 as $1(\cos 540° + i \sin 540°)$. Using DeMoivre's theorem, we have

$$(-1)^{1/3} = 1^{1/3}(\cos \tfrac{1}{3} \cdot 540° + i \sin \tfrac{1}{3} \cdot 540°)$$
$$= \cos 180° + i \sin 180° = -1.$$

We have found the answer we originally anticipated.

Angles of 180° and 900° also have the same terminal side, so we try

$$(-1)^{1/3} = 1^{1/3}(\cos \tfrac{1}{3} \cdot 900° + i \sin \tfrac{1}{3} \cdot 900°)$$
$$= \cos 300° + i \sin 300° = 0.500 - 0.866i.$$

Checking this, we find that it is also a correct root. We may try 1260°, but $\frac{1}{3}(1260°) = 420°$, which has the same functional values as 60° and would give us the answer $0.500 + 0.866i$ again.

8-6 Products, Quotients, Powers, and Roots of Complex Numbers

We have found, therefore, three cube roots of -1: -1, $0.500 + 0.866i$, and $0.500 - 0.866i$. When this is generalized, it can be proved that there are n nth roots of any complex number. The method for finding the n roots is to use θ to find one root and then add $360°$ to θ, $n - 1$ times, in order to find the other roots.

Example G
Find the square roots of i.

We must first properly write i in polar form so that we may use DeMoivre's theorem to find the roots. In polar form, i is

$$i = 1(\cos 90° + i \sin 90°).$$

To find the square roots, we apply DeMoivre's theorem with $n = \frac{1}{2}$:

$$i^{1/2} = 1^{1/2}\left(\cos \frac{90°}{2} + i \sin \frac{90°}{2}\right)$$
$$= \cos 45° + i \sin 45° = 0.707 + 0.707i.$$

To find the other square root, we must write i in polar form as

$$i = 1[\cos(90° + 360°) + i \sin(90° + 360°)]$$
$$= 1(\cos 450° + i \sin 450°).$$

Applying DeMoivre's theorem to i in this form, we have

$$i^{1/2} = 1^{1/2}\left(\cos \frac{450°}{2} + i \sin \frac{450°}{2}\right)$$
$$= \cos 225° + i \sin 225° = -0.707 - 0.707i.$$

Thus, the two square roots of i are $0.707 + 0.707i$ and $-0.707 - 0.707i$.

Example H
Find the six sixth roots of 1.

Here we shall use directly the method for finding the roots of a number, as outlined at the end of Example F.

$$1 = 1(\cos 0° + i \sin 0°)$$

First root: $1^{1/6} = 1^{1/6} \left(\cos \frac{0°}{6} + i \sin \frac{0°}{6} \right)$
$= \cos 0° + i \sin 0° = 1$

Second root: $1^{1/6} = 1^{1/6} \left(\cos \frac{0° + 360°}{6} + i \sin \frac{0° + 360°}{6} \right)$
$= \cos 60° + i \sin 60° = \frac{1}{2} + i \frac{\sqrt{3}}{2}$

Third root: $1^{1/6} = 1^{1/6} \left(\cos \frac{0° + 720°}{6} + i \sin \frac{0° + 720°}{6} \right)$
$= \cos 120° + i \sin 120° = -\frac{1}{2} + i \frac{\sqrt{3}}{2}$

Fourth root: $1^{1/6} = 1^{1/6} \left(\cos \frac{0° + 1080°}{6} + i \sin \frac{0° + 1080°}{6} \right)$
$= \cos 180° + i \sin 180° = -1$

Fifth root: $1^{1/6} = 1^{1/6} \left(\cos \frac{0° + 1440°}{6} + i \sin \frac{0° + 1440°}{6} \right)$
$= \cos 240° + i \sin 240° = -\frac{1}{2} - i \frac{\sqrt{3}}{2}$

Sixth root: $1^{1/6} = 1^{1/6} \left(\cos \frac{0° + 1800°}{6} + i \sin \frac{0° + 1800°}{6} \right)$
$= \cos 300° + i \sin 300° = \frac{1}{2} - i \frac{\sqrt{3}}{2}$

At this point we can see advantages for the various forms of writing complex numbers. Rectangular form lends itself best to addition and subtraction. Polar form is generally used for multiplying, dividing, raising to powers, and finding roots. Exponential form is used for theoretical purposes (e.g., deriving DeMoivre's theorem).

Exercises 8-6

In Exercises 1 through 12 perform the indicated operations. Leave the result in polar form.

1. $[4(\cos 60° + i \sin 60°][2(\cos 20° + i \sin 20°)]$
2. $[3(\cos 120° + i \sin 120°][5(\cos 45° + i \sin 45°)]$
3. $[0.5(\cos 140° + i \sin 140°)][6(\cos 110° + i \sin 110°)]$
4. $[0.4(\cos 320° + i \sin 320°)][5.5(\cos 150° + i \sin 150°)]$
5. $\dfrac{8(\cos 100° + i \sin 100°)}{4(\cos 65° + i \sin 65°)}$
6. $\dfrac{9(\cos 230° + i \sin 230°)}{3(\cos 80° + i \sin 80°)}$
7. $\dfrac{12(\cos 320° + i \sin 320°)}{5(\cos 210° + i \sin 210°)}$
8. $\dfrac{2(\cos 90° + i \sin 90°)}{4(\cos 75° + i \sin 75°)}$
9. $[2(\cos 35° + i \sin 35°)]^3$
10. $[3(\cos 120° + i \sin 120°)]^4$
11. $[2(\cos 135° + i \sin 135°)]^8$
12. $(\cos 142° + i \sin 142°)^{10}$

In Exercises 13 through 18 change each number to polar form and then perform the indicated operations. Express the final result in rectangular and polar forms. Check by performing the same operation in rectangular form.

13. $(3 + 4i)(5 - 12i)$
14. $(-2 + 5i)(-1 - i)$
15. $\dfrac{3 + 4i}{5 - 12i}$
16. $\dfrac{-2 + 5i}{-1 - i}$
17. $(3 + 4i)^4$
18. $(-1 - i)^8$

In Exercises 19 through 26 use DeMoivre's theorem to find the indicated roots. Be sure to find all roots.

19. $\sqrt{4(\cos 60° + i \sin 60°)}$
20. $\sqrt[3]{27(\cos 120° + i \sin 120°)}$
21. $\sqrt[3]{3 - 4i}$
22. $\sqrt{-5 + 12i}$
23. $\sqrt[4]{1}$
24. $\sqrt[3]{8}$
25. $\sqrt[3]{-i}$
26. $\sqrt[4]{i}$

Exercises for Chapter 8

In Exercises 1 through 10 perform the indicated operations, expressing all answers in the simplest rectangular form.

1. $(6 - 2i) + (4 + i)$
2. $(12 + 7i) + (-8 + 6i)$
3. $(18 - 3i) - (12 - 5i)$
4. $(-4 - 2i) - (-6 - 7i)$
5. $(2 + i)(4 - i)$
6. $(-5 + 3i)(8 - 4i)$
7. $(2i)(6 - 3i)(4 + 3i)$
8. $i(3 - 2i) - (i^3)(5 + i)$
9. $\dfrac{i(6 - 4i)}{7 - 2i}$
10. $\dfrac{(7 - i)(8 - i)}{6 + i}$

In Exercises 11 and 12 find the values of x and y for which the equations are true.

11. $3x - 2i = yi - 2$
12. $2xi - 2y = (y + 3)i - 3$

In Exercises 13 through 16 perform the indicated operations graphically; check them algebraically.

13. $(-1 + 5i) + (4 + 6i)$
14. $(7 - 2i) + (5 + 4i)$
15. $(9 + 2i) - (5 - 6i)$
16. $(1 + 4i) - (-3 - 3i)$

8-6 Products, Quotients, Powers, and Roots of Complex Numbers

In Exercises 17 through 20 give the polar and exponential forms of each of the complex numbers.

17. $1 - i$
18. $4 + 3i$
19. $-2 - 7i$
20. $-4i$

In Exercises 21 through 28 give the rectangular form of each of the complex numbers.

21. $2(\cos 225° + i \sin 225°)$
22. $4(\cos 60° + i \sin 60°)$
23. $5(\cos 123° + i \sin 123°)$
24. $2(\cos 296° + i \sin 296°)$
25. $2e^{0.25i}$
26. $e^{3.62i}$
27. $5e^{1.90i}$
28. $4e^{6.04i}$

In Exercises 29 through 36 perform the indicated operations. Leave the result in polar form.

29. $[3(\cos 32° + i \sin 32°)][5(\cos 52° + i \sin 52°)]$
30. $[2.5(\cos 162° + i \sin 162°)][8(\cos 115° + i \sin 115°)]$
31. $\dfrac{24(\cos 165° + i \sin 165°)}{3(\cos 106° + i \sin 106°)}$
32. $\dfrac{18(\cos 403° + i \sin 403°)}{4(\cos 192° + i \sin 192°)}$
33. $[2(\cos 16° + i \sin 16°)]^{10}$
34. $[3(\cos 36° + i \sin 36°)]^{6}$
35. $[3(\cos 54° + i \sin 54°)]^{4}$
36. $[2(\cos 100° + i \sin 100°)]^{3}$

In Exercises 37 through 40 change each number to polar form and then perform the indicated operations. Express the final result in rectangular and polar forms. Check by performing the same operation in rectangular form.

37. $(1 - i)^{10}$
38. $(\sqrt{3} + i)^{8}(1 + i)^{5}$
39. $\dfrac{(5 + 5i)^{4}}{(-1 - i)^{6}}$
40. $(\sqrt{3} - i)^{-8}$

In Exercises 41 through 44 use DeMoivre's theorem to find the indicated roots. Be sure to find all roots.

41. $\sqrt[3]{-8}$
42. $\sqrt[3]{1}$
43. $\sqrt[4]{-i}$
44. $\sqrt[5]{-32}$

Appendix A

Exponents and Radicals

A-1 Integral Exponents

In our study of trigonometry we find a need for exponents and radicals. In this appendix we discuss their meaning and use.

In multiplication we often encounter a number which is to be multiplied by itself several times. Rather than writing this number over and over repeatedly, we use the notation a^n, where a is the number being considered and n is the number of times it appears in the product. **The number a is called the *base*, the number n is called the *exponent*, and, in words, the expression is read as the "nth power of a."**

Example A

$4 \cdot 4 \cdot 4 \cdot 4 \cdot 4 = 4^5$ (the fifth power of 4)
$(-2)(-2)(-2)(-2) = (-2)^4$ (the fourth power of -2)
$a \cdot a = a^2$ (the second power of a, called "a squared")
$(\frac{1}{5})(\frac{1}{5})(\frac{1}{5}) = (\frac{1}{5})^3$ (the third power of $\frac{1}{5}$, called "$\frac{1}{5}$ cubed")
$8 \cdot 8 \cdot 8 \cdot 8 \cdot 8 \cdot 8 \cdot 8 \cdot 8 \cdot 8 = 8^9$

Appendix A: Exponents and Radicals

Certain important operations with exponents will now be stated symbolically. These operations are developed here for positive integers as exponents; they will be verified and illustrated by the examples which follow.

(A-1) $$a^m \cdot a^n = a^{m+n}$$

(A-2) $$\frac{a^m}{a^n} = a^{m-n} \quad (m > n,\ a \neq 0)$$

$$\frac{a^m}{a^n} = \frac{1}{a^{n-m}} \quad (m < n,\ a \neq 0)$$

(A-3) $$(a^m)^n = a^{mn}$$

(A-4) $$(ab)^n = a^n b^n \qquad \left(\frac{a}{b}\right)^n = \frac{a^n}{b^n} \quad (b \neq 0)$$

In applying Equations (A-1) and (A-2), the base a must be the same for the exponents to be added or subtracted. When a problem involves a product of different bases, only exponents of the same base may be combined. The following three examples illustrate the use of Equations (A-1) to (A-4).

Example B

$a^3 \cdot a^5 = a^{3+5} = a^8 \qquad$ since $\qquad a^3 \cdot a^5 = (a \cdot a \cdot a)(a \cdot a \cdot a \cdot a \cdot a) = a^8$

$\dfrac{a^5}{a^3} = a^{5-3} = a^2 \qquad$ since $\qquad \dfrac{a^5}{a^3} = \dfrac{\not{a} \cdot \not{a} \cdot \not{a} \cdot a \cdot a}{\not{a} \cdot \not{a} \cdot \not{a}} = a^2$

$\dfrac{a^3}{a^5} = \dfrac{1}{a^{5-3}} = \dfrac{1}{a^2} \qquad$ since $\qquad \dfrac{a^3}{a^5} = \dfrac{\not{a} \cdot \not{a} \cdot \not{a}}{\not{a} \cdot \not{a} \cdot \not{a} \cdot a \cdot a} = \dfrac{1}{a^2}$

Example C

$(a^5)^3 = a^{5(3)} = a^{15} \qquad$ since $\qquad (a^5)^3 = (a^5)(a^5)(a^5)$
$\qquad\qquad\qquad\qquad\qquad\qquad\qquad\qquad = a^{5+5+5} = a^{15}$

$(ab)^3 = a^3 b^3 \qquad$ since $\qquad (ab)^3 = (ab)(ab)(ab) = a^3 b^3$

$\left(\dfrac{a}{b}\right)^3 = \dfrac{a^3}{b^3} \qquad$ since $\qquad \left(\dfrac{a}{b}\right)^3 = \left(\dfrac{a}{b}\right)\left(\dfrac{a}{b}\right)\left(\dfrac{a}{b}\right) = \dfrac{a^3}{b^3}$

A-1 Integral Exponents

Example D

$$\frac{(3 \cdot 2)^4}{(3 \cdot 5)^3} = \frac{3^4 \, 2^4}{3^3 \, 5^3} = \frac{3 \cdot 2^4}{5^3}$$

$$ax^2 \, (ax)^3 = ax^2 \, (a^3 \, x^3) = a^4 \, x^5$$

$$\frac{(ry^3)^2}{r(y^2)^4} = \frac{r^2 \, y^6}{ry^8} = \frac{r}{y^2}$$

We have developed Equations (A-1) through (A-4) such that they are valid for positive integers as exponents. We shall now extend their use so that zero and the negative integers may also be used as exponents.

In Equation (A-2), if $n = m$ we would have $a^m/a^m = a^{m-m} = a^0$ Also, $a^m/a^m = 1$, since any nonzero quantity divided by itself equals 1. Therefore, for Equation (A-2) to hold when $m = n$, we have:

(A-5) $\qquad\qquad a^0 = 1 \qquad (a \neq 0)$

Equation (A-5) shows the definition of zero as an exponent. Since a has not been specified, this equation states that any nonzero algebraic expression raised to the zero power is 1. Also, the other laws of exponents are valid for this definition.

Example E

Equation (A-1) states that $a^m \cdot a^n = a^{m+n}$. If $n = 0$, we have $a^m \cdot a^0 = a^{m+0} = a^m$. Since $a^0 = 1$, this equation could be written as $a^m (1) = a^m$. This provides further verification for the validity of Equation (A-5).

Example F

$$5^0 = 1 \qquad (2x)^0 = 1 \qquad (ax + b)^0 = 1$$

$$(a^2 \, xb^4)^0 = 1 \qquad (a^2 \, b^0 \, c)^2 = a^4 \, b^0 \, c^2 = a^4 \, c^2$$

If we apply the first form of Equation (A-2) to the case where $n > m$, the resulting exponent is negative. This leads us to the definition of a negative exponent.

Example G
Applying the first form of Equation (A-2) to a^2/a^7, we have

$$\frac{a^2}{a^7} = a^{2-7} = a^{-5}$$

Applying the second form of Equation (A-2) to the same fraction leads to

$$\frac{a^2}{a^7} = \frac{1}{a^{7-2}} = \frac{1}{a^5}$$

For these results to be consistent, it must be true that

$$a^{-5} = \frac{1}{a^5}$$

Following the reasoning in Example C, if we define

$$(A-6) \qquad a^{-n} = \frac{1}{a^n} \qquad (a \neq 0)$$

then all the laws of exponents will hold for negative integers.

Example H

$$3^{-1} = \frac{1}{3} \qquad 4^{-2} = \frac{1}{16} \qquad \frac{1}{a^{-3}} = a^3 \qquad a^4 = \frac{1}{a^{-4}}$$

Example I

$$(a^0 b^2 c)^{-2} = \frac{1}{(b^2 c)^2} = \frac{1}{b^4 c^2}$$

$$\left(\frac{a^3 t}{b^2 x}\right)^{-2} = \frac{(a^3 t)^{-2}}{(b^2 x)^{-2}} = \frac{(b^2 x)^2}{(a^3 t)^2} = \frac{b^4 x^2}{a^6 t^2}$$

From Equation (A-6) and Examples H and I, we see that when a factor is moved from the demoninator to the numerator of a fraction, or conversely, the *sign* of the *exponent* is changed. We should heed the word "factor"; this rule does not apply to moving terms in the numerator or denominator.

Example J

$$\frac{x^{-1} + y^{-1}}{x^2 - y^2} = \frac{\frac{1}{x} + \frac{1}{y}}{x^2 - y^2} = \frac{\frac{y+x}{xy}}{(x+y)(x-y)}$$

$$= \frac{y+x}{xy} \cdot \frac{1}{(x+y)(x-y)} = \frac{y+x}{xy(x+y)(x-y)} = \frac{1}{xy(x-y)}$$

Note that in this example the x^{-1} and y^{-1} in the numerator could not be moved directly to the denominator with positive exponents because they are only terms of the original numerator.

Exercises A-1

In the following exercises express each of the given expressions in the simplest form which uses only positive exponents.

1. $x^3 x^4$
2. $y^2 y^7$
3. $2b^4 b^2$
4. $3k(k^5)$
5. $\dfrac{m^5}{m^3}$
6. $\dfrac{x^8}{x}$
7. $\dfrac{n^5}{n^9}$
8. $\dfrac{s}{s^4}$
9. $(2n)^3$
10. $(ax)^5$
11. $(a^2)^4$
12. $(x^3)^3$
13. $(-t^2)^7$
14. $(-y^3)^5$
15. $\left(\dfrac{2}{b}\right)^3$
16. $\left(\dfrac{x}{y}\right)^7$
17. $(2x^2)^6$
18. $(-c^4)^4$
19. $(-8gs^3)^2$
20. $ax^2(-a^2 x)^2$
21. $(8a)^0$
22. $(3x^2)^0$
23. $2a^{-2}$
24. $(2a)^{-2}$
25. $3x^0 c^{-2}$
26. $5^0 x^{-2} z$
27. $(a+b)^{-1}$
28. $a^{-1} + b^{-1}$
29. $(4xa^{-2})^0$
30. $3(a+b)^0$

31. $b^5 b^{-3}$
32. $2c^4 c^{-7}$
33. $\dfrac{15a^2 n^5}{3an^6}$
34. $\dfrac{(ab^2)^3}{a^2 b^8}$
35. $\dfrac{a(a^2 y^3)^2}{y^7}$
36. $\dfrac{-ax^4(-ax)^4}{(-a^4 x)^4}$
37. $\dfrac{a^2 b^5}{a^{-4} b}$
38. $\dfrac{3a^{-1} y^3}{a^5 y^{-1}}$
39. $(5^0 x^2 a^{-1})^{-1}$
40. $(3m^{-2} n^4)^{-2}$
41. $\left(\dfrac{4a}{x}\right)^{-3}$
42. $\left(\dfrac{2b^2}{y^5}\right)^{-2}$
43. $\left(\dfrac{4x^{-1}}{y^3}\right)^{-2}$
44. $\left(\dfrac{2b^{-1}}{c^{-5}}\right)^{-3}$
45. $(3x^2)^{-4}(2c)^0$
46. $3x^{-4}(6y)^2$
47. $(5t^{-6})(4t^{-7})$
48. $(5x^0 b)^0 7^{-2}$
49. $\left(\dfrac{3a^2}{4b}\right)^{-3}\left(\dfrac{4}{a}\right)^{-5}$
50. $\left(\dfrac{a^{-2}}{b^2}\right)^{-3}\left(\dfrac{a^{-3}}{b^5}\right)^2$
51. $(x^2 y^{-1})^2 - x^{-4}$
52. $3(a^{-1} z^2)^{-3} + c^{-2} z^{-1}$
53. $(a^{-1} + b^{-1})^{-1}$
54. $(2^{-3} - 4^{-2})^2$
55. $\dfrac{x - y^{-1}}{x^{-1} - y}$
56. $\dfrac{x^{-2} - y^{-2}}{x^{-1} - y^{-1}}$
57. $\dfrac{ax^{-2} + a^{-2} x}{a^{-1} + x^{-1}}$
58. $\dfrac{2x^{-2} - 2y^{-2}}{(xy)^{-3}}$
59. $(x - 1)^{-1} + (x + 1)^{-1}$
60. $4(2x - 1)(x + 2)^{-1} - (2x - 1)^2 (x + 2)^{-2}$

A-2 Scientific Notation

We often encounter numbers which are either very large or very small in magnitude. Illustrations of such numbers are given in the following example.

Example A
Television signals travel at about 30,000,000,000 cm/sec. The mass of the earth is about 6,600,000,000,000,000,000,000 tons. A typical

A-2 Scientific Notation

protective coating used on aluminum is about 0.0005 in. thick. The wavelength of some X rays is about 0.000000095 cm.

Writing numbers such as these is inconvenient in ordinary notation, particularly when the number of zeros needed for the proper location of the decimal point is excessive. Therefore, a convenient notation, known as *scientific notation*, is normally used to represent such numbers.

A number written in scientific notation is expressed as the product of a number between 1 and 10 and a power of 10. Symbolically this can be written as

$$P \times 10^k$$

where $1 \leq P < 10$ and k can take on any integral value. The following example illustrates how numbers are written in scientific notation.

Example B

$$340,000 = 3.4(100,000) = 3.4 \times 10^5$$

$$0.017 = \frac{1.7}{100} = 1.7 \times 10^{-2}$$

$$0.000503 = \frac{5.03}{10,000} = 5.03 \times 10^{-4}$$

$$6.82 = 6.82(1) = 6.82 \times 10^0$$

From Example B we can establish a method for changing numbers from ordinary notation to scientific notation. The decimal point is moved so that only one nonzero digit is to its left. The number of places moved is the value of k. It is positive if the decimal point is moved to the left and negative if it is moved to the right. Consider the illustrations in the following example.

Example C

$340000 = 3.4 \times 10^5$ $\quad$ $0.017 = 1.7 \times 10^{-2}$
5 places $\quad\quad\quad\quad\quad\quad\quad$ 2 places

$0.000503 = 5.03 \times 10^{-4}$ $\quad$ $6.82 = 6.82 \times 10^0$
4 places $\quad\quad\quad\quad\quad\quad\quad\quad$ 0 places

To change a number from scientific notation to ordinary notation, the procedure is reversed. The following example illustrates the procedure.

Example D

To change 5.83×10^6 to ordinary notation, we must move the decimal point six places to the right. Therefore, additional zeros must be included for the proper location of the decimal point. Thus, $5.83 \times 10^6 = 5,830,000$.

To change 8.06×10^{-3} to ordinary notation, we must move the decimal point three places to the left. Again, additional zeros must be included. Thus, $8.06 \times 10^{-3} = 0.00806$.

Scientific notation provides a practical way to handle calculations involving numbers of very large or very small magnitude. If all numbers are expressed in scientific notation, the laws of exponents are used to find the decimal point in the result. It is proper to leave the result in scientific notation.

Example E

In determining the result of

$$\frac{95,600,000,000}{0.0286}$$

we may estimate the result as

$$\frac{9 \times 10^{10}}{3 \times 10^{-2}} = 3 \times 10^{12}.$$

Actual calculations would give us

$$\frac{9.56 \times 10^{10}}{2.86 \times 10^{-2}} = \frac{9.56}{2.86} \times 10^{12} = 3.34 \times 10^{12}$$

to three significant digits. Thus, the only division which must be made is $9.56 \div 2.86$. The power of 10 here is sufficiently large that we would generally leave the answer in this form.

Exercises A-2

In Exercises 1 through 8 change the numbers from scientific notation to ordinary notation.

1. 4.5×10^4
2. 6.8×10^7
3. 2.01×10^{-3}
4. 9.61×10^{-5}
5. 3.23×10^0
6. 8.40×10^0
7. 1.86×10
8. 5.44×10^{-1}

In Exercises 9 through 16 change the given numbers from ordinary notation to scientific notation.

9. 40000
10. 5600000
11. 0.0087
12. 0.702
13. 6.89
14. 1.09
15. 0.063
16. 0.0000908

In Exercises 17 through 24 perform the indicated calculations, determining the location of the decimal point by the use of scientific notation.

17. (67000)(3040)
18. (56200)(0.00632)
19. (1280)(86500)(43.8)
20. (0.0000659)(0.00486)(31900)
21. $\dfrac{87400}{0.00895}$
22. $\dfrac{0.00728}{670000}$
23. $\dfrac{(0.0732)(6700)}{0.00134}$
24. $\dfrac{(2430)(97000)}{0.00452}$

In Exercises 25 through 32 change ordinary notation to scientific notation and vice versa.

25. The stress on a certain structure is 22,500 lb/in.^{2}.
26. The distance from the earth to the sun is 93,000,000 mi.
27. The pressure of a certain gas is 6.1×10^{-4} atmospheres.

28. Some computers can perform an addition in 1.5×10^{-6} sec.

29. The age of the earth's crust is 5×10^9 years.

30. The diameter of the sun is 8.64×10^5 mi.

31. The radius of the smallest visible particle is about 0.0001 in.

32. The wavelength of yellow light is about 0.00000059 m.

A-3 Radicals

Another problem often encountered is this: What number multiplied by itself n times gives another specified number? For example, we may ask: What number squared is 9? The answer to this question is the *square root* of 9, which is denoted by $\sqrt{9}$.

The general notation for the nth root of a is $\sqrt[n]{a}$. (When $n = 2$, it is common practice not to put the 2 where n appears.) The $\sqrt{}$ sign is called a *radical sign*.

Example A

$\sqrt{2}$ (the square root of 2)

$\sqrt[3]{2}$ (the cube root of 2)

$\sqrt[4]{2}$ (the fourth root of 2)

$\sqrt[7]{6}$ (the seventh root of 6)

$\sqrt[3]{8}$ (the cube root of 8, which also equals 2)

In considering the question "what number squared is 9?" we can easily see that either +3 or −3 gives a proper result. This implies that both these values equal $\sqrt{9}$. To avoid this ambiguity, **we define the *principal nth root* of a to be positive if a is positive and to be negative if a is negative and n is odd.** This means that $\sqrt{9} = 3$ and not −3 and that $-\sqrt{9} = -3$.

A-3 Radicals

Example B

$\sqrt{4} = 2$ ($\sqrt{4} \neq -2$) $\sqrt{169} = 13$ ($\sqrt{169} \neq -13$)

$-\sqrt{64} = -8$ $-\sqrt{81} = -9$ $\sqrt[4]{256} = 4$

$\sqrt[3]{27} = 3$ $\sqrt[3]{-27} = -3$ $-\sqrt[3]{27} = -(+3) = -3$

To find the square root of a negative number, it is necessary to define a new kind of number. Thus, we define $(bi)^2 = -b^2$, where bi is called an *imaginary number*. (This term is simply the name of the number; these numbers are not imaginary in the usual sense of the word.) By this definition it can be seen that:

(A-7) $\qquad i^2 = -1 \quad$ or $\quad \sqrt{-1} = i$

A more detailed discussion of imaginary numbers is found in Chapter 8. For now, it should be emphasized that although the square root of a negative number gives an imaginary number, the cube root of a negative number gives a negative real number.

Example C

$\sqrt{-4} = 2i$

$\sqrt[3]{-8} = -2$

$\sqrt{64} = 8$ $\sqrt{-64} = 8i$ $\sqrt[3]{64} = 4$ $\sqrt[3]{-64} = -4$

We shall now define the basic operations with radicals so that these definitions are consistent with the laws of exponents. These definitions are as follows:

(A-8) $\qquad \sqrt[n]{a^n} = (\sqrt[n]{a})^n = a$

(A-9) $\qquad \sqrt[n]{a}\sqrt[n]{b} = \sqrt[n]{ab}$

(A-10) $\qquad \sqrt[m]{\sqrt[n]{a}} = \sqrt[mn]{a}$

(A-11) $\qquad \dfrac{\sqrt[n]{a}}{\sqrt[n]{b}} = \sqrt[n]{\dfrac{a}{b}} \quad (b \neq 0)$

The number under the radical sign is called the *radicand*, and the number indicating the root being taken is called the *order* of the radical. At this point these definitions are made for nonnegative radicands.

Example D

$$\sqrt{5^2} = 5$$
$$\sqrt[3]{3}\sqrt[3]{4} = \sqrt[3]{12}$$
$$\sqrt[4]{\sqrt{2}} = \sqrt[8]{2}$$
$$\frac{\sqrt{2}}{\sqrt{7}} = \sqrt{\frac{2}{7}}$$

Equation (A-9) is useful in simplifying radicals. Consider the illustrations in the following example.

Example E

$$\sqrt{8} = \sqrt{(4)(2)} = \sqrt{4}\sqrt{2} = 2\sqrt{2}$$
$$\sqrt{75} = \sqrt{(25)(3)} = \sqrt{25}\sqrt{3} = 5\sqrt{3}$$
$$\sqrt[3]{16} = \sqrt[3]{(8)(2)} = \sqrt[3]{8}\sqrt[3]{2} = 2\sqrt[3]{2}$$

An important operation used when simplifying radicals is that of *rationalizing the denominator*. Generally, if a radical appears in the denominator of an expression, the final result is expressed such that there is no radical in the denominator. To do this we multiply the numerator and denominator of the radicand by the simplest number that will make the denominator a perfect nth power.

Example F

$$\sqrt{\frac{5}{7}} = \sqrt{\frac{5 \cdot 7}{7 \cdot 7}} = \frac{\sqrt{35}}{\sqrt{49}} = \frac{\sqrt{35}}{7}$$

$$\frac{3}{\sqrt{8}} = \frac{3\sqrt{2}}{\sqrt{8 \cdot 2}} = \frac{3\sqrt{2}}{\sqrt{16}} = \frac{3\sqrt{2}}{4}$$

$$\sqrt[3]{\frac{2}{3}} = \sqrt[3]{\frac{2 \cdot 9}{3 \cdot 9}} = \sqrt[3]{\frac{18}{27}} = \frac{\sqrt[3]{18}}{\sqrt[3]{27}} = \frac{\sqrt[3]{18}}{3}$$

Exercises A-3

In the following exercises write each expression in simplest radical form.

1. $\sqrt{25}$
2. $\sqrt{196}$
3. $-\sqrt{121}$
4. $-\sqrt{225}$
5. $\sqrt[3]{125}$
6. $\sqrt[3]{216}$
7. $\sqrt[3]{-125}$
8. $\sqrt[3]{-216}$
9. $\sqrt[4]{16}$
10. $\sqrt[5]{32}$
11. $-\sqrt[3]{-8}$
12. $-\sqrt[5]{-32}$
13. $(\sqrt{5})^2$
14. $(\sqrt{17})^2$
15. $(\sqrt[3]{3})^3$
16. $(\sqrt[5]{4})^5$
17. $\sqrt{24}$
18. $\sqrt{150}$
19. $\sqrt{45}$
20. $\sqrt{98}$
21. $\sqrt{80}$
22. $\sqrt{63}$
23. $\sqrt{128}$
24. $\sqrt{243}$
25. $\sqrt[3]{16}$
26. $\sqrt[4]{48}$
27. $\sqrt[5]{96}$
28. $\sqrt[3]{-16}$
29. $\sqrt{\sqrt{2}}$
30. $\sqrt[3]{\sqrt{5}}$
31. $\sqrt[5]{\sqrt[3]{9}}$
32. $\sqrt{\sqrt[8]{2}}$
33. $\sqrt{\dfrac{3}{2}}$
34. $\sqrt{\dfrac{6}{5}}$
35. $\sqrt{\dfrac{8}{3}}$
36. $\sqrt{\dfrac{18}{5}}$
37. $\sqrt{\dfrac{5}{8}}$
38. $\sqrt{\dfrac{7}{18}}$
39. $\sqrt{\dfrac{8}{45}}$
40. $\sqrt{\dfrac{27}{20}}$
41. $\sqrt[3]{\dfrac{3}{4}}$
42. $\sqrt[4]{\dfrac{2}{5}}$
43. $\sqrt[4]{\dfrac{7}{8}}$
44. $\sqrt[3]{\dfrac{5}{9}}$

A-4 Fractional Exponents

In Section A-1 we defined the laws of exponents as being valid for the integers. In this section we show how these definitions can be extended to include the rational numbers as well.

Equation (A-3) states that $(a^m)^n = a^{mn}$. If we were to let $m = \frac{1}{2}$ and $n = 2$, we would have $(a^{1/2})^2 = a^1$. However, we already have a way of writing a quantity which when squared equals a.

Appendix A: Exponents and Radicals

This is written as $\sqrt{a}$. To be consistent with previous definitions and to allow the laws of exponents to hold, we define:

(A-12) $$a^{1/n} = \sqrt[n]{a}$$

So that Equations (A-3) and (A-12) may hold at the same time, we define:

(A-13) $$a^{m/n} = \sqrt[n]{a^m}$$

It can be shown that these definitions are valid for all the laws of exponents.

Example A
We shall verify here that Equation (A-1) holds for the preceding definitions:
$$a^{1/4} a^{1/4} a^{1/4} a^{1/4} = a^{(1/4)+(1/4)+(1/4)+(1/4)} = a^1.$$
Now $a^{1/4} = \sqrt[4]{a}$ by definition. Also, by definition $\sqrt[4]{a}\sqrt[4]{a}\sqrt[4]{a}\sqrt[4]{a} = a$. Equation (A-1) is thereby verified for $n = 4$. Equation (A-3) is verified by the following:
$$a^{1/4} a^{1/4} a^{1/4} a^{1/4} = a = \sqrt[4]{a^4}.$$

We may interpret Equation (A-13) as "the mth power of the nth root of a," as well as the way in which it is written, which is "the nth root of the mth power of a." This is illustrated in the following example.

Example B
$$(\sqrt[3]{a})^2 = \sqrt[3]{a^2} = a^{2/3}$$
$$8^{2/3} = (\sqrt[3]{8})^2 = (2)^2 = 4 \quad \text{or} \quad 8^{2/3} = \sqrt[3]{8^2} = \sqrt[3]{64} = 4$$

Although both interpretations of Equation (A-13) are possible as indicated in Example B, in evaluating numerical expressions involving fractional exponents it is almost always best to find the root first, as indicated by the denominator of the fractional exponent. This will allow us to find the root of the smaller number, which is normally easier to find.

A-4 Fractional Exponents

Example C
To evaluate $(64)^{5/2}$, we should proceed as follows:

$$(64)^{5/2} = [(64)^{1/2}]^5 = 8^5 = 32{,}768.$$

If we raised 64 to the fifth power first, we would have

$$(64)^{5/2} = (64^5)^{1/2} = (1{,}073{,}741{,}824)^{1/2}.$$

We would now have to evaluate the indicated square root. This demonstrates why it is preferable to find the indicated root first.

Example D

$$(16)^{3/4} = (16^{1/4})^3 = 2^3 = 8$$

$$4^{-1/2} = \frac{1}{4^{1/2}} = \frac{1}{2}$$

$$9^{3/2} = (9^{1/2})^3 = 3^3 = 27$$

The question may arise as to why we use fractional exponents, since we have already defined expressions which are equivalent to their meanings. The answer is that fractional exponents are often easier to handle in more complex expressions involving roots, and therefore any expression involving radicals can be solved by use of fractional exponents.

Example E

$$(8a^2 b^4)^{1/3} = [(8^{1/3})(a^2)^{1/3}(b^4)^{1/3}] = 2a^{2/3} b^{4/3}$$

$$a^{3/4} a^{4/5} = a^{3/4 + 4/5} = a^{31/20}$$

$$(25a^{-2} c^4)^{3/2} = \left(\frac{(25)^{1/2}(c^4)^{1/2}}{(a^2)^{1/2}}\right)^3 = \left(\frac{5c^2}{a}\right)^3 = \frac{125c^6}{a^3}$$

Example F

$$\left(\frac{4^{-3/2} x^{2/3} y^{-7/4}}{2^{3/2} x^{-1/3} y^{3/4}}\right)^{2/3} = \left(\frac{x^{2/3 + 1/3}}{2^{3/2} 4^{3/2} y^{3/4 + 7/4}}\right)^{2/3}$$

$$= \frac{x^{(1)(2/3)}}{2^{(3/2)(2/3)} 4^{(3/2)(2/3)} y^{(10/4)(2/3)}} = \frac{x^{2/3}}{8 y^{5/3}}$$

Exercises A-4

In Exercises 1 through 20 evaluate the given expressions.

1. $(25)^{1/2}$
2. $(49)^{1/2}$
3. $(27)^{1/3}$
4. $(81)^{1/4}$
5. $8^{4/3}$
6. $(125)^{2/3}$
7. $(100)^{25/2}$
8. $(16)^{5/4}$
9. $8^{-1/3}$
10. $16^{-1/4}$
11. $(64)^{-2/3}$
12. $(32)^{-4/5}$
13. $5^{1/2}\,5^{3/2}$
14. $8^{1/3}\,4^{1/2}$
15. $(4^4)^{3/2}$
16. $(3^6)^{2/3}$
17. $\dfrac{(-27)^{1/3}}{6}$
18. $\dfrac{(-8)^{2/3}}{-2}$

19. $(125)^{-2/3} - (100)^{-3/2}$

20. $\dfrac{4^{-1}}{(36)^{-1/2}} - \dfrac{5^{-1/2}}{5^{1/2}}$

In Exercises 21 through 40 use the laws of exponents to simplify the given expressions. Express all answers with positive exponents.

21. $a^{2/3}\,a^{1/2}$
22. $x^{5/6}\,x^{-1/3}$
23. $\dfrac{y^{-1/2}}{y^{2/5}}$
24. $\dfrac{2r^{4/5}}{r^{-1}}$
25. $\dfrac{s^{1/4}\,s^{2/3}}{s^{-1}}$
26. $\dfrac{x^{3/10}}{x^{-1/5}\,x^2}$
27. $(8a^3\,b^6)^{1/3}$
28. $(8b^{-4}\,c^2)^{2/3}$
29. $\left(\dfrac{9t^{-2}}{16}\right)^{3/2}$
30. $\left(\dfrac{a^{5/7}}{a^{2/3}}\right)^{7/4}$
31. $\left(\dfrac{4a^{5/6}\,b^{-1/5}}{a^{2/3}\,b^2}\right)^{-1/2}$
32. $\left(\dfrac{a^0\,b^8\,c^{-1/8}}{ab^{63/64}}\right)^{32/3}$
33. $\dfrac{6x^{-1/2}\,y^{2/3}}{18x^{-1}} \cdot \dfrac{2y^{1/4}}{x^{1/3}}$
34. $\dfrac{3^{-1}\,a^{1/2}}{4^{-1/2}\,b} \div \dfrac{9^{1/2}\,a^{-1/3}}{2b^{-1/4}}$
35. $(x^{-1} + 2x^{-2})^{-1/2}$
36. $(a^{-2} - a^{-4})^{-1/4}$
37. $(a^3)^{-4/3} + a^{-2}$
38. $(4x^6)^{-1/2} - 2x^{-1}$
39. $[(a^{1/2} - a^{-1/2})^2 + 4]^{1/2}$
40. $(3x-1)^{-2/3}(1-x) - (3x-1)^{1/3}$

Appendix B

Approximate Numbers and Significant Digits

When we perform calculations on numbers, we must consider the accuracy of these numbers, since this affects the accuracy of the results obtained. Most of the numbers involved in technical and scientific work are *approximate*, having been arrived at through some process of measurement. However, certain other numbers are *exact*, having been arrived at through some definition or counting process. We can determine whether or not a number is approximate or exact if we know how the number was determined.

Example A
If we measure the length of a rope to be 15.3 ft, we know that 15.3 is approximate. A more precise measuring device may cause us to determine the length as 15.28 ft. However, regardless of the method of measurement used, we cannot determine this length exactly.

If a voltage shown on a voltmeter is read as 116 volts, the 116 is approximate. A more precise voltmeter may show the voltage as 115.7 volts. However, this voltage cannot be determined exactly.

Example B
If a computer counts the cards it has processed and prints this number as 837, this 837 is exact. We know the number of cards was not 836 or 838. Since 837 was determined through a counting process, it is exact.

When we say that 60 sec = 1 min, the 60 is exact, since this is a definition. By this definition there are exactly 60 seconds in 1 minute.

When we are writing approximate numbers we often have to include some zeros so that the decimal point will be properly located. However, except for these zeros, all other digits are considered to be *significant digits*. When we make computations with approximate numbers, we must know the number of significant digits. The following example illustrates how we determine this.

Example C
All numbers in this example are assumed to be approximate.

34.7 has three significant digits.

8900 has two significant digits. We assume that the two zeros are place holders (unless we have specific knowledge to the contrary).

0.039 has two significant digits. The zeros are for proper location of the decimal point.

706.1 has four significant digits. The zero is not used for the location of the decimal point. It shows specifically the number of tens in the number.

5.90 has three significant digits. The zero is not necessary as a placeholder and should not be written unless it is significant.

Other approximate numbers with the proper number of significant digits are listed below:

96000	two	0.0709	three	1.070	four
30900	three	6.000	four	700.00	five
4.006	four	0.0005	one	20008	five

Note from Example C that all nonzero digits are significant. Zeros, other than those used as placeholders for proper positioning of the decimal point, are also significant.

In computations involving approximate numbers, the position of the decimal point as well as the number of significant digits is important. The *precision* of a number refers directly to the decimal position of the last significant digit, whereas the *accuracy* of a number refers to the number of significant digits in the number. Consider the illustrations in the following example.

Example D
Suppose that you are measuring an electric current with two ammeters. One ammeter reads 0.031 amp and the other reads 0.0312 amp. The second reading is more precise, in that the last significant digit is the number of ten-thousandths and the first reading is expressed only to thousandths. The second reading is also more accurate, since it has three significant digits rather than two.

A machine part is measured to be 2.5 cm long. It is coated with a film 0.025 cm thick. The thickness of the film has been measured to a greater precision, although the two measurements have the same accuracy: two significant digits.

A segment of a newly completed highway is 9270 ft long. The concrete surface is 0.8 ft thick. Of these two numbers, 9270 is more accurate, since it contains three significant digits, and 0.8 is more precise, since it is expressed to tenths.

The last significant digit of an approximate number is known not to be completely accurate. It has usually been determined by estimation or *rounding off*. However, we do know that it is at most in error by half a unit in its place value.

Example E
When we measure the rope in Example A to be 15.3 ft, we are saying that the length is at least 15.25 ft and no longer than 15.35 ft. Any value between these two, rounded off to tenths, would be expressed as 15.3 ft.

In converting the fraction $\frac{2}{3}$ to the decimal form 0.667, we are saying that the value is between 0.6665 and 0.6675.

The principle of rounding off a number is to write the closest approximation, with the last significant digit in a specified position or with a specified number of significant digits. We shall now formalize the process of rounding off as follows: If we want a certain number of significant digits, we examine the digit in the next place to the right. If this digit is less than 5, we accept the digit in the last place. If the next digit is 5 or greater, we increase the digit in the last place by 1, and this resulting digit becomes the final significant digit of the approximation. If necessary, we use zeros to replace other digits in order to locate the decimal point properly. Except when the next digit is a 5, and no other nonzero digits are discarded, we have the closest possible approximation with the desired number of significant digits.

Example F
70360 rounded off to three significant digits is 70400. 70430 rounded off to three significant digits is 70400. 187.35 rounded off to four significant digits is 187.4. 71500 rounded off to two significant digits is 72000.

Appendix B: Approximate Numbers and Significant Digits

With the advent of computers and calculators, another method of reducing numbers to a specified number of significant digits is used. This is the process of *truncation*, in which the digits beyond a certain place are discarded. For example, 3.17482 truncated to thousandths is 3.174. For our purposes in this text, when working with approximate numbers we have used only rounding off.

When performing arithmetic operations on approximate numbers, we must be careful not to express the result to a precision or accuracy which is not warranted. Therefore, we use the following rules when we perform the basic operations on approximate numbers:

1. When approximate numbers are added or subtracted, the result is expressed with the precision of the least precise number.
2. When approximate numbers are multipled or divided, the result is expressed with the accuracy of the least accurate number.
3. When the root of an approximate number is found, the result is accurate to the accuracy of the number.
4. Before and during the calculation, all numbers except the least precise or least accurate may be rounded off to one place beyond that of the least precise or least accurate. This procedure is helpful when a calculator is not used. It need not be followed when a calculator is used.

Example G
Add the approximate numbers 73.2, 8.0627, 93.57, 66.296.

The least precise of these numbers is 73.2. Therefore, before performing the addition we may round off the other numbers to hundredths. If we are using a calculator, this need not be done. In either case, after the addition we round off the result to tenths. This leads to

73.2		73.2
8.06		8.0627
93.57	or	93.57
66.30		66.296
241.13		241.1287

Therefore, the final result is 241.1.

Example H
If we multiply 2.4832 by 30.5, we obtain 75.7376. However, since 30.5 has only three significant digits, we express the product and result as $(2.4832)(30.5) = 75.7$.

To find the square root of 3.7, we may express the result only to two significant digits. Thus, $\sqrt{3.7} = 1.9$.

The rules stated here are usually valid for the computational work we may encounter. They are intended only as good practical rules for working with approximate numbers, since the last significant digit is subject to error. If an exact number is included in a calculation, there is no limitation to the number of decimal positions it may take on. The accuracy of the result is limited only by the approximate numbers involved.

Exercises for Appendix B

In Exercises 1 through 8 determine whether the numbers given are exact or approximate.

1. There are 24 hr in one day.
2. The velocity of light is 186,000 mi/sec.
3. The three-stage rocket took 74.6 hr to reach the moon.
4. A man bought 5 lb of nails for $1.56.
5. The melting point of gold is 1063°C.
6. The 21 students had an average test grade of 81.6.
7. A building lot 100 ft by 200 ft cost $3200.
8. In a certain city 5 percent of the people have their money in a bank that pays 5 percent interest.

Appendix B: Approximate Numbers and Significant Digits

In Exercises 9 through 16 determine the number of significant digits in the given approximate numbers.

9. 37.2; 6844
10. 3600; 730
11. 107; 3004
12. 0.8735; 0.0075
13. 6.80; 6.08
14. 90050; 105040
15. 30000; 30000.0
16. 1.00; 0.01

In Exercises 17 through 24 determine which of each pair of approximate numbers is (a) more precise and (b) more accurate.

17. 3.764; 2.81
18. 0.041; 7.673
19. 30.8; 0.01
20. 70,370; 50,400
21. 0.1; 78.0
22. 7040; 37.1
23. 7000; 0.004
24. 50.060; 8.914

In Exercises 25 through 32 round off each of the given approximate numbers (a) to three significant digits and (b) to two significant digits.

25. 4.933
26. 80.53
27. 57893
28. 30490
29. 861.29
30. 9555
31. 0.30505
32. 0.7350

In Exercises 33 through 36 add the given approximate numbers.

33. 3.8
     0.154
    47.26

34.   26
       5.806
     147.29

35. 0.36294
    0.086
    0.5056
    0.74

36. 56.1
      3.0645
    127.38
      0.055

In Exercises 37 through 40 subtract the given approximate numbers.

37. 468.14
     36.7

38. 1.03964
    0.69

39. 57.348
    26.5

40. 8.93
    6.8947

In Exercises 41 through 44 multiply the given approximate numbers.

41. (3.64)(17.06)   42. (0.025)(70.1)
43. (704.6)(0.38)   44. (0.003040)(6079.52)

In Exercises 45 through 48 divide the given approximate numbers.

45. $608 \div 3.9$ 46. $0.4962 \div 827$

47. $\dfrac{596000}{22}$ 48. $\dfrac{53.267}{0.3002}$

In Exercises 49 through 52 find the indicated square roots of the given approximate numbers.

49. $\sqrt{32}$ 50. $\sqrt{6.5}$
51. $\sqrt{19.3}$ 52. $\sqrt{0.0694}$

Appendix C

A Note Regarding Units of Measurement

Presently there are two major systems of measurement, the English system and the metric system. However, the major countries which use the English system have started the conversion to the metric system. There is a great deal of evidence of this transition in American industry, and Congress has passed legislation coordinating the conversion in the United States. The measurements primarily affected by this change are those of length, area, volume, and weight. Therefore, since we are in this state of transition, about half the applied problems in this text dealing with these measurements are in the English system and the other half are in the metric system.

For the information and convenience of the reader, we present here a very brief summary of the basic units of the English system and the metric system, along with a few basic conversions between the systems. The indicated abbreviations are used in the text.

English

Length: 1 yard (yd) = 3 feet (ft)
12 inches (in.) = 1 ft
1 mile (mi) = 5280 ft

Volume: 4 quarts (qt) = 1 gallon (gal)

Weight: 16 ounces (oz) = 1 pound (lb)
1 ton = 2000 lb

Metric

Length: 100 centimeters (cm) = 1 meter (m)
1000 millimeters (mm) = 1 m
1000m = 1 kilometer (km)

Volume: 1000 milliliters (ml) = 1 liter (l)
1000 l = 1 kiloliter (kl)

Weight: (mass)   1000 milligrams (mg) = 1 gram (g)
1000 g = 1 kilogram (kg)

Basic Conversions

1 in. = 2.54 cm	1 l = 1.06 qt
1 m = 39.4 in.	1 kg = 2.2 lb

It can be seen that 1 cm is somewhat less than half an inch and that a meter is a little longer than a yard. A liter is almost the same as a quart, and a kilogram is a little more than 2 lb. These conversion factors should give the reader a sense of the magnitude of some of the basic metric units.

When using the metric system, temperature is normally measured in degrees Celsius. It is probable that °C will become universally used with the metric system. Briefly, 0°C is the temperature at which ice melts and 100°C is the temperature at which water boils. These are equivalent to 32°F and 212°F (Fahrenheit), respectively.

Appendix D

Tables

Table 1 in Appendix D may be used to find the square root of any number with three significant digits. Below are two examples for finding the square root of a number. Further examples are given in Table 1.

Example A
Find $\sqrt{12700}$.

First, approximate the square root of 12,700 by dividing the digits into pairs starting at the decimal point and then finding the first digit of the square root of the first number or pair of numbers.

$$\sqrt{\overset{1}{1} | 27 | 00.}$$

We see that the first digit in the square root of 12700 is 1 and there are two other digits before the decimal point. (Note that there is always one digit above each pair of digits under the radical sign.)

We now go to Table 1 and find the digits 127 in the column labeled n. We go across the row and look at the numbers in the column labeled $\sqrt{n}$ and $\sqrt{10n}$. The number we need starts with a 1 so we pick 1.12694 and copy these digits into the spaces above $\sqrt{12700}$. The position of the decimal in our answer has been determined, and we only need to round the answer to the proper place.

$$\sqrt{\overset{1}{1} | \overset{1}{27} | \overset{2.6}{00.}} \quad \text{or} \quad \sqrt{12700} = 113$$

Example B
Find $\sqrt{0.00697}$

We first divide the digits into pairs starting at the decimal point. The first nonzero pair of numbers is 69. We know that the first significant digit in our answer must be the first digit in $\sqrt{69}$ which is 8. We place the 8 above the 69 and place a zero above the first pair of zeros to hold the decimal point.

$$\sqrt{\overset{.0\ |\ 8\ |}{0.00\ |\ 69\ |\ 70}}$$

We now go to Table 1 and find the digits 697 in the n column and find the digits 8.34865 in the $\sqrt{10n}$ column. We copy these digits into our answer and round the result to the proper number of significant digits. To three significant digits $\sqrt{0.00697} = 0.0835$.

Explanations on the use of the other tables are found in the text.

Table 1. Squares and Square Roots

n	n^2	$\sqrt{n}$	$\sqrt{10n}$	n	n^2	$\sqrt{n}$	$\sqrt{10n}$	n	n^2	$\sqrt{n}$	$\sqrt{10n}$
1.00	1.0000	1.00000	3.16228	1.50	2.2500	1.22474	3.87298	2.00	4.0000	1.41421	4.47214
1.01	1.0201	1.00499	3.17805	1.51	2.2801	1.22882	3.88587	2.01	4.0401	1.41774	4.48330
1.02	1.0404	1.00995	3.19374	1.52	2.3104	1.23288	3.89872	2.02	4.0804	1.42127	4.49444
1.03	1.0609	1.01489	3.20936	1.53	2.3409	1.23693	3.91152	2.03	4.1209	1.42478	4.50555
1.04	1.0816	1.01980	3.22490	1.54	2.3716	1.24097	3.92428	2.04	4.1616	1.42829	4.51664
1.05	1.1025	1.02470	3.24037	1.55	2.4025	1.24499	3.93700	2.05	4.2025	1.43178	4.52769
1.06	1.1236	1.02956	3.25576	1.56	2.4336	1.24900	3.94968	2.06	4.2436	1.43527	4.53872
1.07	1.1449	1.03441	3.27109	1.57	2.4649	1.25300	3.96232	2.07	4.2849	1.43875	4.54973
1.08	1.1664	1.03923	3.28634	1.58	2.4964	1.25698	3.97492	2.08	4.3264	1.44222	4.56070
1.09	1.1881	1.04403	3.30151	1.59	2.5281	1.26095	3.98748	2.09	4.3681	1.44568	4.57165
1.10	1.2100	1.04881	3.31662	1.60	2.5600	1.26491	4.00000	2.10	4.4100	1.44914	4.58258
1.11	1.2321	1.05357	3.33167	1.61	2.5921	1.26886	4.01248	2.11	4.4521	1.45258	4.59347
1.12	1.2544	1.05830	3.34664	1.62	2.6244	1.27279	4.02492	2.12	4.4944	1.45602	4.60435
1.13	1.2769	1.06301	3.36155	1.63	2.6569	1.27671	4.03733	2.13	4.5369	1.45945	4.61519
1.14	1.2996	1.06771	3.37639	1.64	2.6896	1.28062	4.04969	2.14	4.5796	1.46287	4.62601
1.15	1.3225	1.07238	3.39116	1.65	2.7225	1.28452	4.06202	2.15	4.6225	1.46629	4.63681
1.16	1.3456	1.07703	3.40588	1.66	2.7556	1.28841	4.07431	2.16	4.6656	1.46969	4.64758
1.17	1.3689	1.08167	3.42053	1.67	2.7889	1.29228	4.08656	2.17	4.7089	1.47309	4.65833
1.18	1.3924	1.08628	3.43511	1.68	2.8224	1.29615	4.09878	2.18	4.7524	1.47648	4.66905
1.19	1.4161	1.09087	3.44964	1.69	2.8561	1.30000	4.11096	2.19	4.7961	1.47986	4.67974
1.20	1.4400	1.09545	3.46410	1.70	2.8900	1.30384	4.12311	2.20	4.8400	1.48324	4.69042
1.21	1.4641	1.10000	3.47851	1.71	2.9241	1.30767	4.13521	2.21	4.8841	1.48661	4.70106
1.22	1.4884	1.10454	3.49285	1.72	2.9584	1.31149	4.14729	2.22	4.9284	1.48997	4.71169
1.23	1.5129	1.10905	3.50714	1.73	2.9929	1.31529	4.15933	2.23	4.9729	1.49332	4.72229
1.24	1.5376	1.11355	3.52136	1.74	3.0276	1.31909	4.17133	2.24	5.0176	1.49666	4.73286
1.25	1.5625	1.11803	3.53553	1.75	3.0625	1.32288	4.18330	2.25	5.0625	1.50000	4.74342
1.26	1.5876	1.12250	3.54965	1.76	3.0976	1.32665	4.19524	2.26	5.1076	1.50333	4.75395
1.27	1.6129	1.12694	3.56371	1.77	3.1329	1.33041	4.20714	2.27	5.1529	1.50665	4.76445
1.28	1.6384	1.13137	3.57771	1.78	3.1684	1.33417	4.21900	2.28	5.1984	1.50997	4.77493
1.29	1.6641	1.13578	3.59166	1.79	3.2041	1.33791	4.23084	2.29	5.2441	1.51327	4.78539
1.30	1.6900	1.14018	3.60555	1.80	3.2400	1.34164	4.24264	2.30	5.2900	1.51658	4.79583
1.31	1.7161	1.14455	3.61939	1.81	3.2761	1.34536	4.25441	2.31	5.3361	1.51987	4.80625
1.32	1.7424	1.14891	3.63318	1.82	3.3124	1.34907	4.26615	2.32	5.3824	1.52315	4.81664
1.33	1.7689	1.15326	3.64692	1.83	3.3489	1.35277	4.27785	2.33	5.4289	1.52643	4.82701
1.34	1.7956	1.15758	3.66060	1.84	3.3856	1.35647	4.28952	2.34	5.4756	1.52971	4.83735
1.35	1.8225	1.16190	3.67423	1.85	3.4225	1.36015	4.30116	2.35	5.5225	1.53297	4.84768
1.36	1.8496	1.16619	3.68782	1.86	3.4596	1.36382	4.31277	2.36	5.5696	1.53623	4.85798
1.37	1.8769	1.17047	3.70135	1.87	3.4969	1.36748	4.32435	2.37	5.6169	1.53948	4.86826
1.38	1.9044	1.17473	3.71484	1.88	3.5344	1.37113	4.33590	2.38	5.6644	1.54272	4.87852
1.39	1.9321	1.17898	3.72827	1.89	3.5721	1.37477	4.34741	2.39	5.7121	1.54596	4.88876
1.40	1.9600	1.18322	3.74166	1.90	3.6100	1.37840	4.35890	2.40	5.7600	1.54919	4.89898
1.41	1.9881	1.18743	3.75500	1.91	3.6481	1.38203	4.37035	2.41	5.8081	1.55242	4.90918
1.42	2.0164	1.19164	3.76829	1.92	3.6864	1.38564	4.38178	2.42	5.8564	1.55563	4.91935
1.43	2.0449	1.19583	3.78153	1.93	3.7249	1.38924	4.39318	2.43	5.9049	1.55885	4.92950
1.44	2.0736	1.20000	3.79473	1.94	3.7636	1.39284	4.40454	2.44	5.9536	1.56205	4.93964
1.45	2.1025	1.20416	3.80789	1.95	3.8025	1.39642	4.41588	2.45	6.0025	1.56525	4.94975
1.46	2.1316	1.20830	3.82099	1.96	3.8416	1.40000	4.42719	2.46	6.0516	1.56844	4.95984
1.47	2.1609	1.21244	3.83406	1.97	3.8809	1.40357	4.43847	2.47	6.1009	1.57162	4.96991
1.48	2.1904	1.21655	3.84708	1.98	3.9204	1.40712	4.44972	2.48	6.1504	1.57480	4.97996
1.49	2.2201	1.22066	3.86005	1.99	3.9601	1.41067	4.46094	2.49	6.2001	1.57797	4.98999
1.50	2.2500	1.22474	3.87298	2.00	4.0000	1.41421	4.47214	2.50	6.2500	1.58114	5.00000

Illustration for $n = 1.27$:

$1.27^2 = 1.6129 \qquad 0.127^2 = 0.016129$

$12.7^2 = 161.29 \qquad 0.0127^2 = 0.00016129$

$127^2 = 16129 \qquad 0.00127^2 = 0.0000016129$

$1270^2 = 1612900$

Table 1. Continued

n	n^2	$\sqrt{n}$	$\sqrt{10n}$	n	n^2	$\sqrt{n}$	$\sqrt{10n}$	n	n^2	$\sqrt{n}$	$\sqrt{10n}$
2.50	6.2500	1.58114	5.00000	3.00	9.0000	1.73205	5.47723	3.50	12.2500	1.87083	5.91608
2.51	6.3001	1.58430	5.00999	3.01	9.0601	1.73494	5.48635	3.51	12.3201	1.87350	5.92453
2.52	6.3504	1.58745	5.01996	3.02	9.1204	1.73781	5.49545	3.52	12.3904	1.87617	5.93296
2.53	6.4009	1.59060	5.02991	3.03	9.1809	1.74069	5.50454	3.53	12.4609	1.87883	5.94138
2.54	6.4516	1.59374	5.03984	3.04	9.2416	1.74356	5.51362	3.54	12.5316	1.88149	5.94979
2.55	6.5025	1.59687	5.04975	3.05	9.3025	1.74642	5.52268	3.55	12.6025	1.88414	5.95819
2.56	6.5536	1.60000	5.05964	3.06	9.3636	1.74929	5.53173	3.56	12.6736	1.88680	5.96657
2.57	6.6049	1.60312	5.06952	3.07	9.4249	1.75214	5.54076	3.57	12.7449	1.88944	5.97495
2.58	6.6564	1.60624	5.07937	3.08	9.4864	1.75499	5.54977	3.58	12.8164	1.89209	5.98331
2.59	6.7081	1.60935	5.08920	3.09	9.5481	1.75784	5.55878	3.59	12.8881	1.89473	5.99166
2.60	6.7600	1.61245	5.09902	3.10	9.6100	1.76068	5.56776	3.60	12.9600	1.89737	6.00000
2.61	6.8121	1.61555	5.10882	3.11	9.6721	1.76352	5.57674	3.61	13.0321	1.90000	6.00833
2.62	6.8644	1.61864	5.11859	3.12	9.7344	1.76635	5.58570	3.62	13.1044	1.90263	6.01664
2.63	6.9169	1.62173	5.12835	3.13	9.7969	1.76918	5.59464	3.63	13.1769	1.90526	6.02495
2.64	6.9696	1.62481	5.13809	3.14	9.8596	1.77200	5.60357	3.64	13.2496	1.90788	6.03324
2.65	7.0225	1.62788	5.14782	3.15	9.9225	1.77482	5.61249	3.65	13.3225	1.91050	6.04152
2.66	7.0756	1.63095	5.15752	3.16	9.9856	1.77764	5.62139	3.66	13.3956	1.91311	6.04979
2.67	7.1289	1.63401	5.16720	3.17	10.0489	1.78045	5.63028	3.67	13.4689	1.91572	6.05805
2.68	7.1824	1.63707	5.17687	3.18	10.1124	1.78326	5.63915	3.68	13.5424	1.91833	6.06630
2.69	7.2361	1.64012	5.18652	3.19	10.1761	1.78606	5.64801	3.69	13.6161	1.92094	6.07454
2.70	7.2900	1.64317	5.19615	3.20	10.2400	1.78885	5.65685	3.70	13.6900	1.92354	6.08276
2.71	7.3441	1.64621	5.20577	3.21	10.3041	1.79165	5.66569	3.71	13.7641	1.92614	6.09098
2.72	7.3984	1.64924	5.21536	3.22	10.3684	1.79444	5.67450	3.72	13.8384	1.92873	6.09918
2.73	7.4529	1.65227	5.22494	3.23	10.4329	1.79722	5.68331	3.73	13.9129	1.93132	6.10737
2.74	7.5076	1.65529	5.23450	3.24	10.4976	1.80000	5.69210	3.74	13.9876	1.93391	6.11555
2.75	7.5625	1.65831	5.24404	3.25	10.5625	1.80278	5.70088	3.75	14.0625	1.93649	6.12372
2.76	7.6176	1.66132	5.25357	3.26	10.6276	1.80555	5.70964	3.76	14.1376	1.93907	6.13188
2.77	7.6729	1.66433	5.26308	3.27	10.6929	1.80831	5.71839	3.77	14.2129	1.94165	6.14003
2.78	7.7284	1.66733	5.27257	3.28	10.7584	1.81108	5.72713	3.78	14.2884	1.94422	6.14817
2.79	7.7841	1.67033	5.28205	3.29	10.8241	1.81384	5.73585	3.79	14.3641	1.94679	6.15630
2.80	7.8400	1.67332	5.29150	3.30	10.8900	1.81659	5.74456	3.80	14.4400	1.94936	6.16441
2.81	7.8961	1.67631	5.30094	3.31	10.9561	1.81934	5.75326	3.81	14.5161	1.95192	6.17252
2.82	7.9524	1.67929	5.31037	3.32	11.0224	1.82209	5.76194	3.82	14.5924	1.95448	6.18061
2.83	8.0089	1.68226	5.31977	3.33	11.0889	1.82483	5.77062	3.83	14.6689	1.95704	6.18870
2.84	8.0656	1.68523	5.32917	3.34	11.1556	1.82757	5.77927	3.84	14.7456	1.95959	6.19677
2.85	8.1225	1.68819	5.33854	3.35	11.2225	1.83030	5.78792	3.85	14.8225	1.96214	6.20484
2.86	8.1796	1.69115	5.34790	3.36	11.2896	1.83303	5.79655	3.86	14.8996	1.96469	6.21289
2.87	8.2369	1.69411	5.35724	3.37	11.3569	1.83576	5.80517	3.87	14.9769	1.96723	6.22093
2.88	8.2944	1.69706	5.36656	3.38	11.4244	1.83848	5.81378	3.88	15.0544	1.96977	6.22896
2.89	8.3521	1.70000	5.37587	3.39	11.4921	1.84120	5.82237	3.89	15.1321	1.97231	6.23699
2.90	8.4100	1.70294	5.38516	3.40	11.5600	1.84391	5.83095	3.90	15.2100	1.97484	6.24500
2.91	8.4681	1.70587	5.39444	3.41	11.6281	1.84662	5.83952	3.91	15.2881	1.97737	6.25300
2.92	8.5264	1.70880	5.40370	3.42	11.6964	1.84932	5.84808	3.92	15.3664	1.97990	6.26099
2.93	8.5849	1.71172	5.41295	3.43	11.7649	1.85203	5.85662	3.93	15.4449	1.98242	6.26897
2.94	8.6436	1.71464	5.42218	3.44	11.8336	1.85472	5.86515	3.94	15.5236	1.98494	6.27694
2.95	8.7025	1.71756	5.43139	3.45	11.9025	1.85742	5.87367	3.95	15.6025	1.98746	6.28490
2.96	8.7616	1.72047	5.44059	3.46	11.9716	1.86011	5.88218	3.96	15.6816	1.98997	6.29285
2.97	8.8209	1.72337	5.44977	3.47	12.0409	1.86279	5.89067	3.97	15.7609	1.99249	6.30079
2.98	8.8804	1.72627	5.45894	3.48	12.1104	1.86548	5.89915	3.98	15.8404	1.99499	6.30872
2.99	8.9401	1.72916	5.46809	3.49	12.1801	1.86815	5.90762	3.99	15.9201	1.99750	6.31664
3.00	9.0000	1.73205	5.47723	3.50	12.2500	1.87083	5.91608	4.00	16.0000	2.00000	6.32456

Illustration for $n = 1.27$:

$\sqrt{1.27} = 1.12694$ $\sqrt{0.127} = 0.356371$

$\sqrt{12.7} = 3.56371$ $\sqrt{0.0127} = 0.112694$

$\sqrt{127} = 11.2694$ $\sqrt{0.00127} = 0.0356371$

$\sqrt{1270} = 35.6371$ $\sqrt{0.000127} = 0.0112694$

Table 1. Continued

n	n^2	$\sqrt{n}$	$\sqrt{10n}$	n	n^2	$\sqrt{n}$	$\sqrt{10n}$	n	n^2	$\sqrt{n}$	$\sqrt{10n}$
4.00	16.0000	2.00000	6.32456	4.50	20.2500	2.12132	6.70820	5.00	25.0000	2.23607	7.07107
4.01	16.0801	2.00250	6.33246	4.51	20.3401	2.12368	6.71565	5.01	25.1001	2.23830	7.07814
4.02	16.1604	2.00499	6.34035	4.52	20.4304	2.12603	6.72309	5.02	25.2004	2.24054	7.08520
4.03	16.2409	2.00749	6.34823	4.53	20.5209	2.12838	6.73053	5.03	25.3009	2.24277	7.09225
4.04	16.3216	2.00998	6.35610	4.54	20.6116	2.13073	6.73795	5.04	25.4016	2.24499	7.09930
4.05	16.4025	2.01246	6.36396	4.55	20.7025	2.13307	6.74537	5.05	25.5025	2.24722	7.10634
4.06	16.4836	2.01494	6.37181	4.56	20.7936	2.13542	6.75278	5.06	25.6036	2.24944	7.11337
4.07	16.5649	2.01742	6.37966	4.57	20.8849	2.13776	6.76018	5.07	25.7049	2.25167	7.12039
4.08	16.6464	2.01990	6.38749	4.58	20.9764	2.14009	6.76757	5.08	25.8064	2.25389	7.12741
4.09	16.7281	2.02237	6.39531	4.59	21.0681	2.14243	6.77495	5.09	25.9081	2.25610	7.13442
4.10	16.8100	2.02485	6.40312	4.60	21.1600	2.14476	6.78233	5.10	26.0100	2.25832	7.14143
4.11	16.8921	2.02731	6.41093	4.61	21.2521	2.14709	6.78970	5.11	26.1121	2.26053	7.14843
4.12	16.9744	2.02978	6.41872	4.62	21.3444	2.14942	6.79706	5.12	26.2144	2.26274	7.15542
4.13	17.0569	2.03224	6.42651	4.63	21.4369	2.15174	6.80441	5.13	26.3169	2.26495	7.16240
4.14	17.1396	2.03470	6.43428	4.64	21.5296	2.15407	6.81175	5.14	26.4196	2.26716	7.16938
4.15	17.2225	2.03715	6.44205	4.65	21.6225	2.15639	6.81909	5.15	26.5225	2.26936	7.17635
4.16	17.3056	2.03961	6.44981	4.66	21.7156	2.15870	6.82642	5.16	26.6256	2.27156	7.18331
4.17	17.3889	2.04206	6.45755	4.67	21.8089	2.16102	6.83374	5.17	26.7289	2.27376	7.19027
4.18	17.4724	2.04450	6.46529	4.68	21.9024	2.16333	6.84105	5.18	26.8324	2.27596	7.19722
4.19	17.5561	2.04695	6.47302	4.69	21.9961	2.16564	6.84836	5.19	26.9361	2.27816	7.20417
4.20	17.6400	2.04939	6.48074	4.70	22.0900	2.16795	6.85565	5.20	27.0400	2.28035	7.21110
4.21	17.7241	2.05183	6.48845	4.71	22.1841	2.17025	6.86294	5.21	27.1441	2.28254	7.21803
4.22	17.8084	2.05426	6.49615	4.72	22.2784	2.17256	6.87023	5.22	27.2484	2.28473	7.22496
4.23	17.8929	2.05670	6.50384	4.73	22.3729	2.17486	6.87750	5.23	27.3529	2.28692	7.23187
4.24	17.9776	2.05913	6.51153	4.74	22.4676	2.17715	6.88477	5.24	27.4576	2.28910	7.23878
4.25	18.0625	2.06155	6.51920	4.75	22.5625	2.17945	6.89202	5.25	27.5625	2.29129	7.24569
4.26	18.1476	2.06398	6.52687	4.76	22.6576	2.18174	6.89928	5.26	27.6676	2.29347	7.25259
4.27	18.2329	2.06640	6.53452	4.77	22.7529	2.18403	6.90652	5.27	27.7729	2.29565	7.25948
4.28	18.3184	2.06882	6.54217	4.78	22.8484	2.18632	6.91375	5.28	27.8784	2.29783	7.26636
4.29	18.4041	2.07123	6.54981	4.79	22.9441	2.18861	6.92098	5.29	27.9841	2.30000	7.27324
4.30	18.4900	2.07364	6.55744	4.80	23.0400	2.19089	6.92820	5.30	28.0900	2.30217	7.28011
4.31	18.5761	2.07605	6.56506	4.81	23.1361	2.19317	6.93542	5.31	28.1961	2.30434	7.28697
4.32	18.6624	2.07846	6.57267	4.82	23.2324	2.19545	6.94262	5.32	28.3024	2.30651	7.29383
4.33	18.7489	2.08087	6.58027	4.83	23.3289	2.19773	6.94982	5.33	28.4089	2.30868	7.30068
4.34	18.8356	2.08327	6.58787	4.84	23.4256	2.20000	6.95701	5.34	28.5156	2.31084	7.30753
4.35	18.9225	2.08567	6.59545	4.85	23.5225	2.20227	6.96419	5.35	28.6225	2.31301	7.31437
4.36	19.0096	2.08806	6.60303	4.86	23.6196	2.20454	6.97137	5.36	28.7296	2.31517	7.32120
4.37	19.0969	2.09045	6.61060	4.87	23.7169	2.20681	6.97854	5.37	28.8369	2.31733	7.32803
4.38	19.1844	2.09284	6.61816	4.88	23.8144	2.20907	6.98570	5.38	28.9444	2.31948	7.33485
4.39	19.2721	2.09523	6.62571	4.89	23.9121	2.21133	6.99285	5.39	29.0521	2.32164	7.34166
4.40	19.3600	2.09762	6.63325	4.90	24.0100	2.21359	7.00000	5.40	29.1600	2.32379	7.34847
4.41	19.4481	2.10000	6.64078	4.91	24.1081	2.21585	7.00714	5.41	29.2681	2.32594	7.35527
4.42	19.5364	2.10238	6.64831	4.92	24.2064	2.21811	7.01427	5.42	29.3764	2.32809	7.36206
4.43	19.6249	2.10476	6.65582	4.93	24.3049	2.22036	7.02140	5.43	29.4849	2.33024	7.36885
4.44	19.7136	2.10713	6.66333	4.94	24.4036	2.22261	7.02851	5.44	29.5936	2.33238	7.37564
4.45	19.8025	2.10950	6.67083	4.95	24.5025	2.22486	7.03562	5.45	29.7025	2.33452	7.38241
4.46	19.8916	2.11187	6.67832	4.96	24.6016	2.22711	7.04273	5.46	29.8116	2.33666	7.38918
4.47	19.9809	2.11424	6.68581	4.97	24.7009	2.22935	7.04982	5.47	29.9209	2.33880	7.39594
4.48	20.0704	2.11660	6.69328	4.98	24.8004	2.23159	7.05691	5.48	30.0304	2.34094	7.40270
4.49	20.1601	2.11896	6.70075	4.99	24.9001	2.23383	7.06399	5.49	30.1401	2.34307	7.40945
4.50	20.2500	2.12132	6.70820	5.00	25.0000	2.23607	7.07107	5.50	30.2500	2.34521	7.41620

Illustration for $n = 4.61$:

$4.61^2 = 21.2521 \qquad 0.461^2 = 0.212521$

$46.1^2 = 2125.21 \qquad 0.0461^2 = 0.00212521$

$461^2 = 212521 \qquad 0.00461^2 = 0.0000212521$

$4610^2 = 21252100$

Table 1. Continued

n	n^2	$\sqrt{n}$	$\sqrt{10n}$	n	n^2	$\sqrt{n}$	$\sqrt{10n}$	n	n^2	$\sqrt{n}$	$\sqrt{10n}$
5.50	30.2500	2.34521	7.41620	6.00	36.0000	2.44949	7.74597	6.50	42.2500	2.54951	8.06226
5.51	30.3601	2.34734	7.42294	6.01	36.1201	2.45153	7.75242	6.51	42.3801	2.55147	8.06846
5.52	30.4704	2.34947	7.42967	6.02	36.2404	2.45357	7.75887	6.52	42.5104	2.55343	8.07465
5.53	30.5809	2.35160	7.43640	6.03	36.3609	2.45561	7.76531	6.53	42.6409	2.55539	8.08084
5.54	30.6916	2.35372	7.44312	6.04	36.4816	2.45764	7.77174	6.54	42.7716	2.55734	8.08703
5.55	30.8025	2.35584	7.44983	6.05	36.6025	2.45967	7.77817	6.55	42.9025	2.55930	8.09321
5.56	30.9136	2.35797	7.45654	6.06	36.7236	2.46171	7.78460	6.56	43.0336	2.56125	8.09938
5.57	31.0249	2.36008	7.46324	6.07	36.8449	2.46374	7.79102	6.57	43.1649	2.56320	8.10555
5.58	31.1364	2.36220	7.46994	6.08	36.9664	2.46577	7.79744	6.58	43.2964	2.56515	8.11172
5.59	31.2481	2.36432	7.47663	6.09	37.0881	2.46779	7.80385	6.59	43.4281	2.56710	8.11788
5.60	31.3600	2.36643	7.48331	6.10	37.2100	2.46982	7.81025	6.60	43.5600	2.56905	8.12404
5.61	31.4721	2.36854	7.48999	6.11	37.3321	2.47184	7.81665	6.61	43.6921	2.57099	8.13019
5.62	31.5844	2.37065	7.49667	6.12	37.4544	2.47386	7.82304	6.62	43.8244	2.57294	8.13634
5.63	31.6969	2.37276	7.50333	6.13	37.5769	2.47588	7.82943	6.63	43.9569	2.57488	8.14248
5.64	31.8096	2.37487	7.50999	6.14	37.6996	2.47790	7.83582	6.64	44.0896	2.57682	8.14862
5.65	31.9225	2.37697	7.51665	6.15	37.8225	2.47992	7.84219	6.65	44.2225	2.57876	8.15475
5.66	32.0356	2.37908	7.52330	6.16	37.9456	2.48193	7.84857	6.66	44.3556	2.58070	8.16088
5.67	32.1489	2.38118	7.52994	6.17	38.0689	2.48395	7.85493	6.67	44.4889	2.58263	8.16701
5.68	32.2624	2.38328	7.53658	6.18	38.1924	2.48596	7.86130	6.68	44.6224	2.58457	8.17313
5.69	32.3761	2.38537	7.54321	6.19	38.3161	2.48797	7.86766	6.69	44.7561	2.58650	8.17924
5.70	32.4900	2.38747	7.54983	6.20	38.4400	2.48998	7.87401	6.70	44.8900	2.58844	8.18535
5.71	32.6041	2.38956	7.55645	6.21	38.5641	2.49199	7.88036	6.71	45.0241	2.59037	8.19146
5.72	32.7184	2.39165	7.56307	6.22	38.6884	2.49399	7.88670	6.72	45.1584	2.59230	8.19756
5.73	32.8329	2.39374	7.56968	6.23	38.8129	2.49600	7.89303	6.73	45.2929	2.59422	8.20366
5.74	32.9476	2.39583	7.57628	6.24	38.9376	2.49800	7.89937	6.74	45.4276	2.59615	8.20975
5.75	33.0625	2.39792	7.58288	6.25	39.0625	2.50000	7.90569	6.75	45.5625	2.59808	8.21584
5.76	33.1776	2.40000	7.58947	6.26	39.1876	2.50200	7.91202	6.76	45.6976	2.60000	8.22192
5.77	33.2929	2.40208	7.59605	6.27	39.3129	2.50400	7.91833	6.77	45.8329	2.60192	8.22800
5.78	33.4084	2.40416	7.60263	6.28	39.4384	2.50599	7.92465	6.78	45.9684	2.60384	8.23408
5.79	33.5241	2.40624	7.60920	6.29	39.5641	2.50799	7.93095	6.79	46.1041	2.60576	8.24015
5.80	33.6400	2.40832	7.61577	6.30	39.6900	2.50998	7.93725	6.80	46.2400	2.60768	8.24621
5.81	33.7561	2.41039	7.62234	6.31	39.8161	2.51197	7.94355	6.81	46.3761	2.60960	8.25227
5.82	33.8724	2.41247	7.62889	6.32	39.9424	2.51396	7.94984	6.82	46.5124	2.61151	8.25833
5.83	33.9889	2.41454	7.63544	6.33	40.0689	2.51595	7.95613	6.83	46.6489	2.61343	8.26438
5.84	34.1056	2.41661	7.64199	6.34	40.1956	2.51794	7.96241	6.84	46.7856	2.61534	8.27043
5.85	34.2225	2.41868	7.64853	6.35	40.3225	2.51992	7.96869	6.85	46.9225	2.61725	8.27647
5.86	34.3396	2.42074	7.65506	6.36	40.4496	2.52190	7.97496	6.86	47.0596	2.61916	8.28251
5.87	34.4569	2.42281	7.66159	6.37	40.5769	2.52389	7.98123	6.87	47.1969	2.62107	8.28855
5.88	34.5744	2.42487	7.66812	6.38	40.7044	2.52587	7.98749	6.88	47.3344	2.62298	8.29458
5.89	34.6921	2.42693	7.67463	6.39	40.8321	2.52784	7.99375	6.89	47.4721	2.62488	8.30060
5.90	34.8100	2.42899	7.68115	6.40	40.9600	2.52982	8.00000	6.90	47.6100	2.62679	8.30662
5.91	34.9281	2.43105	7.68765	6.41	41.0881	2.53180	8.00625	6.91	47.7481	2.62869	8.31264
5.92	35.0464	2.43311	7.69415	6.42	41.2164	2.53377	8.01249	6.92	47.8864	2.63059	8.31865
5.93	35.1649	2.43516	7.70065	6.43	41.3449	2.53574	8.01873	6.93	48.0249	2.63249	8.32466
5.94	35.2836	2.43721	7.70714	6.44	41.4736	2.53772	8.02496	6.94	48.1636	2.63439	8.33067
5.95	35.4025	2.43926	7.71362	6.45	41.6025	2.53969	8.03119	6.95	48.3025	2.63629	8.33667
5.96	35.5216	2.44131	7.72010	6.46	41.7316	2.54165	8.03741	6.96	48.4416	2.63818	8.34266
5.97	35.6409	2.44336	7.72658	6.47	41.8609	2.54362	8.04363	6.97	48.5809	2.64008	8.34865
5.98	35.7604	2.44540	7.73305	6.48	41.9904	2.54558	8.04984	6.98	48.7204	2.64197	8.35464
5.99	35.8801	2.44745	7.73951	6.49	42.1201	2.54755	8.05605	6.99	48.8601	2.64386	8.36062
6.00	36.0000	2.44949	7.74597	6.50	42.2500	2.54951	8.06226	7.00	49.0000	2.64575	8.36660

Illustration for $n = 4.61$:

$\sqrt{4.61} = 2.14709$ $\sqrt{0.461} = 0.678970$

$\sqrt{46.1} = 6.78970$ $\sqrt{0.0461} = 0.214709$

$\sqrt{461} = 21.4709$ $\sqrt{0.00461} = 0.0678970$

$\sqrt{4610} = 67.8970$ $\sqrt{0.000461} = 0.0214709$

Table 1. Continued

n	n^2	$\sqrt{n}$	$\sqrt{10n}$	n	n^2	$\sqrt{n}$	$\sqrt{10n}$	n	n^2	$\sqrt{n}$	$\sqrt{10n}$
7.00	49.0000	2.64575	8.36660	7.50	56.2500	2.73861	8.66025	8.00	64.0000	2.82843	8.94427
7.01	49.1401	2.64764	8.37257	7.51	56.4001	2.74044	8.66603	8.01	64.1601	2.83019	8.94986
7.02	49.2804	2.64953	8.37854	7.52	56.5504	2.74226	8.67179	8.02	64.3204	2.83196	8.95545
7.03	49.4209	2.65141	8.38451	7.53	56.7009	2.74408	8.67756	8.03	64.4809	2.83373	8.96103
7.04	49.5616	2.65330	8.39047	7.54	56.8516	2.74591	8.68332	8.04	64.6416	2.83549	8.96660
7.05	49.7025	2.65518	8.39643	7.55	57.0025	2.74773	8.68907	8.05	64.8025	2.83725	8.97218
7.06	49.8436	2.65707	8.40238	7.56	57.1536	2.74955	8.69483	8.06	64.9636	2.83901	8.97775
7.07	49.9849	2.65895	8.40833	7.57	57.3049	2.75136	8.70057	8.07	65.1249	2.84077	8.98332
7.08	50.1264	2.66083	8.41427	7.58	57.4564	2.75318	8.70632	8.08	65.2864	2.84253	8.98888
7.09	50.2681	2.66271	8.42021	7.59	57.6081	2.75500	8.71206	8.09	65.4481	2.84429	8.99444
7.10	50.4100	2.66458	8.42615	7.60	57.7600	2.75681	8.71780	8.10	65.6100	2.84605	9.00000
7.11	50.5521	2.66646	8.43208	7.61	57.9121	2.75862	8.72353	8.11	65.7721	2.84781	9.00555
7.12	50.6944	2.66833	8.43801	7.62	58.0644	2.76043	8.72926	8.12	65.9344	2.84956	9.01110
7.13	50.8369	2.67021	8.44393	7.63	58.2169	2.76225	8.73499	8.13	66.0969	2.85132	9.01665
7.14	50.9796	2.67208	8.44985	7.64	58.3696	2.76405	8.74071	8.14	66.2596	2.85307	9.02219
7.15	51.1225	2.67395	8.45577	7.65	58.5225	2.76586	8.74643	8.15	66.4225	2.85482	9.02774
7.16	51.2656	2.67582	8.46168	7.66	58.6756	2.76767	8.75214	8.16	66.5856	2.85657	9.03327
7.17	51.4089	2.67769	8.46759	7.67	58.8289	2.76948	8.75785	8.17	66.7489	2.85832	9.03881
7.18	51.5524	2.67955	8.47349	7.68	58.9824	2.77128	8.76356	8.18	66.9124	2.86007	9.04434
7.19	51.6961	2.68142	8.47939	7.69	59.1361	2.77308	8.76926	8.19	67.0761	2.86182	9.04986
7.20	51.8400	2.68328	8.48528	7.70	59.2900	2.77489	8.77496	8.20	67.2400	2.86356	9.05539
7.21	51.9841	2.68514	8.49117	7.71	59.4441	2.77669	8.78066	8.21	67.4041	2.86531	9.06091
7.22	52.1284	2.68701	8.49706	7.72	59.5984	2.77849	8.78635	8.22	67.5684	2.86705	9.06642
7.23	52.2729	2.68887	8.50294	7.73	59.7529	2.78029	8.79204	8.23	67.7329	2.86880	9.07193
7.24	52.4176	2.69072	8.50882	7.74	59.9076	2.78209	8.79773	8.24	67.8976	2.87054	9.07744
7.25	52.5625	2.69258	8.51469	7.75	60.0625	2.78388	8.80341	8.25	68.0625	2.87228	9.08295
7.26	52.7076	2.69444	8.52056	7.76	60.2176	2.78568	8.80909	8.26	68.2276	2.87402	9.08845
7.27	52.8529	2.69629	8.52643	7.77	60.3729	2.78747	8.81476	8.27	68.3929	2.87576	9.09395
7.28	52.9984	2.69815	8.53229	7.78	60.5284	2.78927	8.82043	8.28	68.5584	2.87750	9.09945
7.29	53.1441	2.70000	8.53815	7.79	60.6841	2.79106	8.82610	8.29	68.7241	2.87924	9.10494
7.30	53.2900	2.70185	8.54400	7.80	60.8400	2.79285	8.83176	8.30	68.8900	2.88097	9.11043
7.31	53.4361	2.70370	8.54985	7.81	60.9961	2.79464	8.83742	8.31	69.0561	2.88271	9.11592
7.32	53.5824	2.70555	8.55570	7.82	61.1524	2.79643	8.84308	8.32	69.2224	2.88444	9.12140
7.33	53.7289	2.70740	8.56154	7.83	61.3089	2.79821	8.84873	8.33	69.3889	2.88617	9.12688
7.34	53.8756	2.70924	8.56738	7.84	61.4656	2.80000	8.85438	8.34	69.5556	2.88791	9.13236
7.35	54.0225	2.71109	8.57321	7.85	61.6225	2.80179	8.86002	8.35	69.7225	2.88964	9.13783
7.36	54.1696	2.71293	8.57904	7.86	61.7796	2.80357	8.86566	8.36	69.8896	2.89137	9.14330
7.37	54.3169	2.71477	8.58487	7.87	61.9369	2.80535	8.87130	8.37	70.0569	2.89310	9.14877
7.38	54.4644	2.71662	8.59069	7.88	62.0944	2.80713	8.87694	8.38	70.2244	2.89482	9.15423
7.39	54.6121	2.71846	8.59651	7.89	62.2521	2.80891	8.88257	8.39	70.3921	2.89655	9.15969
7.40	54.7600	2.72029	8.60233	7.90	62.4100	2.81069	8.88819	8.40	70.5600	2.89828	9.16515
7.41	54.9081	2.72213	8.60814	7.91	62.5681	2.81247	8.89382	8.41	70.7281	2.90000	9.17061
7.42	55.0564	2.72397	8.61394	7.92	62.7264	2.81425	8.89944	8.42	70.8964	2.90172	9.17606
7.43	55.2049	2.72580	8.61974	7.93	62.8849	2.81603	8.90505	8.43	71.0649	2.90345	9.18150
7.44	55.3536	2.72764	8.62554	7.94	63.0436	2.81780	8.91067	8.44	71.2336	2.90517	9.18695
7.45	55.5025	2.72947	8.63134	7.95	63.2025	2.81957	8.91628	8.45	71.4025	2.90689	9.19239
7.46	55.6516	2.73130	8.63713	7.96	63.3616	2.82135	8.92188	8.46	71.5716	2.90861	9.19783
7.47	55.8009	2.73313	8.64292	7.97	63.5209	2.82312	8.92749	8.47	71.7409	2.91033	9.20326
7.48	55.9504	2.73496	8.64870	7.98	63.6804	2.82489	8.93308	8.48	71.9104	2.91204	9.20869
7.49	56.1001	2.73679	8.65448	7.99	63.8401	2.82666	8.93868	8.49	72.0801	2.91376	9.21412
7.50	56.2500	2.73861	8.66025	8.00	64.0000	2.82843	8.94427	8.50	72.2500	2.91548	9.21954

Illustration for $n = 8.39$:

$8.39^2 = 70.3921$ $0.839^2 = 0.703921$

$83.9^2 = 7039.21$ $0.0839^2 = 0.00703921$

$839^2 = 703921$ $0.00839^2 = 0.0000703921$

$8390^2 = 70392100$

Table 1. Continued

n	n^2	$\sqrt{n}$	$\sqrt{10n}$	n	n^2	$\sqrt{n}$	$\sqrt{10n}$	n	n^2	$\sqrt{n}$	$\sqrt{10n}$
8.50	72.2500	2.91548	9.21954	9.00	81.0000	3.00000	9.48683	9.50	90.2500	3.08221	9.74679
8.51	72.4201	2.91719	9.22497	9.01	81.1801	3.00167	9.49210	9.51	90.4401	3.08383	9.75192
8.52	72.5904	2.91890	9.23038	9.02	81.3604	3.00333	9.49737	9.52	90.6304	3.08545	9.75705
8.53	72.7609	2.92062	9.23580	9.03	81.5409	3.00500	9.50263	9.53	90.8209	3.08707	9.76217
8.54	72.9316	2.92233	9.24121	9.04	81.7216	3.00666	9.50789	9.54	91.0116	3.08869	9.76729
8.55	73.1025	2.92404	9.24662	9.05	81.9025	3.00832	9.51315	9.55	91.2025	3.09031	9.77241
8.56	73.2736	2.92575	9.25203	9.06	82.0836	3.00998	9.51840	9.56	91.3936	3.09192	9.77753
8.57	73.4449	2.92746	9.25743	9.07	82.2649	3.01164	9.52365	9.57	91.5849	3.09354	9.78264
8.58	73.6164	2.92916	9.26283	9.08	82.4464	3.01330	9.52890	9.58	91.7764	3.09516	9.78775
8.59	73.7881	2.93087	9.26823	9.09	82.6281	3.01496	9.53415	9.59	91.9681	3.09677	9.79285
8.60	73.9600	2.93258	9.27362	9.10	82.8100	3.01662	9.53939	9.60	92.1600	3.09839	9.79796
8.61	74.1321	2.93428	9.27901	9.11	82.9921	3.01828	9.54463	9.61	92.3521	3.10000	9.80306
8.62	74.3044	2.93598	9.28440	9.12	83.1744	3.01993	9.54987	9.62	92.5444	3.10161	9.80816
8.63	74.4769	2.93769	9.28978	9.13	83.3569	3.02159	9.55510	9.63	92.7369	3.10322	9.81326
8.64	74.6496	2.93939	9.29516	9.14	83.5396	3.02324	9.56033	9.64	92.9296	3.10483	9.81835
8.65	74.8225	2.94109	9.30054	9.15	83.7225	3.02490	9.56556	9.65	93.1225	3.10644	9.82344
8.66	74.9956	2.94279	9.30591	9.16	83.9056	3.02655	9.57079	9.66	93.3156	3.10805	9.82853
8.67	75.1689	2.94449	9.31128	9.17	84.0889	3.02820	9.57601	9.67	93.5089	3.10966	9.83362
8.68	75.3424	2.94618	9.31665	9.18	84.2724	3.02985	9.58123	9.68	93.7024	3.11127	9.83870
8.69	75.5161	2.94788	9.32202	9.19	84.4561	3.03150	9.58645	9.69	93.8961	3.11288	9.84378
8.70	75.6900	2.94958	9.32738	9.20	84.6400	3.03315	9.59166	9.70	94.0900	3.11448	9.84886
8.71	75.8641	2.95127	9.33274	9.21	84.8241	3.03480	9.59687	9.71	94.2841	3.11609	9.85393
8.72	76.0384	2.95296	9.33809	9.22	85.0084	3.03645	9.60208	9.72	94.4784	3.11769	9.85901
8.73	76.2129	2.95466	9.34345	9.23	85.1929	3.03809	9.60729	9.73	94.6729	3.11929	9.86408
8.74	76.3876	2.95635	9.34880	9.24	85.3776	3.03974	9.61249	9.74	94.8676	3.12090	9.86914
8.75	76.5625	2.95804	9.35414	9.25	85.5625	3.04138	9.61769	9.75	95.0625	3.12250	9.87421
8.76	76.7376	2.95973	9.35949	9.26	85.7476	3.04302	9.62289	9.76	95.2576	3.12410	9.87927
8.77	76.9129	2.96142	9.36483	9.27	85.9329	3.04467	9.62808	9.77	95.4529	3.12570	9.88433
8.78	77.0884	2.96311	9.37017	9.28	86.1184	3.04631	9.63328	9.78	95.6484	3.12730	9.88939
8.79	77.2641	2.96479	9.37550	9.29	86.3041	3.04795	9.63846	9.79	95.8441	3.12890	9.89444
8.80	77.4400	2.96648	9.38083	9.30	86.4900	3.04959	9.64365	9.80	96.0400	3.13050	9.89949
8.81	77.6161	2.96816	9.38616	9.31	86.6761	3.05123	9.64883	9.81	96.2361	3.13209	9.90454
8.82	77.7924	2.96985	9.39149	9.32	86.8624	3.05287	9.65401	9.82	96.4324	3.13369	9.90959
8.83	77.9689	2.97153	9.39681	9.33	87.0489	3.05450	9.65919	9.83	96.6289	3.13528	9.91464
8.84	78.1456	2.97321	9.40213	9.34	87.2356	3.05614	9.66437	9.84	96.8256	3.13688	9.91968
8.85	78.3225	2.97489	9.40744	9.35	87.4225	3.05778	9.66954	9.85	97.0225	3.13847	9.92472
8.86	78.4996	2.97658	9.41276	9.36	87.6096	3.05941	9.67471	9.86	97.2196	3.14006	9.92975
8.87	78.6769	2.97825	9.41807	9.37	87.7969	3.06105	9.67988	9.87	97.4169	3.14166	9.93479
8.88	78.8544	2.97993	9.42338	9.38	87.9844	3.06268	9.68504	9.88	97.6144	3.14325	9.93982
8.89	79.0321	2.98161	9.42868	9.39	88.1721	3.06431	9.69020	9.89	97.8121	3.14484	9.94485
8.90	79.2100	2.98329	9.43398	9.40	88.3600	3.06594	9.69536	9.90	98.0100	3.14643	9.94987
8.91	79.3881	2.98496	9.43928	9.41	88.5481	3.06757	9.70052	9.91	98.2081	3.14802	9.95490
8.92	79.5664	2.98664	9.44458	9.42	88.7364	3.06920	9.70567	9.92	98.4064	3.14960	9.95992
8.93	79.7449	2.98831	9.44987	9.43	88.9249	3.07083	9.71082	9.93	98.6049	3.15119	9.96494
8.94	79.9236	2.98998	9.45516	9.44	89.1136	3.07246	9.71597	9.94	98.8036	3.15278	9.96995
8.95	80.1025	2.99166	9.46044	9.45	89.3025	3.07409	9.72111	9.95	99.0025	3.15436	9.97497
8.96	80.2816	2.99333	9.46573	9.46	89.4916	3.07571	9.72625	9.96	99.2016	3.15595	9.97998
8.97	80.4609	2.99500	9.47101	9.47	89.6809	3.07734	9.73139	9.97	99.4009	3.15753	9.98499
8.98	80.6404	2.99666	9.47629	9.48	89.8704	3.07896	9.73653	9.98	99.6004	3.15911	9.98999
8.99	80.8201	2.99833	9.48156	9.49	90.0601	3.08058	9.74166	9.99	99.8001	3.16070	9.99500
9.00	81.0000	3.00000	9.48683	9.50	90.2500	3.08221	9.74679	10.00	100.000	3.16228	10.0000

Illustration for $n = 8.39$:

$\sqrt{8.39} = 2.89655$ $\sqrt{0.839} = 0.915969$

$\sqrt{83.9} = 9.15969$ $\sqrt{0.0839} = 0.289655$

$\sqrt{839} = 28.9655$ $\sqrt{0.00839} = 0.0915969$

$\sqrt{8390} = 91.5969$ $\sqrt{0.000839} = 0.0289655$

Table 2. Four-Place Logarithms of Numbers

n	0	1	2	3	4	5	6	7	8	9
10	0000	0043	0086	0128	0170	0212	0253	0294	0334	0374
11	0414	0453	0492	0531	0569	0607	0645	0682	0719	0755
12	0792	0828	0864	0899	0934	0969	1004	1038	1072	1106
13	1139	1173	1206	1239	1271	1303	1335	1367	1399	1430
14	1461	1492	1523	1553	1584	1614	1644	1673	1703	1732
15	1761	1790	1818	1847	1875	1903	1931	1959	1987	2014
16	2041	2068	2095	2122	2148	2175	2201	2227	2253	2279
17	2304	2330	2355	2380	2405	2430	2455	2480	2504	2529
18	2553	2577	2601	2625	2648	2672	2695	2718	2742	2765
19	2788	2810	2833	2856	2878	2900	2923	2945	2967	2989
20	3010	3032	3054	3075	3096	3118	3139	3160	3181	3201
21	3222	3243	3263	3284	3304	3324	3345	3365	3385	3404
22	3424	3444	3464	3483	3502	3522	3541	3560	3579	3598
23	3617	3636	3655	3674	3692	3711	3729	3747	3766	3784
24	3802	3820	3838	3856	3874	3892	3909	3927	3945	3962
25	3979	3997	4014	4031	4048	4065	4082	4099	4116	4133
26	4150	4166	4183	4200	4216	4232	4249	4265	4281	4298
27	4314	4330	4346	4362	4378	4393	4409	4425	4440	4456
28	4472	4487	4502	4518	4533	4548	4564	4579	4594	4609
29	4624	4639	4654	4669	4683	4698	4713	4728	4742	4757
30	4771	4786	4800	4814	4829	4843	4857	4871	4886	4900
31	4914	4928	4942	4955	4969	4983	4997	5011	5024	5038
32	5051	5065	5079	5092	5105	5119	5132	5145	5159	5172
33	5185	5198	5211	5224	5237	5250	5263	5276	5289	5302
34	5315	5328	5340	5353	5366	5378	5391	5403	5416	5428
35	5441	5453	5465	5478	5490	5502	5514	5527	5539	5551
36	5563	5575	5587	5599	5611	5623	5635	5647	5658	5670
37	5682	5694	5705	5717	5729	5740	5752	5763	5775	5786
38	5798	5809	5821	5832	5843	5855	5866	5877	5888	5899
39	5911	5922	5933	5944	5955	5966	5977	5988	5999	6010
40	6021	6031	6042	6053	6064	6075	6085	6096	6107	6117
41	6128	6138	6149	6160	6170	6180	6191	6201	6212	6222
42	6232	6243	6253	6263	6274	6284	6294	6304	6314	6325
43	6335	6345	6355	6365	6375	6385	6395	6405	6415	6425
44	6435	6444	6454	6464	6474	6484	6493	6503	6513	6522
45	6532	6542	6551	6561	6571	6580	6590	6599	6609	6618
46	6628	6637	6646	6656	6665	6675	6684	6693	6702	6712
47	6721	6730	6739	6749	6758	6767	6776	6785	6794	6803
48	6812	6821	6830	6839	6848	6857	6866	6875	6884	6893
49	6902	6911	6920	6928	6937	6946	6955	6964	6972	6981
50	6990	6998	7007	7016	7024	7033	7042	7050	7059	7067
51	7076	7084	7093	7101	7110	7118	7126	7135	7143	7152
52	7160	7168	7177	7185	7193	7202	7210	7218	7226	7235
53	7243	7251	7259	7267	7275	7284	7292	7300	7308	7316
54	7324	7332	7340	7348	7356	7364	7372	7380	7388	7396

Table 2. Continued

n	0	1	2	3	4	5	6	7	8	9
55	7404	7412	7419	7427	7435	7443	7451	7459	7466	7474
56	7482	7490	7497	7505	7513	7520	7528	7536	7543	7551
57	7559	7566	7574	7582	7589	7597	7604	7612	7619	7627
58	7634	7642	7649	7657	7664	7672	7679	7686	7694	7701
59	7709	7716	7723	7731	7738	7745	7752	7760	7767	7774
60	7782	7789	7796	7803	7810	7818	7825	7832	7839	7846
61	7853	7860	7868	7875	7882	7889	7896	7903	7910	7917
62	7924	7931	7938	7945	7952	7959	7966	7973	7980	7987
63	7993	8000	8007	8014	8021	8028	8035	8041	8048	8055
64	8062	8069	8075	8082	8089	8096	8102	8109	8116	8122
65	8129	8136	8142	8149	8156	8162	8169	8176	8182	8189
66	8195	8202	8209	8215	8222	8228	8235	8241	8248	8254
67	8261	8267	8274	8280	8287	8293	8299	8306	8312	8319
68	8325	8331	8338	8344	8351	8357	8363	8370	8376	8382
69	8388	8395	8401	8407	8414	8420	8426	8432	8439	8445
70	8451	8457	8463	8470	8476	8482	8488	8494	8500	8506
71	8513	8519	8525	8531	8537	8543	8549	8555	8561	8567
72	8573	8579	8585	8591	8597	8603	8609	8615	8621	8627
73	8633	8639	8645	8651	8657	8663	8669	8675	8681	8686
74	8692	8698	8704	8710	8716	8722	8727	8733	8739	8745
75	8751	8756	8762	8768	8774	8779	8785	8791	8797	8802
76	8808	8814	8820	8825	8831	8837	8842	8848	8854	8859
77	8865	8871	8876	8882	8887	8893	8899	8904	8910	8915
78	8921	8927	8932	8938	8943	8949	8954	8960	8965	8971
79	8976	8982	8987	8993	8998	9004	9009	9015	9020	9025
80	9031	9036	9042	9047	9053	9058	9063	9069	9074	9079
81	9085	9090	9096	9101	9106	9112	9117	9122	9128	9133
82	9138	9143	9149	9154	9159	9165	9170	9175	9180	9186
83	9191	9196	9201	9206	9212	9217	9222	9227	9232	9238
84	9243	9248	9253	9258	9263	9269	9274	9279	9284	9289
85	9294	9299	9304	9309	9315	9320	9325	9330	9335	9340
86	9345	9350	9355	9360	9365	9370	9375	9380	9385	9390
87	9395	9400	9405	9410	9415	9420	9425	9430	9435	9440
88	9445	9450	9455	9460	9465	9469	9474	9479	9484	9489
89	9494	9499	9504	9509	9513	9518	9523	9528	9533	9538
90	9542	9547	9552	9557	9562	9566	9571	9576	9581	9586
91	9590	9595	9600	9605	9609	9614	9619	9624	9628	9633
92	9638	9643	9647	9652	9657	9661	9666	9671	9675	9680
93	9685	9689	9694	9699	9703	9708	9713	9717	9722	9727
94	9731	9736	9741	9745	9750	9754	9759	9763	9768	9773
95	9777	9782	9786	9791	9795	9800	9805	9809	9814	9818
96	9823	9827	9832	9836	9841	9845	9850	9854	9859	9863
97	9868	9872	9877	9881	9886	9890	9894	9899	9903	9908
98	9912	9917	9921	9926	9930	9934	9939	9943	9948	9952
99	9956	9961	9965	9969	9974	9978	9983	9987	9991	9996

Table 3. Four-Place Values of Functions and Radians

Degrees	Radians	sin θ	cos θ	tan θ	cot θ	sec θ	csc θ		
0° 00'	.0000	.0000	1.0000	.0000	—	1.000	—	1.5708	90° 00'
10	.0029	.0029	1.0000	.0029	343.8	1.000	343.8	1.5679	50
20	.0058	.0058	1.0000	.0058	171.9	1.000	171.9	1.5650	40
30	.0087	.0087	1.0000	.0087	114.6	1.000	114.6	1.5621	30
40	.0116	.0116	.9999	.0116	85.94	1.000	85.95	1.5592	20
50	.0145	.0145	.9999	.0145	68.75	1.000	68.76	1.5563	10
1° 00'	.0175	.0175	.9998	.0175	57.29	1.000	57.30	1.5533	89° 00'
10	.0204	.0204	.9998	.0204	49.10	1.000	49.11	1.5504	50
20	.0233	.0233	.9997	.0233	42.96	1.000	42.98	1.5475	40
30	.0262	.0262	.9997	.0262	38.19	1.000	38.20	1.5446	30
40	.0291	.0291	.9996	.0291	34.37	1.000	34.38	1.5417	20
50	.0320	.0320	.9995	.0320	31.24	1.001	31.26	1.5388	10
2° 00'	.0349	.0349	.9994	.0349	28.64	1.001	28.65	1.5359	88° 00'
10	.0378	.0378	.9993	.0378	26.43	1.001	26.45	1.5330	50
20	.0407	.0407	.9992	.0407	24.54	1.001	24.56	1.5301	40
30	.0436	.0436	.9990	.0437	22.90	1.001	22.93	1.5272	30
40	.0465	.0465	.9989	.0466	21.47	1.001	21.49	1.5243	20
50	.0495	.0494	.9988	.0495	20.21	1.001	20.23	1.5213	10
3° 00'	.0524	.0523	.9986	.0524	19.08	1.001	19.11	1.5184	87° 00'
10	.0553	.0552	.9985	.0553	18.07	1.002	18.10	1.5155	50
20	.0582	.0581	.9983	.0582	17.17	1.002	17.20	1.5126	40
30	.0611	.0610	.9981	.0612	16.35	1.002	16.38	1.5097	30
40	.0640	.0640	.9980	.0641	15.60	1.002	15.64	1.5068	20
50	.0669	.0669	.9978	.0670	14.92	1.002	14.96	1.5039	10
4° 00'	.0698	.0698	.9976	.0699	14.30	1.002	14.34	1.5010	86° 00'
10	.0727	.0727	.9974	.0729	13.73	1.003	13.76	1.4981	50
20	.0756	.0756	.9971	.0758	13.20	1.003	13.23	1.4952	40
30	.0785	.0785	.9969	.0787	12.71	1.003	12.75	1.4923	30
40	.0814	.0814	.9967	.0816	12.25	1.003	12.29	1.4893	20
50	.0844	.0843	.9964	.0846	11.83	1.004	11.87	1.4864	10
5° 00'	.0873	.0872	.9962	.0875	11.43	1.004	11.47	1.4835	85° 00'
10	.0902	.0901	.9959	.0904	11.06	1.004	11.10	1.4806	50
20	.0931	.0929	.9957	.0934	10.71	1.004	10.76	1.4777	40
30	.0960	.0958	.9954	.0963	10.39	1.005	10.43	1.4748	30
40	.0989	.0987	.9951	.0992	10.08	1.005	10.13	1.4719	20
50	.1018	.1016	.9948	.1022	9.788	1.005	9.839	1.4690	10
6° 00'	.1047	.1045	.9945	.1051	9.514	1.006	9.567	1.4661	84° 00'
		cos θ	sin θ	cot θ	tan θ	csc θ	sec θ	Radians	Degrees

Table 3. Continued

Degrees	Radians	sin θ	cos θ	tan θ	cot θ	sec θ	csc θ		
6° 00'	.1047	.1045	.9945	.1051	9.514	1.006	9.567	1.4661	84° 00'
10	.1076	.1074	.9942	.1080	9.255	1.006	9.309	1.4632	50
20	.1105	.1103	.9939	.1110	9.010	1.006	9.065	1.4603	40
30	.1134	.1132	.9936	.1139	8.777	1.006	8.834	1.4573	30
40	.1164	.1161	.9932	.1169	8.556	1.007	8.614	1.4544	20
50	.1193	.1190	.9929	.1198	8.345	1.007	8.405	1.4515	10
7° 00'	.1222	.1219	.9925	.1228	8.144	1.008	8.206	1.4486	83° 00'
10	.1251	.1248	.9922	.1257	7.953	1.008	8.016	1.4457	50
20	.1280	.1276	.9918	.1287	7.770	1.008	7.834	1.4428	40
30	.1309	.1305	.9914	.1317	7.596	1.009	7.661	1.4399	30
40	.1338	.1334	.9911	.1346	7.429	1.009	7.496	1.4370	20
50	.1367	.1363	.9907	.1376	7.269	1.009	7.337	1.4341	10
8° 00'	.1396	.1392	.9903	.1405	7.115	1.010	7.185	1.4312	82° 00'
10	.1425	.1421	.9899	.1435	6.968	1.010	7.040	1.4283	50
20	.1454	.1449	.9894	.1465	6.827	1.011	6.900	1.4254	40
30	.1484	.1478	.9890	.1495	6.691	1.011	6.765	1.4224	30
40	.1513	.1507	.9886	.1524	6.561	1.012	6.636	1.4195	20
50	.1542	.1536	.9881	.1554	6.435	1.012	6.512	1.4166	10
9° 00'	.1571	.1564	.9877	.1584	6.314	1.012	6.392	1.4137	81° 00'
10	.1600	.1593	.9872	.1614	6.197	1.013	6.277	1.4108	50
20	.1629	.1622	.9868	.1644	6.084	1.013	6.166	1.4079	40
30	.1658	.1650	.9863	.1673	5.976	1.014	6.059	1.4050	30
40	.1687	.1679	.9858	.1703	5.871	1.014	5.955	1.4021	20
50	.1716	.1708	.9853	.1733	5.769	1.015	5.855	1.3992	10
10° 00'	.1745	.1736	.9848	.1763	5.671	1.015	5.759	1.3963	80° 00'
10	.1774	.1765	.9843	.1793	5.576	1.016	5.665	1.3934	50
20	.1804	.1794	.9838	.1823	5.485	1.016	5.575	1.3904	40
30	.1833	.1822	.9833	.1853	5.396	1.017	5.487	1.3875	30
40	.1862	.1851	.9827	.1883	5.309	1.018	5.403	1.3846	20
50	.1891	.1880	.9822	.1914	5.226	1.018	5.320	1.3817	10
11° 00'	.1920	.1908	.9816	.1944	5.145	1.019	5.241	1.3788	79° 00'
10	.1949	.1937	.9811	.1974	5.066	1.019	5.164	1.3759	50
20	.1978	.1965	.9805	.2004	4.989	1.020	5.089	1.3730	40
30	.2007	.1994	.9799	.2035	4.915	1.020	5.016	1.3701	30
40	.2036	.2022	.9793	.2065	4.843	1.021	4.945	1.3672	20
50	.2065	.2051	.9787	.2095	4.773	1.022	4.876	1.3643	10
12° 00'	.2094	.2079	.9781	.2126	4.705	1.022	4.810	1.3614	78° 00'
		cos θ	sin θ	cot θ	tan θ	csc θ	sec θ	Radians	Degrees

Table 3. Continued

Degrees	Radians	sin θ	cos θ	tan θ	cot θ	sec θ	csc θ		
12° 00'	.2094	.2079	.9781	.2126	4.705	1.022	4.810	1.3614	78° 00'
10	.2123	.2108	.9775	.2156	4.638	1.023	4.745	1.3584	50
20	.2153	.2136	.9769	.2186	4.574	1.024	4.682	1.3555	40
30	.2182	.2164	.9763	.2217	4.511	1.024	4.620	1.3526	30
40	.2211	.2193	.9757	.2247	4.449	1.025	4.560	1.3497	20
50	.2240	.2221	.9750	.2278	4.390	1.026	4.502	1.3468	10
13° 00'	.2269	.2250	.9744	.2309	4.331	1.026	4.445	1.3439	77° 00'
10	.2298	.2278	.9737	.2339	4.275	1.027	4.390	1.3410	50
20	.2327	.2306	.9730	.2370	4.219	1.028	4.336	1.3381	40
30	.2356	.2334	.9724	.2401	4.165	1.028	4.284	1.3352	30
40	.2385	.2363	.9717	.2432	4.113	1.029	4.232	1.3323	20
50	.2414	.2391	.9710	.2462	4.061	1.030	4.182	1.3294	10
14° 00'	.2443	.2419	.9703	.2493	4.011	1.031	4.134	1.3265	76° 00'
10	.2473	.2447	.9696	.2524	3.962	1.031	4.086	1.3235	50
20	.2502	.2476	.9689	.2555	3.914	1.032	4.039	1.3206	40
30	.2531	.2504	.9681	.2586	3.867	1.033	3.994	1.3177	30
40	.2560	.2532	.9674	.2617	3.821	1.034	3.950	1.3148	20
50	.2589	.2560	.9667	.2648	3.776	1.034	3.906	1.3119	10
15° 00'	.2618	.2588	.9659	.2679	3.732	1.035	3.864	1.3090	75° 00'
10	.2647	.2616	.9652	.2711	3.689	1.036	3.822	1.3061	50
20	.2676	.2644	.9644	.2742	3.647	1.037	3.782	1.3032	40
30	.2705	.2672	.9636	.2773	3.606	1.038	3.742	1.3003	30
40	.2734	.2700	.9628	.2805	3.566	1.039	3.703	1.2974	20
50	.2763	.2728	.9621	.2836	3.526	1.039	3.665	1.2945	10
16° 00'	.2793	.2756	.9613	.2867	3.487	1.040	3.628	1.2915	74° 00'
10	.2822	.2784	.9605	.2899	3.450	1.041	3.592	1.2886	50
20	.2851	.2812	.9596	.2931	3.412	1.042	3.556	1.2857	40
30	.2880	.2840	.9588	.2962	3.376	1.043	3.521	1.2828	30
40	.2909	.2868	.9580	.2994	3.340	1.044	3.487	1.2799	20
50	.2938	.2896	.9572	.3026	3.305	1.045	3.453	1.2770	10
17° 00'	.2967	.2924	.9563	.3057	3.271	1.046	3.420	1.2741	73° 00'
10	.2996	.2952	.9555	.3089	3.237	1.047	3.388	1.2712	50
20	.3025	.2979	.9546	.3121	3.204	1.048	3.356	1.2683	40
30	.3054	.3007	.9537	.3153	3.172	1.049	3.326	1.2654	30
40	.3083	.3035	.9528	.3185	3.140	1.049	3.295	1.2625	20
50	.3113	.3062	.9520	.3217	3.108	1.050	3.265	1.2595	10
18° 00'	.3142	.3090	.9511	.3249	3.078	1.051	3.236	1.2566	72° 00'
		cos θ	sin θ	cot θ	tan θ	csc θ	sec θ	Radians	Degrees

Table 3. Continued

Degrees	Radians	sin θ	cos θ	tan θ	cot θ	sec θ	csc θ		
18° 00′	.3142	.3090	.9511	.3249	3.078	1.051	3.236	1.2566	72° 00′
10	.3171	.3118	.9502	.3281	3.047	1.052	3.207	1.2537	50
20	.3200	.3145	.9492	.3314	3.018	1.053	3.179	1.2508	40
30	.3229	.3173	.9483	.3346	2.989	1.054	3.152	1.2479	30
40	.3258	.3201	.9474	.3378	2.960	1.056	3.124	1.2450	20
50	.3287	.3228	.9465	.3411	2.932	1.057	3.098	1.2421	10
19° 00′	.3316	.3256	.9455	.3443	2.904	1.058	3.072	1.2392	71° 00′
10	.3345	.3283	.9446	.3476	2.877	1.059	3.046	1.2363	50
20	.3374	.3311	.9436	.3508	2.850	1.060	3.021	1.2334	40
30	.3403	.3338	.9426	.3541	2.824	1.061	2.996	1.2305	30
40	.3432	.3365	.9417	.3574	2.798	1.062	2.971	1.2275	20
50	.3462	.3393	.9407	.3607	2.773	1.063	2.947	1.2246	10
20° 00′	.3491	.3420	.9397	.3640	2.747	1.064	2.924	1.2217	70° 00′
10	.3520	.3448	.9387	.3673	2.723	1.065	2.901	1.2188	50
20	.3549	.3475	.9377	.3706	2.699	1.066	2.878	1.2159	40
30	.3578	.3502	.9367	.3739	2.675	1.068	2.855	1.2130	30
40	.3607	.3529	.9356	.3772	2.651	1.069	2.833	1.2101	20
50	.3636	.3557	.9346	.3805	2.628	1.070	2.812	1.2072	10
21° 00′	.3665	.3584	.9336	.3839	2.605	1.071	2.790	1.2043	69° 00′
10	.3694	.3611	.9325	.3872	2.583	1.072	2.769	1.2014	50
20	.3723	.3638	.9315	.3906	2.560	1.074	2.749	1.1985	40
30	.3752	.3665	.9304	.3939	2.539	1.075	2.729	1.1956	30
40	.3782	.3692	.9293	.3973	2.517	1.076	2.709	1.1926	20
50	.3811	.3719	.9283	.4006	2.496	1.077	2.689	1.1897	10
22° 00′	.3840	.3746	.9272	.4040	2.475	1.079	2.669	1.1868	68° 00′
10	.3869	.3773	.9261	.4074	2.455	1.080	2.650	1.1839	50
20	.3898	.3800	.9250	.4108	2.434	1.081	2.632	1.1810	40
30	.3927	.3827	.9239	.4142	2.414	1.082	2.613	1.1781	30
40	.3956	.3854	.9228	.4176	2.394	1.084	2.595	1.1752	20
50	.3985	.3881	.9216	.4210	2.375	1.085	2.577	1.1723	10
23° 00′	.4014	.3907	.9205	.4245	2.356	1.086	2.559	1.1694	67° 00′
10	.4043	.3934	.9194	.4279	2.337	1.088	2.542	1.1665	50
20	.4072	.3961	.9182	.4314	2.318	1.089	2.525	1.1636	40
30	.4102	.3987	.9171	.4348	2.300	1.090	2.508	1.1606	30
40	.4131	.4014	.9159	.4383	2.282	1.092	2.491	1.1577	20
50	.4160	.4041	.9147	.4417	2.264	1.093	2.475	1.1548	10
24° 00′	.4189	.4067	.9135	.4452	2.246	1.095	2.459	1.1519	66° 00′
		cos θ	sin θ	cot θ	tan θ	csc θ	sec θ	Radians	Degrees

Table 3. Continued

Degrees	Radians	sin θ	cos θ	tan θ	cot θ	sec θ	csc θ		
24° 00'	.4189	.4067	.9135	.4452	2.246	1.095	2.459	1.1519	66° 00'
10	.4218	.4094	.9124	.4487	2.229	1.096	2.443	1.1490	50
20	.4247	.4120	.9112	.4522	2.211	1.097	2.427	1.1461	40
30	.4276	.4147	.9100	.4557	2.194	1.099	2.411	1.1432	30
40	.4305	.4173	.9088	.4592	2.177	1.100	2.396	1.1403	20
50	.4334	.4200	.9075	.4628	2.161	1.102	2.381	1.1374	10
25° 00'	.4363	.4226	.9063	.4663	2.145	1.103	2.366	1.1345	65° 00'
10	.4392	.4253	.9051	.4699	2.128	1.105	2.352	1.1316	50
20	.4422	.4279	.9038	.4734	2.112	1.106	2.337	1.1286	40
30	.4451	.4305	.9026	.4770	2.097	1.108	2.323	1.1257	30
40	.4480	.4331	.9013	.4806	2.081	1.109	2.309	1.1228	20
50	.4509	.4358	.9001	.4841	2.066	1.111	2.295	1.1199	10
26° 00'	.4538	.4384	.8988	.4877	2.050	1.113	2.281	1.1170	64° 00'
10	.4567	.4410	.8975	.4913	2.035	1.114	2.268	1.1141	50
20	.4596	.4436	.8962	.4950	2.020	1.116	2.254	1.1112	40
30	.4625	.4462	.8949	.4986	2.006	1.117	2.241	1.1083	30
40	.4654	.4488	.8936	.5022	1.991	1.119	2.228	1.1054	20
50	.4683	.4514	.8923	.5059	1.977	1.121	2.215	1.1025	10
27° 00'	.4712	.4540	.8910	.5095	1.963	1.122	2.203	1.0996	63° 00'
10	.4741	.4566	.8897	.5132	1.949	1.124	2.190	1.0966	50
20	.4771	.4592	.8884	.5169	1.935	1.126	2.178	1.0937	40
30	.4800	.4617	.8870	.5206	1.921	1.127	2.166	1.0908	30
40	.4829	.4643	.8857	.5243	1.907	1.129	2.154	1.0879	20
50	.4858	.4669	.8843	.5280	1.894	1.131	2.142	1.0850	10
28° 00'	.4887	.4695	.8829	.5317	1.881	1.133	2.130	1.0821	62° 00'
10	.4916	.4720	.8816	.5354	1.868	1.134	2.118	1.0792	50
20	.4945	.4746	.8802	.5392	1.855	1.136	2.107	1.0763	40
30	.4974	.4772	.8788	.5430	1.842	1.138	2.096	1.0734	30
40	.5003	.4797	.8774	.5467	1.829	1.140	2.085	1.0705	20
50	.5032	.4823	.8760	.5505	1.816	1.142	2.074	1.0676	10
29° 00'	.5061	.4848	.8746	.5543	1.804	1.143	2.063	1.0647	61° 00'
10	.5091	.4874	.8732	.5581	1.792	1.145	2.052	1.0617	50
20	.5120	.4899	.8718	.5619	1.780	1.147	2.041	1.0588	40
30	.5149	.4924	.8704	.5658	1.767	1.149	2.031	1.0559	30
40	.5178	.4950	.8689	.5696	1.756	1.151	2.020	1.0530	20
50	.5207	.4975	.8675	.5735	1.744	1.153	2.010	1.0501	10
30° 00'	.5236	.5000	.8660	.5774	1.732	1.155	2.000	1.0472	60° 00'
		cos θ	sin θ	cot θ	tan θ	csc θ	sec θ	Radians	Degrees

Table 3. Continued

Degrees	Radians	sin θ	cos θ	tan θ	cot θ	sec θ	csc θ		
30° 00'	.5236	.5000	.8660	.5774	1.732	1.155	2.000	1.0472	60° 00'
10	.5265	.5025	.8646	.5812	1.720	1.157	1.990	1.0443	50
20	.5294	.5050	.8631	.5851	1.709	1.159	1.980	1.0414	40
30	.5323	.5075	.8616	.5890	1.698	1.161	1.970	1.0385	30
40	.5352	.5100	.8601	.5930	1.686	1.163	1.961	1.0356	20
50	.5381	.5125	.8587	.5969	1.675	1.165	1.951	1.0327	10
31° 00'	.5411	.5150	.8572	.6009	1.664	1.167	1.942	1.0297	59° 00'
10	.5440	.5175	.8557	.6048	1.653	1.169	1.932	1.0268	50
20	.5469	.5200	.8542	.6088	1.643	1.171	1.923	1.0239	40
30	.5498	.5225	.8526	.6128	1.632	1.173	1.914	1.0210	30
40	.5527	.5250	.8511	.6168	1.621	1.175	1.905	1.0181	20
50	.5556	.5275	.8496	.6208	1.611	1.177	1.896	1.0152	10
32° 00'	.5585	.5299	.8480	.6249	1.600	1.179	1.887	1.0123	58° 00'
10	.5614	.5324	.8465	.6289	1.590	1.181	1.878	1.0094	50
20	.5643	.5348	.8450	.6330	1.580	1.184	1.870	1.0065	40
30	.5672	.5373	.8434	.6371	1.570	1.186	1.861	1.0036	30
40	.5701	.5398	.8418	.6412	1.560	1.188	1.853	1.0007	20
50	.5730	.5422	.8403	.6453	1.550	1.190	1.844	.9977	10
33° 00'	.5760	.5446	.8387	.6494	1.540	1.192	1.836	.9948	57° 00'
10	.5789	.5471	.8371	.6536	1.530	1.195	1.828	.9919	50
20	.5818	.5495	.8355	.6577	1.520	1.197	1.820	.9890	40
30	.5847	.5519	.8339	.6619	1.511	1.199	1.812	.9861	30
40	.5876	.5544	.8323	.6661	1.501	1.202	1.804	.9832	20
50	.5905	.5568	.8307	.6703	1.492	1.204	1.796	.9803	10
34° 00'	.5934	.5592	.8290	.6745	1.483	1.206	1.788	.9774	56° 00'
10	.5963	.5616	.8274	.6787	1.473	1.209	1.781	.9745	50
20	.5992	.5640	.8258	.6830	1.464	1.211	1.773	.9716	40
30	.6021	.5664	.8241	.6873	1.455	1.213	1.766	.9687	30
40	.6050	.5688	.8225	.6916	1.446	1.216	1.758	.9657	20
50	.6080	.5712	.8208	.6959	1.437	1.218	1.751	.9628	10
35° 00'	.6109	.5736	.8192	.7002	1.428	1.221	1.743	.9599	55° 00'
10	.6138	.5760	.8175	.7046	1.419	1.223	1.736	.9570	50
20	.6167	.5783	.8158	.7089	1.411	1.226	1.729	.9541	40
30	.6196	.5807	.8141	.7133	1.402	1.228	1.722	.9512	30
40	.6225	.5831	.8124	.7177	1.393	1.231	1.715	.9483	20
50	.6254	.5854	.8107	.7221	1.385	1.233	1.708	.9454	10
36° 00'	.6283	.5878	.8090	.7265	1.376	1.236	1.701	.9425	54° 00'
		cos θ	sin θ	cot θ	tan θ	csc θ	sec θ	Radians	Degrees

Table 3. Continued

Degrees	Radians	sin θ	cos θ	tan θ	cot θ	sec θ	csc θ		
36° 00′	.6283	.5878	.8090	.7265	1.376	1.236	1.701	.9425	54° 00′
10	.6312	.5901	.8073	.7310	1.368	1.239	1.695	.9396	50
20	.6341	.5925	.8056	.7355	1.360	1.241	1.688	.9367	40
30	.6370	.5948	.8039	.7400	1.351	1.244	1.681	.9338	30
40	.6400	.5972	.8021	.7445	1.343	1.247	1.675	.9308	20
50	.6429	.5995	.8004	.7490	1.335	1.249	1.668	.9279	10
37° 00′	.6458	.6018	.7986	.7536	1.327	1.252	1.662	.9250	53° 00′
10	.6487	.6041	.7969	.7581	1.319	1.255	1.655	.9221	50
20	.6516	.6065	.7951	.7627	1.311	1.258	1.649	.9192	40
30	.6545	.6088	.7934	.7673	1.303	1.260	1.643	.9163	30
40	.6574	.6111	.7916	.7720	1.295	1.263	1.636	.9134	20
50	.6603	.6134	.7898	.7766	1.288	1.266	1.630	.9105	10
38° 00′	.6632	.6157	.7880	.7813	1.280	1.269	1.624	.9076	52° 00′
10	.6661	.6180	.7862	.7860	1.272	1.272	1.618	.9047	50
20	.6690	.6202	.7844	.7907	1.265	1.275	1.612	.9018	40
30	.6720	.6225	.7826	.7954	1.257	1.278	1.606	.8988	30
40	.6749	.6248	.7808	.8002	1.250	1.281	1.601	.8959	20
50	.6778	.6271	.7790	.8050	1.242	1.284	1.595	.8930	10
39° 00′	.6807	.6293	.7771	.8098	1.235	1.287	1.589	.8901	51° 00′
10	.6836	.6316	.7753	.8146	1.228	1.290	1.583	.8872	50
20	.6865	.6338	.7735	.8195	1.220	1.293	1.578	.8843	40
30	.6894	.6361	.7716	.8243	1.213	1.296	1.572	.8814	30
40	.6923	.6383	.7698	.8292	1.206	1.299	1.567	.8785	20
50	.6952	.6406	.7679	.8342	1.199	1.302	1.561	.8756	10
40° 00′	.6981	.6428	.7660	.8391	1.192	1.305	1.556	.8727	50° 00′
10	.7010	.6450	.7642	.8441	1.185	1.309	1.550	.8698	50
20	.7039	.6472	.7623	.8491	1.178	1.312	1.545	.8668	40
30	.7069	.6494	.7604	.8541	1.171	1.315	1.540	.8639	30
40	.7098	.6517	.7585	.8591	1.164	1.318	1.535	.8610	20
50	.7127	.6539	.7566	.8642	1.157	1.322	1.529	.8581	10
41° 00′	.7156	.6561	.7547	.8693	1.150	1.325	1.524	.8552	49° 00′
10	.7185	.6583	.7528	.8744	1.144	1.328	1.519	.8523	50
20	.7214	.6604	.7509	.8796	1.137	1.332	1.514	.8494	40
30	.7243	.6626	.7490	.8847	1.130	1.335	1.509	.8465	30
40	.7272	.6648	.7470	.8899	1.124	1.339	1.504	.8436	20
50	.7301	.6670	.7451	.8952	1.117	1.342	1.499	.8407	10
42° 00′	.7330	.6691	.7431	.9004	1.111	1.346	1.494	.8378	48° 00′
		cos θ	sin θ	cot θ	tan θ	csc θ	sec θ	Radians	Degrees

Table 3. Continued

Degrees	Radians	sin θ	cos θ	tan θ	cot θ	sec θ	csc θ		
42° 00′	.7330	.6691	.7431	.9004	1.111	1.346	1.494	.8378	48° 00′
10	.7359	.6713	.7412	.9057	1.104	1.349	1.490	.8348	50
20	.7389	.6734	.7392	.9110	1.098	1.353	1.485	.8319	40
30	.7418	.6756	.7373	.9163	1.091	1.356	1.480	.8290	30
40	.7447	.6777	.7353	.9217	1.085	1.360	1.476	.8261	20
50	.7476	.6799	.7333	.9271	1.079	1.364	1.471	.8232	10
43° 00′	.7505	.6820	.7314	.9325	1.072	1.367	1.466	.8203	47° 00′
10	.7534	.6841	.7294	.9380	1.066	1.371	1.462	.8174	50
20	.7563	.6862	.7274	.9435	1.060	1.375	1.457	.8145	40
30	.7592	.6884	.7254	.9490	1.054	1.379	1.453	.8116	30
40	.7621	.6905	.7234	.9545	1.048	1.382	1.448	.8087	20
50	.7650	.6926	.7214	.9601	1.042	1.386	1.444	.8058	10
44° 00′	.7679	.6947	.7193	.9657	1.036	1.390	1.440	.8029	46° 00′
10	.7709	.6967	.7173	.9713	1.030	1.394	1.435	.7999	50
20	.7738	.6988	.7153	.9770	1.024	1.398	1.431	.7970	40
30	.7767	.7009	.7133	.9827	1.018	1.402	1.427	.7941	30
40	.7796	.7030	.7112	.9884	1.012	1.406	1.423	.7912	20
50	.7825	.7050	.7092	.9942	1.006	1.410	1.418	.7883	10
45° 00′	.7854	.7071	.7071	1.000	1.000	1.414	1.414	.7854	45° 00′
		cos θ	sin θ	cot θ	tan θ	csc θ	sec θ	Radians	Degrees

Table 4. Four-Place Logarithms of Trigonometric Functions — Angle θ in Degrees
Attach -10 to Logarithms Obtained from This Table

Angle θ	log sin θ	log csc θ	log tan θ	log cot θ	log sec θ	log cos θ	
0° 00′	No value	No value	No value	No value	10.0000	10.0000	90° 00′
10	7.4637	12.5363	7.4637	12.5363	10.0000	10.0000	50
20	7.7648	12.2352	7.7648	12.2352	10.0000	10.0000	40
30	7.9408	12.0592	7.9409	12.0591	10.0000	10.0000	30
40	8.0658	11.9342	8.0658	11.9342	10.0000	10.0000	20
50	8.1627	11.8373	8.1627	11.8373	10.0000	10.0000	10
1° 00′	8.2419	11.7581	8.2419	11.7581	10.0001	9.9999	89° 00′
10	8.3088	11.6912	8.3089	11.6911	10.0001	9.9999	50
20	8.3668	11.6332	8.3669	11.6331	10.0001	9.9999	40
30	8.4179	11.5821	8.4181	11.5819	10.0001	9.9999	30
40	8.4637	11.5363	8.4638	11.5362	10.0002	9.9998	20
50	8.5050	11.4950	8.5053	11.4947	10.0002	9.9998	10
2° 00′	8.5428	11.4572	8.5431	11.4569	10.0003	9.9997	88° 00′
10	8.5776	11.4224	8.5779	11.4221	10.0003	9.9997	50
20	8.6097	11.3903	8.6101	11.3899	10.0004	9.9996	40
30	8.6397	11.3603	8.6401	11.3599	10.0004	9.9996	30
40	8.6677	11.3323	8.6682	11.3318	10.0005	9.9995	20
50	8.6940	11.3060	8.6945	11.3055	10.0005	9.9995	10
3° 00′	8.7188	11.2812	8.7194	11.2806	10.0006	9.9994	87° 00′
10	8.7423	11.2577	8.7429	11.2571	10.0007	9.9993	50
20	8.7645	11.2355	8.7652	11.2348	10.0007	9.9993	40
30	8.7857	11.2143	8.7865	11.2135	10.0008	9.9992	30
40	8.8059	11.1941	8.8067	11.1933	10.0009	9.9991	20
50	8.8251	11.1749	8.8261	11.1739	10.0010	9.9990	10
4° 00′	8.8463	11.1564	8.8446	11.1554	10.0011	9.9989	86° 00′
10	8.8613	11.1387	8.8624	11.1376	10.0011	9.9989	50
20	8.8783	11.1217	8.8795	11.1205	10.0012	9.9988	40
30	8.8946	11.1054	8.8960	11.1040	10.0013	9.9987	30
40	8.9104	11.0896	8.9118	11.0882	10.0014	9.9986	20
50	8.9256	11.0744	8.9272	11.0728	10.0015	9.9985	10
5° 00′	8.9403	11.0597	8.9420	11.0580	10.0017	9.9983	85° 00′
10	8.9545	11.0455	8.9563	11.0437	10.0018	9.9982	50
20	8.9682	11.0318	8.9701	11.0299	10.0019	9.9981	40
30	8.9816	11.0184	8.9836	11.0164	10.0020	9.9980	30
40	8.9945	11.0055	8.9966	11.0034	10.0021	9.9979	20
50	9.0070	10.9930	9.0093	10.9907	10.0023	9.9977	10
6° 00′	9.0192	10.9808	9.0216	10.9784	10.0024	9.9976	84° 00′
	log cos θ	log sec θ	log cot θ	log tan θ	log csc θ	log sin θ	Angle θ

Table 4. Continued
Attach −10 to Logarithms Obtained from This Table

Angle θ	log sin θ	log csc θ	log tan θ	log cot θ	log sec θ	log cos θ	
6° 00′	9.0192	10.9808	9.0216	10.9784	10.0024	9.9976	84° 00′
10	9.0311	10.9689	9.0336	10.9664	10.0025	9.9975	50
20	9.0426	10.9574	9.0453	10.9547	10.0027	9.9973	40
30	9.0539	10.9461	9.0567	10.9433	10.0028	9.9972	30
40	9.0648	10.9352	9.0678	10.9322	10.0029	9.9971	20
50	9.0755	10.9245	9.0786	10.9214	10.0031	9.9969	10
7° 00′	9.0859	10.9141	9.0891	10.9109	10.0032	9.9968	83° 00′
10	9.0961	10.9039	9.0995	10.9005	10.0034	9.9966	50
20	9.1060	10.8940	9.1096	10.8904	10.0036	9.9964	40
30	9.1157	10.8843	9.1194	10.8806	10.0037	9.9963	30
40	9.1252	10.8748	9.1291	10.8709	10.0039	9.9961	20
50	9.1345	10.8655	9.1385	10.8615	10.0041	9.9959	10
8° 00′	9.1436	10.8564	9.1478	10.8522	10.0042	9.9958	82° 00′
10	9.1525	10.8475	9.1569	10.8431	10.0044	9.9956	50
20	9.1612	10.8388	9.1658	10.8342	10.0046	9.9954	40
30	9.1697	10.8303	9.1745	10.8255	10.0048	9.9952	30
40	9.1781	10.8219	9.1831	10.8169	10.0050	9.9950	20
50	9.1863	10.8137	9.1915	10.8085	10.0052	9.9948	10
9° 00′	9.1943	10.8057	9.1997	10.8003	10.0054	9.9946	81° 00′
10	9.2022	10.7978	9.2078	10.7922	10.0056	9.9944	50
20	9.2100	10.7900	9.2158	10.7842	10.0058	9.9942	40
30	9.2176	10.7824	9.2236	10.7764	10.0060	9.9940	30
40	9.2251	10.7749	9.2313	10.7687	10.0062	9.9938	20
50	9.2324	10.7676	9.2389	10.7611	10.0064	9.9936	10
10° 00′	9.2397	10.7603	9.2463	10.7537	10.0066	9.9934	80° 00′
10	9.2468	10.7532	9.2536	10.7464	10.0069	9.9931	50
20	9.2538	10.7462	9.2609	10.7391	10.0071	9.9929	40
30	9.2606	10.7394	9.2680	10.7320	10.0073	9.9927	30
40	9.2674	10.7326	9.2750	10.7250	10.0076	9.9924	20
50	9.2740	10.7260	9.2819	10.7181	10.0078	9.9922	10
11° 00′	9.2806	10.7194	9.2887	10.7113	10.0081	9.9919	79° 00′
10	9.2870	10.7130	9.2953	10.7047	10.0083	9.9917	50
20	9.2934	10.7066	9.3020	10.6980	10.0086	9.9914	40
30	9.2997	10.7003	9.3085	10.6915	10.0088	9.9912	30
40	9.3058	10.6942	9.3149	10.6851	10.0091	9.9909	20
50	9.3119	10.6881	9.3212	10.6788	10.0093	9.9907	10
12° 00′	9.3179	10.6821	9.3275	10.6725	10.0096	9.9904	78° 00′
	log cos θ	log sec θ	log cot θ	log tan θ	log csc θ	log sin θ	Angle θ

Table 4. Continued
Attach −10 to Logarithms Obtained from This Table

Angle θ	log sin θ	log csc θ	log tan θ	log cot θ	log sec θ	log cos θ	
12° 00'	9.3179	10.6821	9.3275	10.6725	10.0096	9.9904	78° 00'
10	9.3238	10.6762	9.3336	10.6664	10.0099	9.9901	50
20	9.3296	10.6704	9.3397	10.6603	10.0101	9.9899	40
30	9.3353	10.6647	9.3458	10.6542	10.0104	9.9896	30
40	9.3410	10.6590	9.3517	10.6483	10.0107	9.9893	20
50	9.3466	10.6534	9.3576	10.6424	10.0110	9.9890	10
13° 00'	9.3521	10.6479	9.3634	10.6366	10.0113	9.9887	77° 00'
10	9.3575	10.6425	9.3691	10.6309	10.0116	9.9884	50
20	9.3629	10.6371	9.3748	10.6252	10.0119	9.9881	40
30	9.3682	10.6318	9.3804	10.6196	10.0122	9.9878	30
40	9.3734	10.6266	9.3859	10.6141	10.0125	9.9875	20
50	9.3786	10.6214	9.3914	10.6086	10.0128	9.9872	10
14° 00'	9.3837	10.6163	9.3968	10.6032	10.0131	9.9869	76° 00'
10	9.3887	10.6113	9.4021	10.5979	10.0134	9.9866	50
20	9.3937	10.6063	9.4074	10.5926	10.0137	9.9863	40
30	9.3986	10.6014	9.4127	10.5873	10.0141	9.9859	30
40	9.4035	10.5965	9.4178	10.5822	10.0144	9.9856	20
50	9.4083	10.5917	9.4230	10.5770	10.0147	9.9853	10
15° 00'	9.4130	10.5870	9.4281	10.5719	10.0151	9.9849	75° 00'
10	9.4177	10.5823	9.4331	10.5669	10.0154	9.9846	50
20	9.4223	10.5777	9.4381	10.5619	10.0157	9.9843	40
30	9.4269	10.5731	9.4430	10.5570	10.0161	9.9839	30
40	9.4314	10.5686	9.4479	10.5521	10.0164	9.9836	20
50	9.4359	10.5641	9.4527	10.5473	10.0168	9.9832	10
16° 00'	9.4403	10.5597	9.4575	10.5425	10.0172	9.9828	74° 00'
10	9.4447	10.5553	9.4622	10.5378	10.0175	9.9825	50
20	9.4491	10.5509	9.4669	10.5331	10.0179	9.9821	40
30	9.4533	10.5467	9.4716	10.5284	10.0183	9.9817	30
40	9.4576	10.5424	9.4762	10.5238	10.0186	9.9814	20
50	9.4618	10.5382	9.4808	10.5192	10.0190	9.9810	10
17° 00'	9.4659	10.5341	9.4853	10.5147	10.0194	9.9806	73° 00'
10	9.4700	10.5300	9.4898	10.5102	10.0198	9.9802	50
20	9.4741	10.5259	9.4943	10.5057	10.0202	9.9798	40
30	9.4781	10.5219	9.4987	10.5013	10.0206	9.9794	30
40	9.4821	10.5179	9.5031	10.4969	10.0210	9.9790	20
50	9.4861	10.5139	9.5075	10.4925	10.0214	9.9786	10
18° 00'	9.4900	10.5100	9.5118	10.4882	10.0218	9.9782	72° 00'
	log cos θ	log sec θ	log cot θ	log tan θ	log csc θ	log sin θ	Angle θ

Table 4. Continued
Attach −10 to Logarithms Obtained from This Table

Angle θ	log sin θ	log csc θ	log tan θ	log cot θ	log sec θ	log cos θ	
18° 00′	9.4900	10.5100	9.5118	10.4882	10.0218	9.9782	72° 00′
10	9.4939	10.5061	9.5161	10.4839	10.0222	9.9778	50
20	9.4977	10.5023	9.5203	10.4797	10.0226	9.9774	40
30	9.5015	10.4985	9.5245	10.4755	10.0230	9.9770	30
40	9.5052	10.4948	9.5287	10.4713	10.0235	9.9765	20
50	9.5090	10.4910	9.5329	10.4671	10.0239	9.9761	10
19° 00′	9.5126	10.4874	9.5370	10.4630	10.0243	9.9757	71° 00′
10	9.5163	10.4837	9.5411	10.4589	10.0248	9.9752	50
20	9.5199	10.4801	9.5451	10.4549	10.0252	9.9748	40
30	9.5235	10.4765	9.5491	10.4509	10.0257	9.9743	30
40	9.5270	10.4730	9.5531	10.4469	10.0261	9.9739	20
50	9.5306	10.4694	9.5571	10.4429	10.0266	9.9734	10
20° 00′	9.5341	10.4659	9.5611	10.4389	10.0270	9.9730	70° 00′
10	9.5375	10.4625	9.5650	10.4350	10.0275	9.9725	50
20	9.5409	10.4591	9.5689	10.4311	10.0279	9.9721	40
30	9.5443	10.4557	9.5727	10.4273	10.0284	9.9716	30
40	9.5477	10.4523	9.5766	10.4234	10.0289	9.9711	20
50	9.5510	10.4490	9.5804	10.4196	10.0294	9.9706	10
21° 00′	9.5543	10.4457	9.5842	10.4158	10.0298	9.9702	69° 00′
10	9.5576	10.4424	9.5879	10.4121	10.0303	9.9697	50
20	9.5609	10.4391	9.5917	10.4083	10.0308	9.9692	40
30	9.5641	10.4359	9.5954	10.4046	10.0313	9.9687	30
40	9.5673	10.4327	9.5991	10.4009	10.0318	9.9682	20
50	9.5704	10.4296	9.6028	10.3972	10.0323	9.9677	10
22° 00′	9.5736	10.4264	9.6064	10.3936	10.0328	9.9672	68° 00′
10	9.5767	10.4233	9.6100	10.3900	10.0333	9.9667	50
20	9.5798	10.4202	9.6136	10.3864	10.0339	9.9661	40
30	9.5828	10.4172	9.6172	10.3828	10.0344	9.9656	30
40	9.5859	10.4141	9.6208	10.3792	10.0349	9.9651	20
50	9.5889	10.4111	9.6243	10.3757	10.0354	9.9646	10
23° 00′	9.5919	10.4081	9.6279	10.3721	10.0360	9.9640	67° 00′
10	9.5948	10.4052	9.6314	10.3686	10.0365	9.9635	50
20	9.5978	10.4022	9.6348	10.3652	10.0371	9.9629	40
30	9.6007	10.3993	9.6383	10.3617	10.0376	9.9624	30
40	9.6036	10.3964	9.6417	10.3583	10.0382	9.9618	20
50	9.6065	10.3935	9.6452	10.3548	10.0387	9.9613	10
24° 00′	9.6093	10.3907	9.6486	10.3514	10.0393	9.9607	66° 00′
	log cos θ	log sec θ	log cot θ	log tan θ	log csc θ	log sin θ	Angle θ

Table 4. Continued
Attach −10 to Logarithms Obtained from This Table

Angle θ	log sin θ	log csc θ	log tan θ	log cot θ	log sec θ	log cos θ	
24° 00′	9.6093	10.3907	9.6486	10.3514	10.0393	9.9607	66° 00′
10	9.6121	10.3879	9.6520	10.3480	10.0398	9.9602	50
20	9.6149	10.3851	9.6553	10.3447	10.0404	9.9596	40
30	9.6177	10.3823	9.6587	10.3413	10.0410	9.9590	30
40	9.6205	10.3795	9.6620	10.3380	10.0416	9.9584	20
50	9.6232	10.3768	9.6654	10.3346	10.0421	9.9579	10
25° 00′	9.6259	10.3741	9.6687	10.3313	10.0427	9.9573	65° 00′
10	9.6286	10.3714	9.6720	10.3280	10.0433	9.9567	50
20	9.6313	10.3687	9.6752	10.3248	10.0439	9.9561	40
30	9.6340	10.3660	9.6785	10.3215	10.0445	9.9555	30
40	9.6366	10.3634	9.6817	10.3183	10.0451	9.9549	20
50	9.6392	10.3608	9.6850	10.3150	10.0457	9.9543	10
26° 00′	9.6418	10.3582	9.6882	10.3118	10.0463	9.9537	64° 00′
10	9.6444	10.3556	9.6914	10.3086	10.0470	9.9530	50
20	9.6470	10.3530	9.6946	10.3054	10.0476	9.9524	40
30	9.6495	10.3505	9.6977	10.3023	10.0482	9.9518	30
40	9.6521	10.3479	9.7009	10.2991	10.0488	9.9512	20
50	9.6546	10.3454	9.7040	10.2960	10.0495	9.9505	10
27° 00′	9.6570	10.3430	9.7072	10.2928	10.0501	9.9499	63° 00′
10	9.6595	10.3405	9.7103	10.2897	10.0508	9.9492	50
20	9.6620	10.3380	9.7134	10.2866	10.0514	9.9486	40
30	9.6644	10.3356	9.7165	10.2835	10.0521	9.9479	30
40	9.6668	10.3332	9.7196	10.2804	10.0527	9.9473	20
50	9.6692	10.3308	9.7226	10.2774	10.0534	9.9466	10
28° 00′	9.6716	10.3284	9.7257	10.2743	10.0541	9.9459	62° 00′
10	9.6740	10.3260	9.7287	10.2713	10.0547	9.9453	50
20	9.6763	10.3237	9.7317	10.2683	10.0554	9.9446	40
30	9.6787	10.3213	9.7348	10.2652	10.0561	9.9439	30
40	9.6810	10.3190	9.7378	10.2622	10.0568	9.9432	20
50	9.6833	10.3167	9.7408	10.2592	10.0575	9.9425	10
29° 00′	9.6856	10.3144	9.7438	10.2562	10.0582	9.9418	61° 00′
10	9.6878	10.3122	9.7467	10.2533	10.0589	9.9411	50
20	9.6901	10.3099	9.7497	10.2503	10.0596	9.9404	40
30	9.6923	10.3077	9.7526	10.2474	10.0603	9.9397	30
40	9.6946	10.3054	9.7556	10.2444	10.0610	9.9390	20
50	9.6968	10.3032	9.7585	10.2415	10.0617	9.9383	10
30° 00′	9.6990	10.3010	9.7614	10.2386	10.0625	9.9375	60° 00′
	log cos θ	log sec θ	log cot θ	log tan θ	log csc θ	log sin θ	Angle θ

Table 4. Continued
Attach −10 to Logarithms Obtained from This Table

Angle θ	log sin θ	log csc θ	log tan θ	log cot θ	log sec θ	log cos θ	
30° 00′	9.6990	10.3010	9.7614	10.2386	10.0625	9.9375	60° 00′
10	9.7012	10.2988	9.7644	10.2356	10.0632	9.9368	50
20	9.7033	10.2967	9.7673	10.2327	10.0639	9.9361	40
30	9.7055	10.2945	9.7701	10.2299	10.0647	9.9353	30
40	9.7076	10.2924	9.7730	10.2270	10.0654	9.9346	20
50	9.7097	10.2903	9.7759	10.2241	10.0662	9.9338	10
31° 00′	9.7118	10.2882	9.7788	10.2212	10.0669	9.9331	59° 00′
10	9.7139	10.2861	9.7816	10.2184	10.0677	9.9323	50
20	9.7160	10.2840	9.7845	10.2155	10.0685	9.9315	40
30	9.7181	10.2819	9.7873	10.2127	10.0692	9.9308	30
40	9.7201	10.2799	9.7902	10.2098	10.0700	9.9300	20
50	9.7222	10.2778	9.7930	10.2070	10.0708	9.9292	10
32° 00′	9.7242	10.2758	9.7958	10.2042	10.0716	9.9284	58° 00′
10	9.7262	10.2738	9.7986	10.2014	10.0724	9.9276	50
20	9.7282	10.2718	9.8014	10.1986	10.0732	9.9268	40
30	9.7302	10.2698	9.8042	10.1958	10.0740	9.9260	30
40	9.7322	10.2678	9.8070	10.1930	10.0748	9.9252	20
50	9.7342	10.2658	9.8097	10.1903	10.0756	9.9244	10
33° 00′	9.7361	10.2639	9.8125	10.1875	10.0764	9.9236	57° 00′
10	9.7380	10.2620	9.8153	10.1847	10.0772	9.9228	50
20	9.7400	10.2600	9.8180	10.1820	10.0781	9.9219	40
30	9.7419	10.2581	9.8208	10.1792	10.0789	9.9211	30
40	9.7438	10.2562	9.8235	10.1765	10.0797	9.9203	20
50	9.7457	10.2543	9.8263	10.1737	10.0806	9.9194	10
34° 00′	9.7476	10.2524	9.8290	10.1710	10.0814	9.9186	56° 00′
10	9.7494	10.2506	9.8317	10.1683	10.0823	9.9177	50
20	9.7513	10.2487	9.8344	10.1656	10.0831	9.9169	40
30	9.7531	10.2469	9.8371	10.1629	10.0840	9.9160	30
40	9.7550	10.2450	9.8398	10.1602	10.0849	9.9151	20
50	9.7568	10.2432	9.8425	10.1575	10.0858	9.9142	10
35° 00′	9.7586	10.2414	9.8452	10.1548	10.0866	9.9134	55° 00′
10	9.7604	10.2396	9.8479	10.1521	10.0875	9.9125	50
20	9.7622	10.2378	9.8506	10.1494	10.0884	9.9116	40
30	9.7640	10.2360	9.8533	10.1467	10.0893	9.9107	30
40	9.7657	10.2343	9.8559	10.1441	10.0902	9.9098	20
50	9.7675	10.2325	9.8586	10.1414	10.0911	9.9089	10
36° 00′	9.7692	10.2308	9.8613	10.1387	10.0920	9.9080	54° 00′
	log cos θ	log sec θ	log cot θ	log tan θ	log csc θ	log sin θ	Angle θ

Table 4. Continued
Attach −10 to Logarithms Obtained from This Table

Angle θ	log sin θ	log csc θ	log tan θ	log cot θ	log sec θ	log cos θ	
36° 00'	9.7692	10.2308	9.8613	10.1387	10.0920	9.9080	54° 00'
10	9.7710	10.2290	9.8639	10.1361	10.0930	9.9070	50
20	9.7727	10.2273	9.8666	10.1334	10.0939	9.9061	40
30	9.7744	10.2256	9.8692	10.1308	10.0948	9.9052	30
40	9.7761	10.2239	9.8718	10.1282	10.0958	9.9042	20
50	9.7778	10.2222	9.8745	10.1255	10.0967	9.9033	10
37° 00'	9.7795	10.2205	9.8771	10.1229	10.0977	9.9023	53° 00'
10	9.7811	10.2189	9.8797	10.1203	10.0986	9.9014	50
20	9.7828	10.2172	9.8824	10.1176	10.0996	9.9004	40
30	9.7844	10.2156	9.8850	10.1150	10.1005	9.8995	30
40	9.7861	10.2139	9:8876	10.1124	10.1015	9.8985	20
50	9.7877	10.2123	9.8902	10.1098	10.1025	9.8975	10
38° 00'	9.7893	10.2107	9.8928	10.1072	10.1035	9.8965	52° 00'
10	9.7910	10.2090	9.8954	10.1046	10.1045	9.8955	50
20	9.7926	10.2074	9.8980	10.1020	10.1055	9.8945	40
30	9.7941	10.2059	9.9006	10.0994	10.1065	9.8935	30
40	9.7957	10.2043	9.9032	10.0968	10.1075	9.8925	20
50	9.7973	10.2027	9.9058	10.0942	10.1085	9.8915	10
39° 00'	9.7989	10.2011	9.9084	10.0916	10.1095	9.8905	51° 00'
10	9.8004	10.1996	9.9110	10.0890	10.1105	9.8895	50
20	9.8020	10.1980	9.9135	10.0865	10.1116	9.8884	40
30	9.8035	10.1965	9.9161	10.0839	10.1126	9.8874	30
40	9.8050	10.1950	9.9187	10.0813	10.1136	9.8864	20
50	9.8066	10.1934	9.9212	10.0788	10.1147	9.8853	10
40° 00'	9.8081	10.1919	9.9238	10.0762	10.1157	9.8843	50° 00'
10	9.8096	10.1904	9.9264	10.0736	10.1168	9.8832	50
20	9.8111	10.1889	9.9289	10.0711	10.1179	9.8821	40
30	9.8125	10.1875	9.9315	10.0685	10.1190	9.8810	30
40	9.8140	10.1860	9.9341	10.0659	10.1200	9.8800	20
50	9.8155	10.1845	9.9366	10.0634	10.1211	9.8789	10
41° 00'	9.8169	10.1831	9.9392	10.0608	10.1222	9.8778	49° 00'
10	9.8184	10.1816	9.9417	10.0583	10.1233	9.8767	50
20	9.8198	10.1802	9.9443	10.0557	10.1244	9.8756	40
30	9.8213	10.1787	9.9468	10.0532	10.1255	9.8745	30
40	9.8227	10.1773	9.9494	10.0506	10.1267	9.8733	20
50	9.8241	10.1759	9.9519	10.0481	10.1278	9.8722	10
42° 00'	9.8255	10.1745	9.9544	10.0456	10.1289	9.8711	48° 00'
	log cos θ	log sec θ	log cot θ	log tan θ	log csc θ	log sin θ	Angle θ

Table 4. Continued

Attach −10 to Logarithms Obtained from This Table

Angle θ	log sin θ	log csc θ	log tan θ	log cot θ	log sec θ	log cos θ	
42° 00'	9.8255	10.1745	9.9544	10.0456	10.1289	9.8711	48° 00'
10	9.8269	10.1731	9.9570	10.0430	10.1301	9.8699	50
20	9.8283	10.1717	9.9595	10.0405	10.1312	9.8688	40
30	9.8297	10.1703	9.9621	10.0379	10.1324	9.8676	30
40	9.8311	10.1689	9.9646	10.0354	10.1335	9.8665	20
50	9.8324	10.1676	9.9671	10.0329	10.1347	9.8653	10
43° 00'	9.8338	10.1662	9.9697	10.0303	10.1359	9.8641	47° 00'
10	9.8351	10.1649	9.9722	10.0278	10.1371	9.8629	50
20	9.8365	10.1635	9.9747	10.0253	10.1382	9.8618	40
30	9.8378	10.1622	9.9772	10.0228	10.1394	9.8606	30
40	9.8391	10.1609	9.9798	10.0202	10.1406	9.8594	20
50	9.8405	10.1595	9.9823	10.0177	10.1418	9.8582	10
44° 00'	9.8418	10.1582	9.9848	10.0152	10.1431	9.8569	46° 00'
10	9.8431	10.1569	9.9874	10.0126	10.1443	9.8557	50
20	9.8444	10.1556	9.9899	10.0101	10.1455	9.8545	40
30	9.8457	10.1543	9.9924	10.0076	10.1468	9.8532	30
40	9.8469	10.1531	9.9949	10.0051	10.1480	9.8520	20
50	9.8482	10.1518	9.9975	10.0025	10.1493	9.8507	10
45° 00'	9.8495	10.1505	10.0000	10.0000	10.1505	9.8495	45° 00'
	log cos θ	log sec θ	log cot θ	log tan θ	log csc θ	log sin θ	Angle θ

Answers to Odd-Numbered Exercises

Exercises 1-1, page 10

1. $A = \pi r^2$ 3. $c = 2\pi r$ 5. $A = 5l$ 7. $A = s^2; s = \sqrt{A}$
9. $C = 77 + 25(t - 3) = 25t + 2$ 11. $I = 200(0.05t) = 10t$ 13. $3, -1$
15. $11, -7$ 17. $-18, 70$ 19. $7, -32$ 21. $\dfrac{a + 2a^2}{4}, 0$

23. $3s^2 + s + 6, 12s^2 - 2s + 6$

25. Square the value of the independent variable, and add 2 to the result.

27. Multiply the independent variable by six, cube the independent variable, and subtract the second result from the first.

29. The domain is all real numbers.
    The range is all real numbers.

31. The domain is all real numbers greater than or equal to -1.
    The range is all real numbers greater than or equal to 0.

33. 54π cu in.   35. $25°, -\dfrac{100°}{9}$

Exercises 1-2, page 14

1. $A(2,1), B(-1,2), C(-2,-3)$

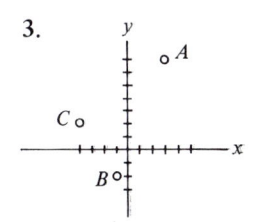

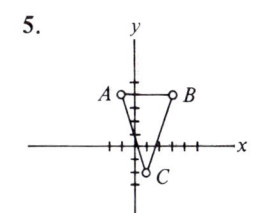

3.

5.

isosceles triangle

7. (5,4)

9. They are on a line parallel to the y-axis and one unit to the right of the y-axis.

11. They are on a line bisecting the first and third quadrants.

13. 0   **15.** They are to the right of the y-axis.

17. They are in the fourth quadrant.   **19.** I, III.

Exercises 1-3, page 19

1. 16.3, 98.6   **3.** 1.33, 9.13   **5.** 0.761, 0.120   **7.** 0.0756, 0.0284
9. 5.00   **11.** 6.66   **13.** 80.0   **15.** 0.472   **17.** 0.693   **19.** 0.056
21. 5   **23.** 9.06   **25.** 7.8 ft   **27.** 243 cm   **29.** 3.4 ft   **31.** 7.8 ft

Exercises 1-4, page 27

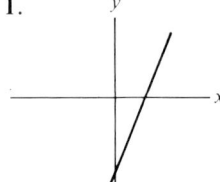

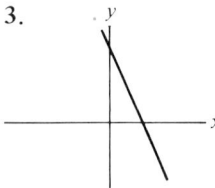

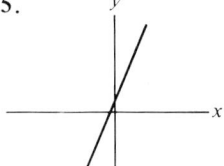

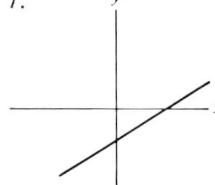

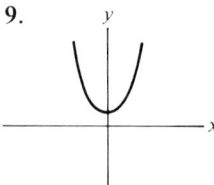

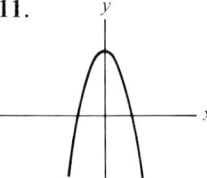

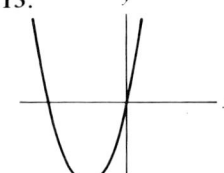

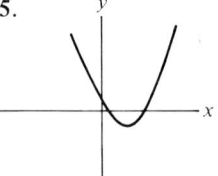

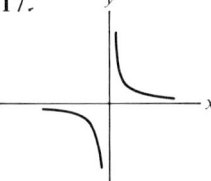

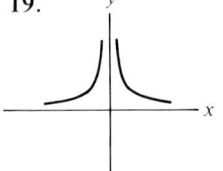

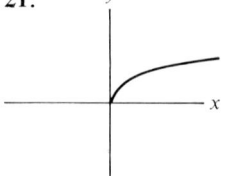

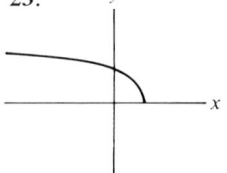

35. The domain is all real numbers except -1.
The range is all real numbers except 1.

37. They are in the first quadrant on a line parallel to the y-axis and one unit to the right of the y-axis.

39. $(6,2)$ **41.** 10.2 ft **43.**

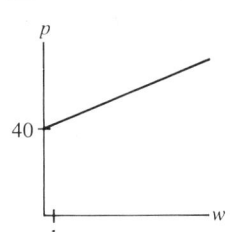

Exercises 2-1, page 37

1.

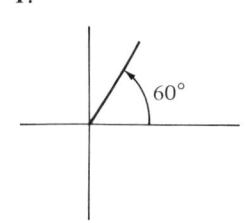

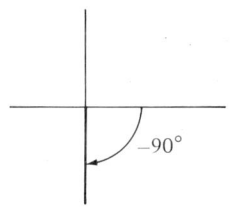

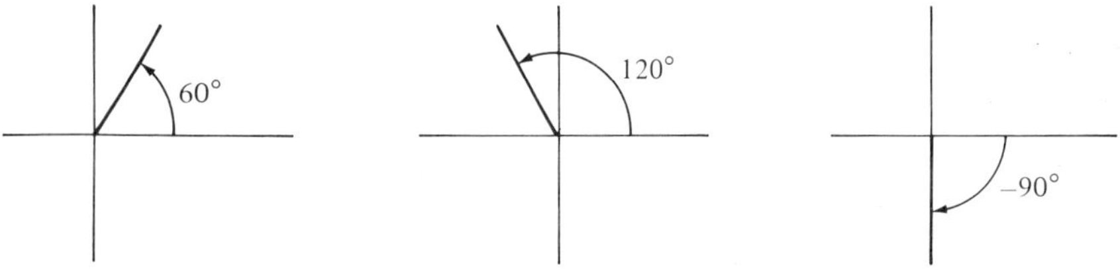

3.

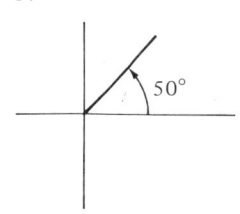

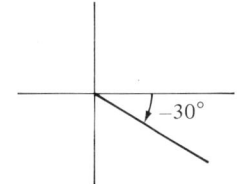

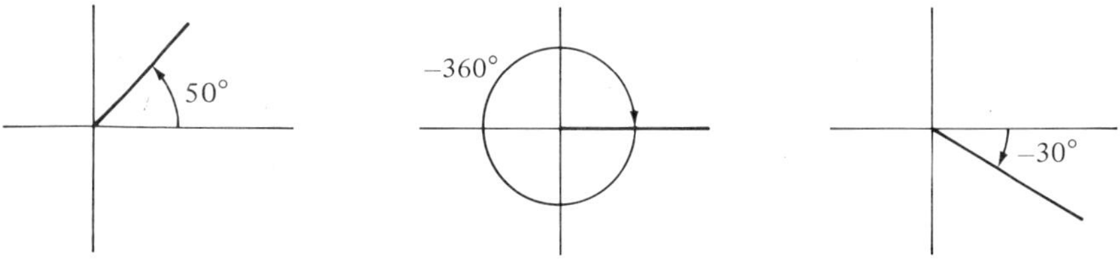

5. $405°, -315°$ **7.** $510°, -210°$ **9.** $430°30', -289°30'$
11. $638.1°, -81.9°$ **13.** $15.2°$ **15.** $86.05°$ **17.** $47°30'$ **19.** $5°37'$

Exercises 2-2, page 41

1. $\sin \theta = \frac{3}{5}$, $\cos \theta = \frac{4}{5}$, $\tan \theta = \frac{3}{4}$, $\cot \theta = \frac{4}{3}$, $\sec \theta = \frac{5}{4}$, $\csc \theta = \frac{5}{3}$

3. $\sin \theta = \frac{8}{17}$, $\cos \theta = \frac{15}{17}$, $\tan \theta = \frac{8}{15}$, $\cot \theta = \frac{15}{8}$, $\sec \theta = \frac{17}{15}$, $\csc \theta = \frac{17}{8}$

5. $\sin \theta = \frac{\sqrt{15}}{4}$, $\cos \theta = \frac{1}{4}$, $\tan \theta = \sqrt{15}$, $\cot \theta = \frac{1}{\sqrt{15}}$, $\sec \theta = 4$, $\csc \theta = \frac{4}{\sqrt{15}}$

Answers to Odd-Numbered Exercises

25. The domain is all real numbers.
    The range is all real numbers.

27. The domain is all real numbers.
    The range is all real numbers greater than or equal to 1.

29. The domain is all real numbers.
    The range is all real numbers greater than or equal to -4.

31. The domain is all real numbers except 0.
    The range is all real numbers except 0.

33.

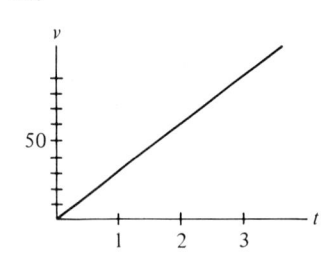

35.

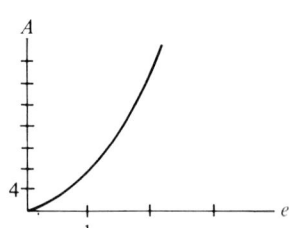

Exercises for Chapter 1, page 28

1. $T = 0.06C$ 3. $A = \frac{l}{2}(1000 - l)$ 5. $16, -47$ 7. $-3, 3$ 9. $u - 9u^?$

11. $6xh + 3h^2 - 2h$ 13. 20.8 15. 6.11 17. 5 19. 8.49

21.

23.

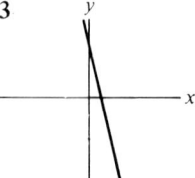

25.

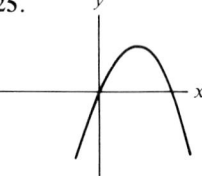

27.

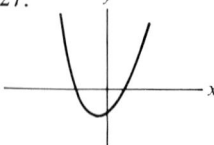

29.

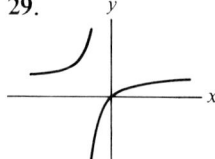

31.

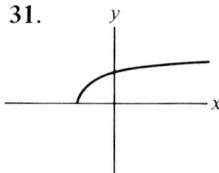

33. The domain is all real numbers.
    The range is all numbers less than or equal to 4.

Answers to Odd-Numbered Exercises   345

7. $\sin\theta = \dfrac{5}{\sqrt{29}}$, $\cos\theta = \dfrac{2}{\sqrt{29}}$, $\tan\theta = \dfrac{5}{2}$, $\cot\theta = \dfrac{2}{5}$, $\sec\theta = \dfrac{\sqrt{29}}{2}$, $\csc\theta = \dfrac{\sqrt{29}}{5}$

9. $\dfrac{1}{\sqrt{2}}, \sqrt{2}$ 11. $\dfrac{\sqrt{5}}{2}, \dfrac{2}{\sqrt{5}}$ 13. $\dfrac{\sqrt{51}}{7}, \dfrac{10}{7}$ 15. $\dfrac{1}{\sqrt{17}}, \dfrac{\sqrt{17}}{4}$

17. $\sin\theta = \tfrac{4}{5}, \tan\theta = \tfrac{4}{3}$ 19. $\sec\theta, \tan\theta$

Exercises 2-3, page 46

1. $\sin 40° = 0.64$, $\cos 40° = 0.77$, $\tan 40° = 0.84$, $\cot 40° = 1.19$, $\sec 40° = 1.30$, $\csc 40° = 1.56$

3. $\sin 15° = 0.26$, $\cos 15° = 0.97$, $\tan 15° = 0.27$, $\cot 15° = 3.73$, $\sec 15° = 1.04$, $\csc 15° = 3.86$

5. 1   7. 2   9. $\dfrac{1}{\sqrt{3}}$ 11. 2   13. 0.3256   15. 2.356   17. 0.9250

19. 1.812   21. 0.5528   23. 0.8930   25. 7.991   27. 0.4490

29. 0.8949   31. 0.9686   33. $5°30'$ 35. $72°10'$ 37. $51°6'$

39. $39°7'$ 41. $30.6°$ 43. $79.3°$ 45. 162 ft/sec

Exercises 2-4, page 52

1.

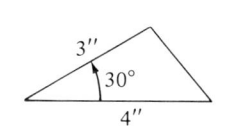

3.

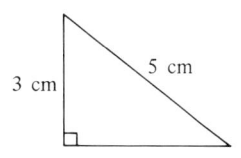

5. $A = 30°, b = 8\sqrt{3}, c = 16$ 7. $A = 45°, B = 45°, c = 7\sqrt{2}$

9. $A = 30°, a = 5, B = 60°$ 11. $B = 30°, b = \dfrac{8}{\sqrt{3}}, c = \dfrac{16}{\sqrt{3}}$

13. $a = 30.1, B = 58°0', b = 48.2$ 15. $A = 52°10', B = 37°50', c = 71.8$

17. $A = 53°0', b = 0.668, c = 1.11$ 19. $A = 58°40', a = 143, B = 31°20'$

21. $B = 12.2°, b = 1450, c = 6850$ 23. $A = 25.8°, B = 64.2°, b = 311$

25. $a = \sqrt{c^2 - b^2}$, $\cos A = \dfrac{b}{c}$, $\sin B = \dfrac{b}{c}$

27. $a = c\sin A, b = c\cos A, B = 90° - A$

Exercises 2-5, page 56

1. 10.2 ft   3. 28.3 cm   5. 9.08 ft   7. 165 m   9. 42°50′
11. 2.28 m   13. 0.242 ft   15. 32.5°   17. 9.42 ft   19. 234 ft
21. $x = 11.8$ ft, $y = 8.25$ ft   23. 595 ft

Exercises for Chapter 2, page 59

1. 377°0′, −343°0′   3. 142.5°, −577.5°   5. 31.9°   7. 38.1°
9. 17°30′   11. 49°42′
13. $\sin \theta = \frac{7}{25}$, $\cos \theta = \frac{24}{25}$, $\tan \theta = \frac{7}{24}$, $\cot \theta = \frac{24}{7}$, $\sec \theta = \frac{25}{24}$, $\csc \theta = \frac{25}{7}$
15. $\sin \theta = \frac{1}{\sqrt{2}}$, $\cos \theta = \frac{1}{\sqrt{2}}$, $\tan \theta = 1$, $\cot \theta = 1$, $\sec \theta = \sqrt{2}$, $\csc \theta = \sqrt{2}$
17. $\cos \theta = \frac{12}{13}$, $\cot \theta = \frac{12}{5}$ 19. $\cos \theta = \frac{1}{\sqrt{5}}$, $\csc \theta = \frac{\sqrt{5}}{2}$ 21. 0.9511
23. 1.829   25. 6.498   27. 0.5609   29. 1.317   31. 1.352
33. 32°10′   35. 62°10′   37. 41°27′   39. 60°4′   41. 18.2°
43. 57.6°   45. $a = 2, B = 60°, b = 2\sqrt{3}$ 47. $A = 30°, a = 3\sqrt{3}, c = 6\sqrt{3}$
49. $a = 1.83, B = 73°0′, c = 6.27$ 51. $A = 51°30′, B = 38°30′, c = 104$
53. $a = 0.626, B = 40°17′, b = 0.530$ 55. $A = 41°49′, B = 48°11′, b = 11.2$
57. $B = 52.5°, b = 15.6, c = 19.7$ 59. $A = 31.2°, a = 3.94, B = 58.8°$
61. 12.6 ft   63. 84.9 m   65. 194 ft   67. 4.36 m   69. 25°10′
71. 32.0 ft   73. 4892 ft   75. 1.69 mi

Exercises 3-1, page 70

1. +, −, −   3. +, +, −   5. +, +, +   7. +, −, +
9. $\sin \theta = \frac{1}{\sqrt{5}}$, $\cos \theta = \frac{2}{\sqrt{5}}$, $\tan \theta = \frac{1}{2}$, $\cot \theta = 2$, $\sec \theta = \frac{\sqrt{5}}{2}$, $\csc \theta = \sqrt{5}$
11. $\sin \theta = \frac{-3}{\sqrt{13}}$, $\cos \theta = \frac{-2}{\sqrt{13}}$, $\tan \theta = \frac{3}{2}$, $\cot \theta = \frac{2}{3}$, $\sec \theta = -\frac{\sqrt{13}}{2}$, $\csc \theta = -\frac{\sqrt{13}}{3}$
13. $\sin \theta = \frac{12}{13}$, $\cos \theta = -\frac{5}{13}$, $\tan \theta = -\frac{12}{5}$, $\cot \theta = -\frac{5}{12}$, $\sec \theta = -\frac{13}{5}$, $\csc \theta = \frac{13}{12}$

Answers to Odd-Numbered Exercises   347

15. $\sin\theta = -\dfrac{2}{\sqrt{29}}$, $\cos\theta = \dfrac{5}{\sqrt{29}}$, $\tan\theta = -\dfrac{2}{5}$, $\cot\theta = -\dfrac{5}{2}$, $\sec\theta = \dfrac{\sqrt{29}}{5}$, $\csc\theta = -\dfrac{\sqrt{29}}{2}$

17. II   19. II

Exercises 3-2, page 77

1. $\sin 20°$; $-\cos 40°$ 3. $-\tan 75°$; $-\csc 58°$ 5. $-\sin 57°$; $-\cot 6°$
7. $\cos 40°$; $-\tan 40°$ 9. $\cos 20°$; $\cot 5°$ 11. $-\sec 32°$; $-\sin 14°$
13. $-\tan 25°$; $\csc 25°$ 15. $-\sin 30°$; $\sec 37°$ 17. -0.2588 19. -0.2756
21. -1.036 23. -2.366 25. -0.6036 27. -1.834 29. $30°, 210°$
31. $237°40', 302°20'$ 33. $252°19', 287°41'$ 35. $125.6°, 305.6°$
37. -0.7002 39. -0.7771
41. $\sin\theta = \frac{12}{13}$, $\tan\theta = \frac{12}{5}$, $\cot\theta = \frac{5}{12}$, $\sec\theta = \frac{13}{5}$, $\csc\theta = \frac{13}{12}$
43. $\sin\theta = -\frac{8}{17}$, $\cos\theta = -\frac{15}{17}$, $\tan\theta = \frac{8}{15}$, $\cot\theta = \frac{15}{8}$, $\csc\theta = -\frac{17}{8}$
45. $<$ 47. $=$ 49. $-\sin 35°, -\tan 70°$ 51. $-\cot 312°, \sec 241°$

Exercises 3-3, page 84

1. $\dfrac{\pi}{12}, \dfrac{5\pi}{6}$ 3. $\dfrac{5\pi}{12}, \dfrac{11\pi}{6}$ 5. $\dfrac{7\pi}{6}, \dfrac{3\pi}{2}$ 7. $\dfrac{8\pi}{9}, \dfrac{13\pi}{9}$ 9. $72°, 270°$
11. $10°, 315°$ 13. $170°, 300°$ 15. $15°, 27°$ 17. 0.401 19. 4.40
21. 5.82 23. 3.11 25. $43.0°$ 27. $172°$ 29. $140°$ 31. $940°$
33. 0.7071 35. 3.732 37. -1.732 39. -0.1219 41. 0.9057
43. 0.5132 45. -0.4797 47. -0.9890 49. $0.3142, 2.8278$
51. $2.9326, 6.0736$ 53. $0.8308, 5.4522$ 55. $2.4432, 3.8408$

Exercises 3-4, page 88

1. 10.5 in.   3. 52.4 sq in.   5. 1.20 rad   7. 2.88 sq in.   9. 0.262 ft
11. 129 km   13. 94.2 in.   15. 1040 mi/hr   17. 11,300 in./min
19. 65.9 sq cm   21. 9.35×10^{-3} rad/hr

Exercises 3-5, page 95

1. $(0,1)$ 3. $(-1,0)$ 5. $(0,-1)$ 7. $(1,0)$ 9. II   11. IV   13. III

15. I 17. + 19. − 21. + 23. + 25. + 27. − 29. − 31. +

33. $\sin\theta = \frac{12}{13}$, $\cos\theta = \frac{5}{13}$, $\tan\theta = \frac{12}{5}$

35. $\sin\theta = \frac{4}{5}$, $\cos\theta = -\frac{3}{5}$, $\tan\theta = -\frac{4}{3}$

37. $\sin\theta = -\frac{8}{17}$, $\cos\theta = -\frac{15}{17}$, $\tan\theta = \frac{8}{15}$

39. $\sin\theta = -\frac{24}{25}$, $\cos\theta = \frac{7}{25}$, $\tan\theta = -\frac{24}{7}$

41. $\sin\theta = \frac{3}{5}$, $\cos\theta = \frac{4}{5}$, $\tan\theta = \frac{3}{4}$

43. $\sin\theta = -\frac{7}{25}$, $\cos\theta = \frac{24}{25}$, $\tan\theta = \frac{-7}{24}$

45. $\sin\theta = -\frac{5}{\sqrt{29}}$, $\cos\theta = -\frac{2}{\sqrt{29}}$, $\tan\theta = \frac{5}{2}$

47. $\sin\theta = -\frac{2}{\sqrt{13}}$, $\cos\theta = \frac{-3}{\sqrt{13}}$, $\tan\theta = \frac{2}{3}$

49. $\cos\theta = \frac{4}{5}$, $\tan\theta = \frac{3}{4}$ 51. $\sin\theta = -\frac{12}{13}$, $\tan\theta = \frac{12}{5}$

53. $\sin\theta = \frac{8}{17}$, $\cos\theta = -\frac{15}{17}$ 55. $\cos\theta = \frac{7}{25}$, $\tan\theta = -\frac{24}{7}$

Exercises for Chapter 3, page 96

1. $\sin\theta = \frac{4}{5}$, $\cos\theta = \frac{3}{5}$, $\tan\theta = \frac{4}{3}$, $\cot\theta = \frac{3}{4}$, $\sec\theta = \frac{5}{3}$, $\csc\theta = \frac{5}{4}$

3. $\sin\theta = \frac{5}{13}$, $\cos\theta = -\frac{12}{13}$, $\tan\theta = -\frac{5}{12}$, $\cot\theta = -\frac{12}{5}$, $\sec\theta = -\frac{13}{12}$, $\csc\theta = \frac{13}{5}$

5. $\sin\theta = -\frac{2}{\sqrt{53}}$, $\cos\theta = \frac{7}{\sqrt{53}}$, $\tan\theta = -\frac{2}{7}$, $\cot\theta = -\frac{7}{2}$, $\sec\theta = \frac{\sqrt{53}}{7}$, $\csc\theta = -\frac{\sqrt{53}}{2}$

7. $\sin\theta = -\frac{3}{\sqrt{13}}$, $\cos\theta = -\frac{2}{\sqrt{13}}$, $\tan\theta = \frac{3}{2}$, $\cot\theta = \frac{2}{3}$, $\sec\theta = -\frac{\sqrt{13}}{2}$, $\csc\theta = -\frac{\sqrt{13}}{3}$

9. $-\cos 48°$, $\tan 14°$ 11. $-\sin 71°$, $\sec 15°$ 13. $\cos 28°$, $-\sin 27°$

15. $-\csc 23°$, $\sin 25°$ 17. $\frac{2\pi}{9}, \frac{17\pi}{20}$ 19. $\frac{4\pi}{15}, \frac{9\pi}{8}$ 21. $252°, 130°$

23. $12°, 330°$ 25. $32.1°$ 27. $206.3°$ 29. 1.74 31. 0.358

33. 4.58 35. 2.38 37. -0.4226 39. 4.230 41. -0.5878

43. 1.195 45. $17°, 163°$ 47. $70°45', 289°15'$ 49. $0.5760, 5.707$

51. $4.1863, 5.2387$

53. $\cos\theta = \frac{3}{5}$, $\tan\theta = -\frac{4}{3}$, $\cot\theta = -\frac{3}{4}$, $\sec\theta = \frac{5}{3}$, $\csc\theta = -\frac{5}{4}$

55. 0.393 ft   57. 36.2 sq cm   59. 188 in./sec   61. 6.91 ft/sec

Answers to Odd-Numbered Exercises

Exercises 4-1, page 110

1.   3.   5.   7.   9.   11.

13. $A_x = 3.22, A_y = 7.97$ 15. $A_x = -62.9, A_y = 44.1$
17. $A_x = 2.11, A_y = -8.79$ 19. $A_x = -8.52, A_y = 13.4$
21. $A_x = 0.5375, \mathbf{A}_y = 0.8432$ 23. $A_x = 62.63, A_y = -26.00$
25. $R = 10.0, \theta_R = 58°50'$ 27. $R = 2.74, \theta_R = 111°0'$
29. $R = 2130, \theta_R = 107.7°$ 31. $R = 1.42, \theta_R = 299.2°$
33. $R = 52.67, \theta_R = 76°25'$ 35. $R = 1.572, \theta_R = 308°6'$
37. $R = 29.2, \theta_R = 10°50'$ 39. $R = 47.0, \theta_R = 101°10'$
41. $R = 27.2, \theta_R = 33.0°$ 43. $R = 12.8, \theta_R = 25.2°$
45. $R = 288.6, \theta_R = 88°27'$ 47. $R = 27.16, \theta_R = 138°2'$

Exercises 4-2, page 116

1. 81.5 lb, 56.5° from the 45.0-lb force
3. 343 kg, 18°0' from the 220-kg force   5. 510 mi, 28.1° south of west
7. 37.7 km, 77.4° south of west   9. $V_H = 80.3$ ft/sec, $V_V = 89.2$ ft/sec
11. 112 m/sec, 26.6° from the direction of the plane
13. 26.5 lb, 48.6° from resultant   15. 5.20 mi/hr, perpendicular to bank

Exercises 4-3, page 123

1. $b = 38.1, C = 66°0', c = 46.1$ 3. $a = 2790, b = 2590, C = 109°0'$
5. $B = 12°0', C = 150°0', c = 7.44$ 7. $A = 125°30', a = 0.0777, c = 0.00583$
9. $A = 99.5°, b = 55.2, c = 24.2$ 11. $A = 68.2°, a = 553, c = 538$
13. $A_1 = 61°30', C_1 = 70°20', c_1 = 5.61$
    $A_2 = 118°30', C_2 = 13°20', c_2 = 1.37$
15. No solution   17. 10.6 ft, 7.83 ft   19. 25,200 m   21. 23.0 mi

Exercises 4-4, page 130

1. $A = 50°20', B = 75°40', c = 6.31$ 3. $A = 70°50', B = 11°10', c = 4750$
5. $A = 34.7°, B = 40.7°, C = 104.6°$ 7. $A = 18.2°, B = 22.2°, C = 139.6°$
9. $A = 6.0°, B = 16.0°, c = 1150$ 11. $A = 82.3°, b = 21.6, C = 11.4°$

13. $A = 38°50', B = 36°40', b = 98.1$ **15.** $A = 46.9°, B = 61.8°, C = 71.3°$
17. $R = 85.2, \theta_R = 31°0'$ **19.** $R = 683, \theta_R = 67°10'$ **21.** 1140 ft
23. $96°50'$ **25.** 8.88 mi/hr at 13.4° from bank   **27.** 42.4°

Exercises 4-5, page 136

1. 9.01   **3.** 45.3   **5.** 104   **7.** 13.8   **9.** 15,500   **11.** 27,900
13. 34.3   **15.** 58,200   **17.** 2130   **19.** 15,100   **21.** 124   **23.** 137,000
25. 265 sq in.   **27.** 20,900 sq cm   **29.** 874 sq yd   **31.** 118,000 sq cm

Exercises for Chapter 4, page 137

1. $A_x = 57.4, A_y = 30.5$ **3.** $A_x = 0.754, A_y = 0.528$ **5.** $965, 8°30'$
7. $26.1, 146°0'$ **9.** $62.9, 105.5°$ **11.** $71.9, 336.4°$
13. (a) $b = 18.1, C = 64°0', c = 17.5$ **(b)** 118
15. (a) $A = 21°10', b = 34.9, c = 51.5$ **(b)** 325
17. (a) $a = 52.0, A = 40°0', C = 30°0'$ **(b)** 989
19. (a) $A = 54°40', a = 12.7, B = 68°40'$ **(b)** 76.2
21. (a) $A = 32.3°, b = 267, C = 17.7°$ **(b)** 7550
23. (a) $A = 148.7°, B = 9.3°, c = 5.67$ **(b)** 3.61
25. (a) $A = 36.8°, B = 25.0°, C = 118.2°$ **(b)** 89.8
27. (a) $A = 20.6°, B = 35.5°, C = 123.9°$ **(b)** 19.2   **29.** 845 mi/hr
31. 27.0 ft/sec, $33°40'$ with ground   **33.** 770 m   **35.** 174 ft
37. 284,000 sq ft   **39.** 8,410 ft   **41.** 27,300 km   **43.** 186 mi

Exercises 5-1, page 147

1. $\log_3 27 = 3$ **3.** $\log_4 256 = 4$ **5.** $\log_4\left(\frac{1}{16}\right) = -2$ **7.** $\log_2\left(\frac{1}{64}\right) = -6$
9. $\log_8 2 = \frac{1}{3}$ **11.** $\log_{\frac{1}{4}}\left(\frac{1}{16}\right) = 2$ **13.** $81 = 3^4$ **15.** $9 = 9^1$ **17.** $5 = 25^{\frac{1}{2}}$
19. $3 = 243^{\frac{1}{5}}$ **21.** $0.01 = 10^{-2}$ **23.** $16 = (0.5)^{-4}$ **25.** 2   **27.** -2
29. 343   **31.** $\frac{1}{4}$ **33.** 9   **35.** $\frac{1}{64}$ **37.** 0.2   **39.** -3

Answers to Odd-Numbered Exercises

Exercises 5-2, page 154

1.

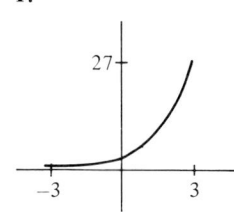

3.

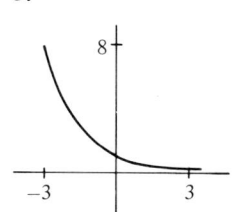

5.

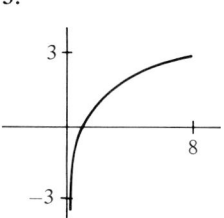

7.

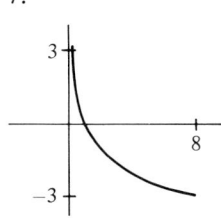

9. $\log_5 x + \log_5 y$ **11.** $\log_3 r - \log_3 s$ **13.** $3\log_2 a$
15. $2 + \log_3 2$ **17.** $-1 - \log_2 3$ **19.** $\frac{1}{2}(1 + \log_3 2)$
21. $4 + 3\log_2 3$ **23.** $3 + \log_{10} 3$ **25.** $\log_b ac$
27. $\log_5 3$ **29.** $\log_b x^{\frac{3}{2}}$ **31.** $\log_b (4n^3)$
33. $y = 2x$ **35.** $y = \frac{49}{x^3}$ **37.** 0.602 **39.** -0.301

Exercises 5-3, page 160

1. 2.7536   **3.** 8.8062 − 10   **5.** 6.9657   **7.** 6.0682 − 10   **9.** 0.0224
11. 9.3773 − 10   **13.** 8.8652   **15.** 8.6512 − 10   **17.** 27400
19. 0.0496   **21.** 2000   **23.** 0.724   **25.** 89.02   **27.** 4.065×10^{-4}
29. 1.427   **31.** 0.005788   **33.** 9.0607   **35.** 8.8751 − 10

Exercises 5-4, page 163

1. 64.58   **3.** 0.03742   **5.** 98.74   **7.** 1.757   **9.** 308.8   **11.** 0.6046
13. 1.323   **15.** 25.33   **17.** 3.740   **19.** 0.003190   **21.** 1011
23. 1.460×10^9

Exercises 5-5, page 167

1. 9.5767 − 10   **3.** 0.2107   **5.** 9.7916 − 10   **7.** 0.1834
9. 9.8982 − 10   **11.** 9.2766 − 10   **13.** 28°10′   **15.** 50°16′
17. 48°56′   **19.** 1°40′ or 1°50′   **21.** $a = 85.35, b = 11.87, B = 7°55′$
23. $b = 9506, C = 42°10′, c = 6703$ **25.** $a = 12.22, C = 68°8′, c = 12.31$
27. $A = 37°58′, B = 52°2′, c = 485.3$ **29.** 98.89 in.   **31.** 13,940 sq m

Exercises 5-6, page 172

1. 3.258   3. 6.447   5. 0.4447   7. 3.822   9. −0.6912   11. −2.957
13. 1.921   15. 6.425   17. 1.602   19. 0.3053

Exercises for Chapter 5, page 173

1. 10,000   3. $\frac{1}{5}$ 5. 6   7. $\frac{5}{3}$ 9. 6   11. 100   13. $2 + \log_2 7$
15. $1 + \frac{1}{2}\log_4 3$ 17. $y = \frac{4}{x}$ 19. $y = \frac{8}{x}$ 21. 9.423   23. 122.8
25. 1.180×10^{15} 27. 2.037   29. 4.771   31. 29.28
33. $b = 99.71, C = 67°50', c = 100.2$ 35. $a = 68.41, B = 41°23', C = 23°22'$
37. 2.181   39. 0.7276   41. $7031

Exercises 6-1, page 185

1. 0, −0.7, −1, −0.7, 0, 0.7, 1, 0.7, 0, −0.7, −1, −0.7, 0, 0.7, 1, 0.7, 0

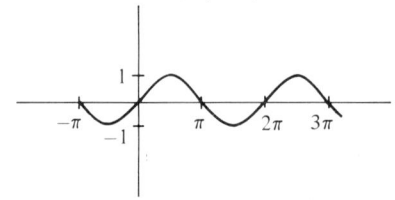

3. −10, −7.0, 0, 7.0, 10, 7.0, 0, −7.0, −10, −7.0, 0, 7.0, 10, 7.0, 0, −7.0, −10

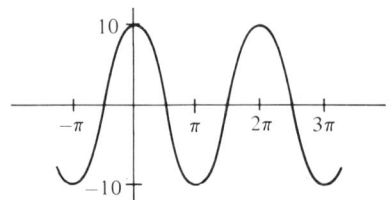

5.

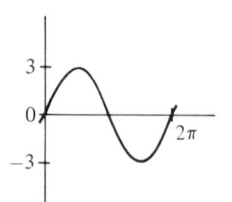

7.

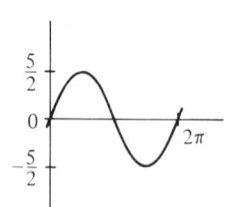

9.

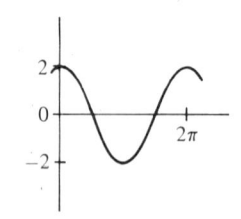

Answers to Odd-Numbered Exercises   353

11. **13.** 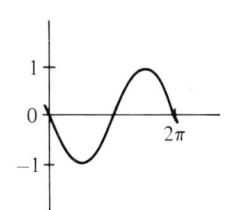 **15.**

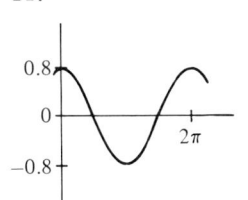

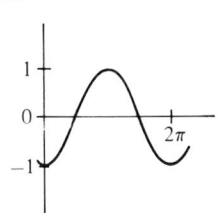

17. $0, 0.84, 0.91, 0.14, -0.76, -0.96,$
$-0.28, 0.66$

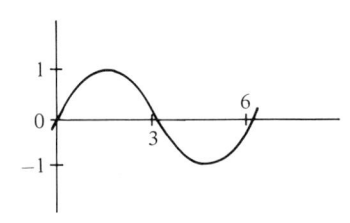

19. $1, 0.54, -0.42, -0.99, -0.65, 0.28,$
$0.96, 0.75$

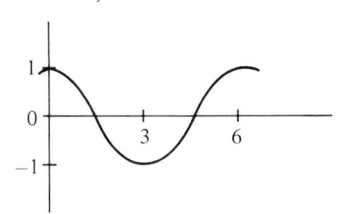

Exercises 6-2, page 190

1. $\dfrac{\pi}{3}$ **3.** $\dfrac{\pi}{4}$ **5.** $\dfrac{\pi}{6}$ **7.** $\dfrac{\pi}{8}$ **9.** 1   **11.** $\dfrac{1}{2}$ **13.** 6π **15.** 3π **17.** 3

19. $\dfrac{2}{\pi}$ **21.**   **23.**

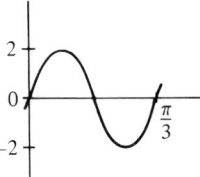

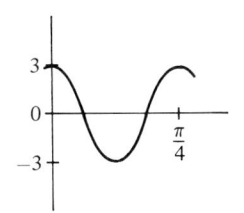

Answers to Odd-Numbered Exercises

25.

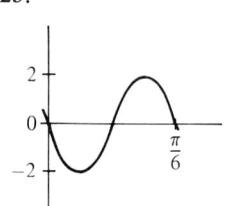

27.

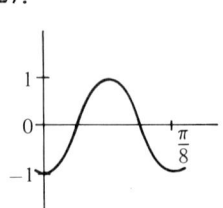

29.

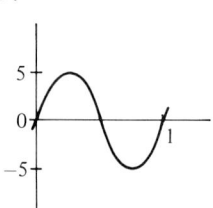

31.

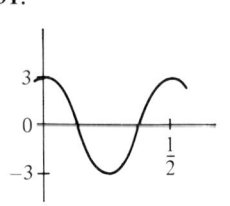

33.

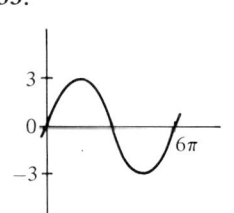

35.

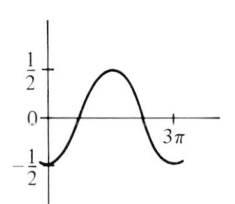

37.

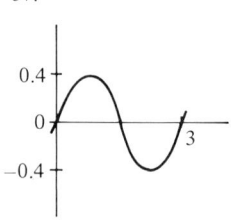

39.

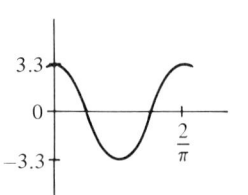

Exercises 6-3, page 195

1. $1, 2\pi, \dfrac{\pi}{6}$ (R)

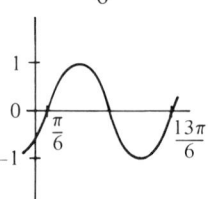

3. $1, 2\pi, \dfrac{\pi}{6}$ (L)

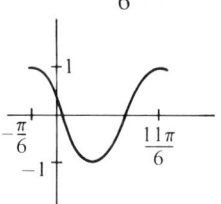

5. $2, \pi, \dfrac{\pi}{4}$ (L)

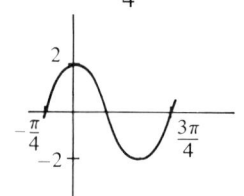

7. $1, \pi, \dfrac{\pi}{2}$ (R)

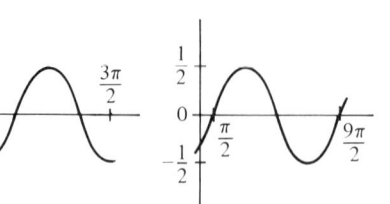

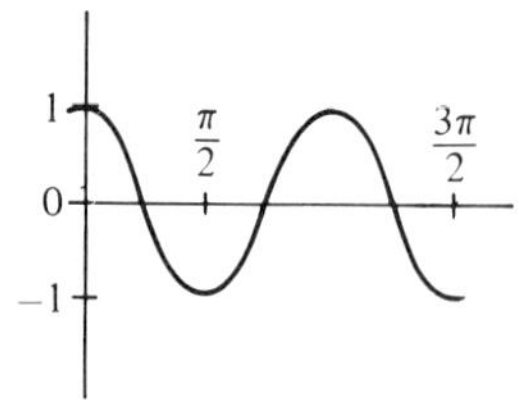

9. $\dfrac{1}{2}, 4\pi, \dfrac{\pi}{2}$ (R)

11. $3, 6\pi, \pi$ (L)

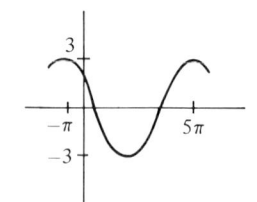

Answers to Odd-Numbered Exercises   355

13. $1, 3, \frac{1}{8}$ (L)   **15.** $\frac{3}{4}, \frac{1}{2}, \frac{1}{20}$ (R) 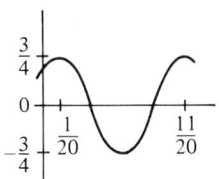   **17.** $0.6, 1, \frac{1}{2\pi}$ (R)

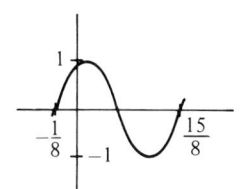

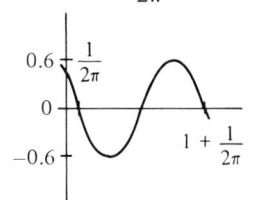

19. $4, \frac{2}{3}, \frac{2}{3\pi}$ (L) 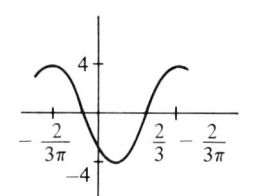   **21.** $1, \frac{2}{\pi}, \frac{1}{\pi}$ (R) 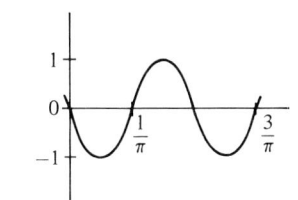   **23.** $2\pi, \frac{\pi}{2}, \frac{1}{4}$ (L)

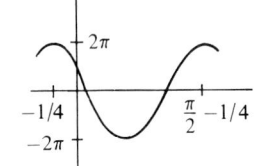

Exercises 6-4, page 202

1. undef., $-1.7, -1, -0.58, 0, 0.58$
   $1, 1.7$, undef., $-1.7, -1, -0.58, 0$

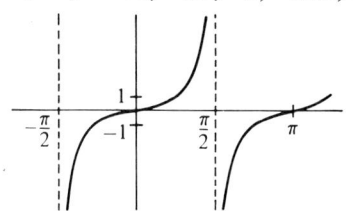

3. undef., $2, 1.4, 1.2, 1, 1.2,$
   $1.4, 2$, undef., $-2, -1.4, -1.2, -1$

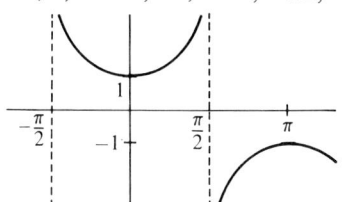

5.

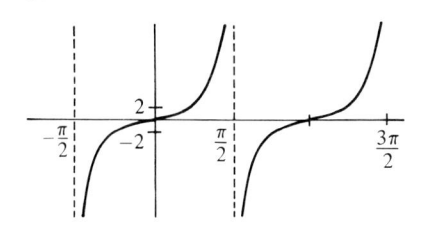

7.

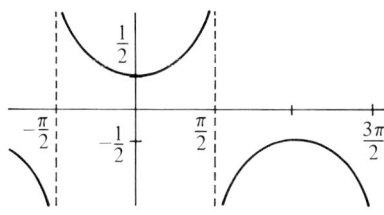

356   Answers to Odd-Numbered Exercises

9.

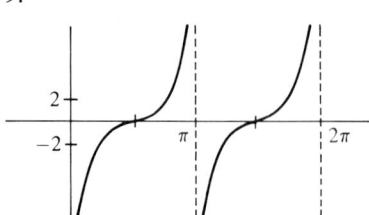

11.

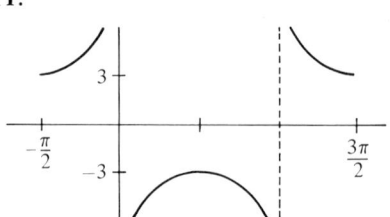

13.

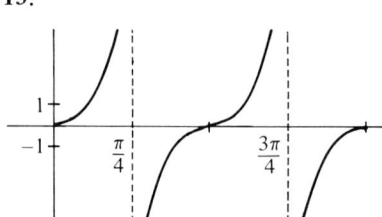

15.

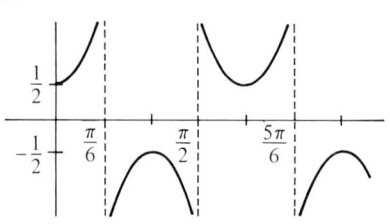

17.

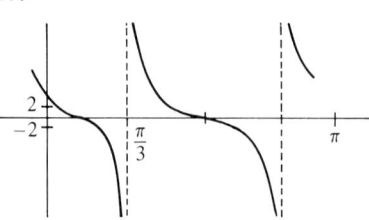

19.

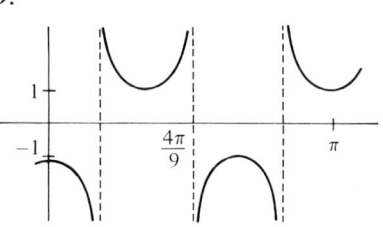

21.

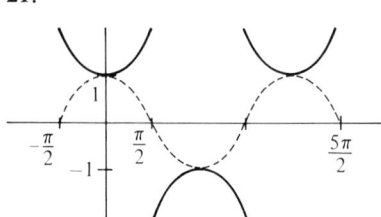

23.

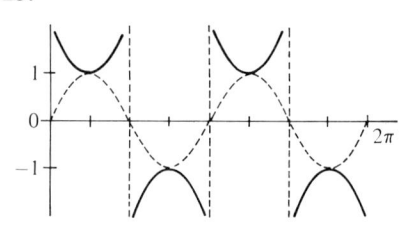

Exercises 6-5, page 206

1.

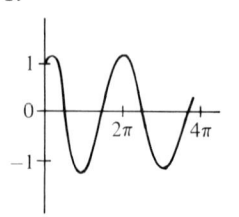

3.

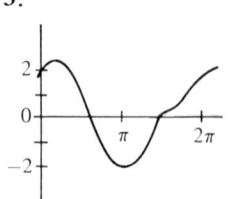

5.

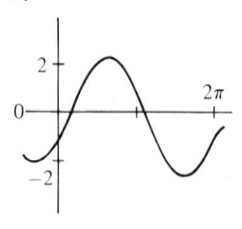

Answers to Odd-Numbered Exercises

7. 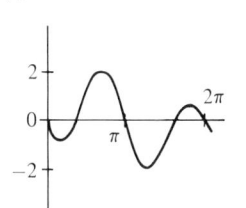
9. 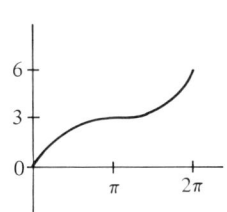
11. 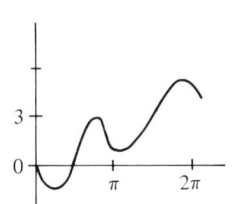
13. 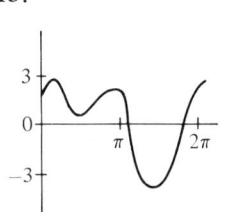
15. 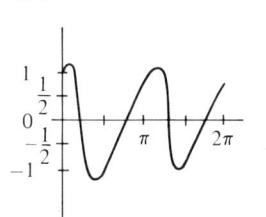
17. 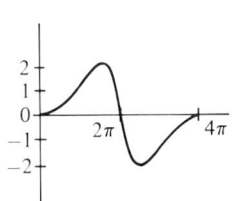
19. 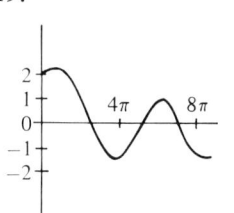
21. 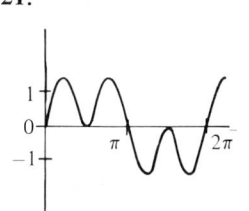
23. 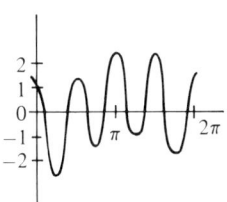
25. 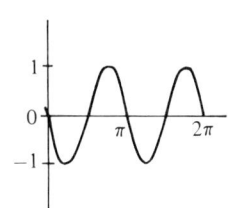
27. 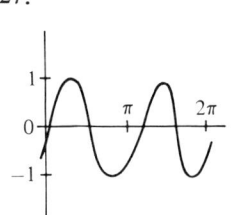
29. 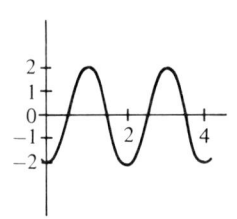
31. 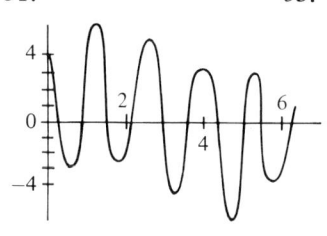
33.

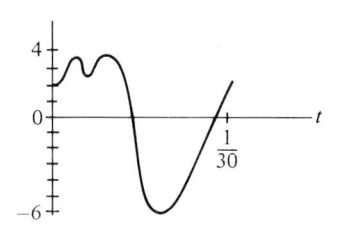

Exercises for Chapter 6, page 209

1.

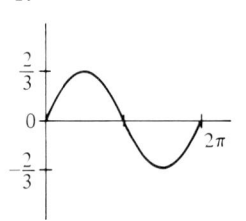

3.

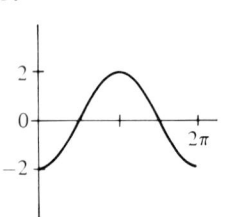

5.

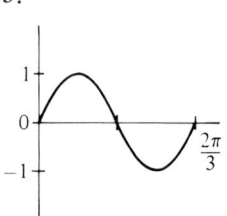

7.

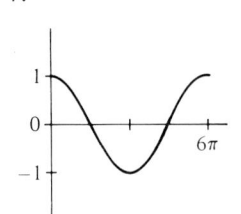

9.

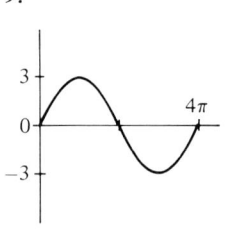

11.

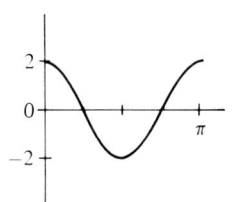

13.

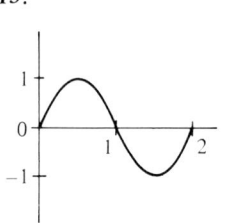

15.

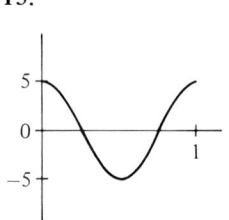

17.

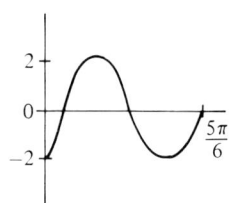

19.

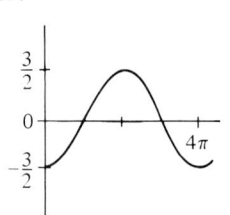

21.

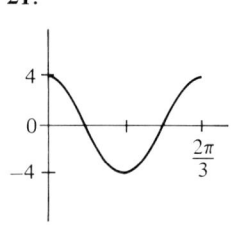

23.

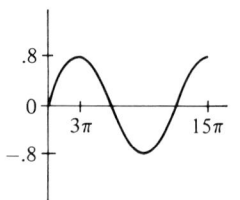

25.

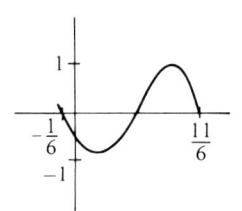

27.

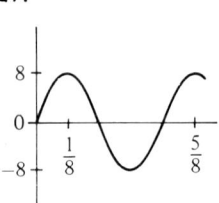

29.

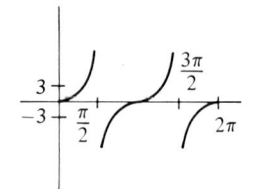

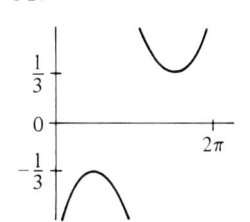

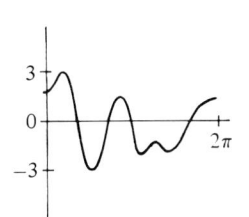

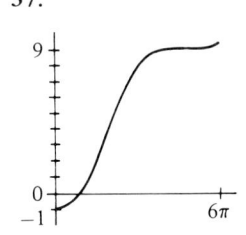

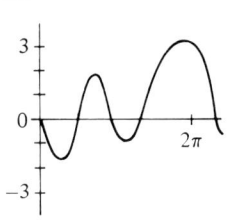

Answers to Odd-Numbered Exercises

31. [graph] 33. [graph] 35. [graph]

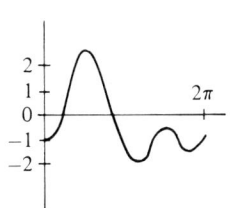

37. [graph] 39. [graph] 41. [graph]

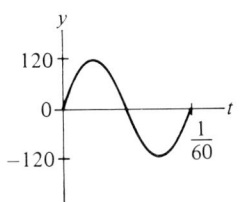

43. [graph]

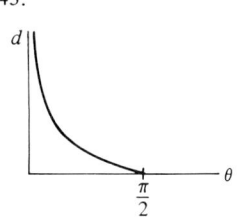

Exercises 7-1, page 220

(Note: "Answers" to trigonometric identities are intermediate steps of suggested reductions of the left member.)

1. $1.483 = \dfrac{1}{0.6745}$ 3. $\left(\dfrac{\sqrt{3}}{2}\right)^2 + \left(-\dfrac{1}{2}\right)^2 = \dfrac{3}{4} + \dfrac{1}{4} = 1$ 5. $\sin\theta\left(\dfrac{\cos\theta}{\sin\theta}\right)$

7. $\left(\dfrac{\sin y}{\cos y}\right) \cdot \csc y = \dfrac{1}{\cos y}$ 9. $\sin x\left(\dfrac{1}{\cos x}\right) = \dfrac{\sin x}{\cos x}$

11. $\dfrac{\cos\theta}{\sin\theta} \cdot \dfrac{1}{\cos\theta} = \dfrac{1}{\sin\theta}$

13. $\sin y\left(\dfrac{\sin y}{\cos y}\right) + \cos y = \dfrac{\sin^2 y + \cos^2 y}{\cos y} = \dfrac{1}{\cos y}$

15. $\dfrac{1}{\cos x} \cdot \dfrac{1}{\sin x} - \dfrac{\cos x}{\sin x} = \dfrac{1 - \cos^2 x}{\cos x \sin x} = \dfrac{\sin^2 x}{\cos x \sin x} = \dfrac{\sin x}{\cos x}$

17. $\csc^2\theta\,(\sin^2\theta)$ 19. $\sin x(\csc^2 x)$ 21. $1 - \sin^2 x$ 23. $1 + \tan^2 y$

25. $\cot\theta\,(\sec^2\theta - 1) = \cot\theta\,(\tan^2\theta)$ 27. $\sin\theta\,(\sec^2\theta - \tan^2\theta)$

29. $\dfrac{\sin x}{\cos x} + \dfrac{\cos x}{\sin x} = \dfrac{\sin^2 x + \cos^2 x}{\cos x \sin x} = \dfrac{1}{\cos x \sin x}$ 31. $(1 - \sin^2 y) - \sin^2 y$

33. $\dfrac{\sin x\,(1 + \cos x)}{1 - \cos^2 x} = \dfrac{1 + \cos x}{\sin x}$

35. $\dfrac{\left(\dfrac{1}{\cos\theta}\right) + \left(\dfrac{1}{\sin\theta}\right)}{1 + \left(\dfrac{\sin\theta}{\cos\theta}\right)} = \dfrac{\left(\dfrac{\sin\theta + \cos\theta}{\cos\theta\sin\theta}\right)}{\left(\dfrac{\cos\theta + \sin\theta}{\cos\theta}\right)} = \dfrac{1}{\sin\theta}$

37. $\dfrac{\sin^2 y}{\cos^2 y}\cdot\cos^2 y + \dfrac{\cos^2 y}{\sin^2 y}\cdot\sin^2 y = \sin^2 y + \cos^2 y$

39. $4\sin x + \dfrac{\sin x}{\cos x} = \sin x\!\left(4 + \dfrac{1}{\cos x}\right)$

41. $\dfrac{1}{\cos\theta} + \dfrac{\sin\theta}{\cos\theta} + \dfrac{\cos\theta}{\sin\theta} = \dfrac{\sin\theta + \sin^2\theta + \cos^2\theta}{\cos\theta\sin\theta}$

43. $(2\sin^2 y - 1)(\sin^2 y - 1)$

45. $\dfrac{(\sec x - 1)^2}{\sec^2 x - 1} = \dfrac{(\sec x - 1)^2}{(\sec x + 1)(\sec x - 1)}$

Exercises 7-2, page 226

1. $\sin 105° = \sin 60° \cos 45° + \cos 60° \sin 45° = \dfrac{\sqrt{3}}{2}\cdot\dfrac{\sqrt{2}}{2} +$
$\dfrac{1}{2}\cdot\dfrac{\sqrt{2}}{2} = \dfrac{\sqrt{6} + \sqrt{2}}{4} = 0.9659$

3. $\tan 15° = \dfrac{\tan 60° - \tan 45°}{1 + \tan 60°\tan 45°} = \dfrac{\sqrt{3} - 1}{1 + (\sqrt{3})(1)} = 0.2679$

5. $\cos 105° = \cos 135°\cos 30° + \sin 135°\sin 30° = -\dfrac{\sqrt{2}}{2}\cdot\dfrac{\sqrt{3}}{2} +$
$\dfrac{\sqrt{2}}{2}\cdot\dfrac{1}{2} = \dfrac{\sqrt{2} - \sqrt{6}}{4} = -0.2588$

7. $\sin 75° = \sin 120°\cos 45° - \cos 120°\sin 45° = \dfrac{\sqrt{3}}{2}\cdot\dfrac{\sqrt{2}}{2} - \dfrac{-1}{2}\cdot\dfrac{\sqrt{2}}{2} =$
$\dfrac{\sqrt{6} + \sqrt{2}}{4} = 0.9659$

9. $\dfrac{-33}{65}$ 11. $\dfrac{-56}{65}$ 13. $\tan 2x$ 15. $\cos 3y$ 17. $\sin(3x - y)$

19. $\cos 2x$

Answers to Odd-Numbered Exercises   361

21. $\sin(270° - x) = \sin 270° \cos x - \cos 270° \sin x = (-1)\cos x - 0(\sin x)$

23. $\cos\left(\dfrac{\pi}{2} - x\right) = \cos\dfrac{\pi}{2}\cos x + \sin\dfrac{\pi}{2}\sin x = 0(\cos x) + 1(\sin x)$

25. $\tan(x + 180°) = \dfrac{\tan x + \tan 180°}{1 - \tan x \tan 180°} = \dfrac{\tan x + 0}{1 - \tan x \cdot 0}$

27. $\sin\left(\dfrac{\pi}{4} + x\right) = \sin\dfrac{\pi}{4}\cos x + \cos\dfrac{\pi}{4}\sin x = \dfrac{\sqrt{2}}{2}\cos x + \dfrac{\sqrt{2}}{2}\sin x$

29. $(\sin x \cos y + \cos x \sin y)(\sin x \cos y - \cos x \sin y) =$
    $\sin^2 x \cos^2 y - \cos^2 x \sin^2 y = \sin^2 x (1 - \sin^2 y) - (1 - \sin^2 x)\sin^2 y$

31. $(\cos x \cos y - \sin x \sin y) + (\cos x \cos y + \sin x \sin y)$

33. Use indicated method. 35. Use indicated method.

37. Use indicated method. 39. Use indicated method.

Exercises 7-3, page 232

1. $\sin 60° = \sin 2(30°) = 2\sin 30° \cos 30° = 2\left(\dfrac{1}{2}\right)\left(\dfrac{\sqrt{3}}{2}\right) = \dfrac{\sqrt{3}}{2}$

3. $\tan 120° = \tan 2(60°) = \dfrac{2\tan 60°}{1 - \tan^2 60°} = \dfrac{2\cdot\sqrt{3}}{1 - (\sqrt{3})^2} = -\sqrt{3}$

5. $\cos 240° = \cos 2(120°) = \cos^2 120° - \sin^2 120° = \left(-\dfrac{1}{2}\right)^2 - \left(\dfrac{\sqrt{3}}{2}\right)^2 = -\dfrac{1}{2}$

7. $\sin 300° = \sin 2(150°) = 2\sin 150° \cos 150° = 2\left(\dfrac{1}{2}\right)\left(\dfrac{-\sqrt{3}}{2}\right) = -\dfrac{\sqrt{3}}{2}$

9. $\dfrac{24}{25}$ 11. $\dfrac{-24}{7}$ 13. $-\dfrac{7}{25}$ 15. $\dfrac{336}{625}$ 17. $2\sin 8x$ 19. $2\sin 6x$

21. $\dfrac{1}{2}\tan 6x$ 23. $-\cos 8x$ 25. $\cos^2\alpha - (1 - \cos^2\alpha)$

27. $\dfrac{1}{\tan 2x} = \dfrac{1 - \tan^2 x}{2\tan x} = \dfrac{1}{2\tan x} - \dfrac{\tan x}{2}$

29. $\sin^2 x + 2\sin x \cos x + \cos^2 x = (\sin^2 x + \cos^2 x) + 2\sin x \cos x$

31. $2\sin x + 2\sin x \cos x = \dfrac{2\sin x(1 + \cos x)(1 - \cos x)}{(1 - \cos x)}$

33. $\dfrac{\sin 3x \cos x - \cos 3x \sin x}{\sin x \cos x} = \dfrac{\sin 2x}{\frac{1}{2}\sin 2x}$

35. $\sin(2x + x) = \sin 2x \cos x + \cos 2x \sin x = (2\sin x \cos x)(\cos x) + (\cos^2 x - \sin^2 x)\sin x$

Exercises 7-4, page 236

1. $\cos 15° = \cos \frac{1}{2}(30°) = \sqrt{\frac{1+\cos 30°}{2}} = \sqrt{\frac{1.8660}{2}} = 0.9659$

3. $\sin 75° = \sin \frac{1}{2}(150°) = \sqrt{\frac{1-\cos 150°}{2}} = \sqrt{\frac{1.8660}{2}} = 0.9659$

5. $\tan 67.5° = \tan \frac{1}{2}(135°) = \sqrt{\frac{1-\cos 135°}{1+\cos 135°}} = \sqrt{\frac{1.7071}{0.2929}} = 2.414$

7. $\sin 165° = \sin \frac{1}{2}(330°) = \sqrt{\frac{1-\cos 330°}{2}} = \sqrt{\frac{0.1340}{2}} = 0.2588$

9. $\sin 3x$ 11. $\tan \alpha$ 13. $2\cos 4\beta$ 15. $4\cos 2x$ 17. $\frac{\sqrt{26}}{26}$ 19. 3

21. $2\left(\sqrt{\frac{1-\cos x}{2}}\right)^2 = \frac{2(1-\cos x)}{2}$ 23. $\frac{\sin \frac{\alpha}{2}}{\cos \frac{\alpha}{2}} = \frac{2\sin \frac{\alpha}{2}\cos \frac{\alpha}{2}}{2\cos^2 \frac{\alpha}{2}}$

25. $\frac{1-\cos \beta}{2\sqrt{\frac{1-\cos \beta}{2}}} = \sqrt{\frac{1-\cos \beta}{2}}$

27. $\tan^2 \frac{\alpha}{2} = \frac{(1-\cos \alpha)(1-\cos \alpha)}{(1+\cos a)(1-\cos a)} = \frac{1-2\cos \alpha + \cos^2 \alpha}{1-\cos^2 \alpha} = \frac{1-2\cos \alpha + \cos^2 \alpha}{\sin^2 \alpha}$

Exercises 7-5, page 241

1. $\frac{\pi}{2}$ 3. $\frac{3\pi}{4}, \frac{7\pi}{4}$ 5. $\frac{\pi}{3}, \frac{2\pi}{3}, \frac{4\pi}{3}, \frac{5\pi}{3}$ 7. $0, \frac{\pi}{6}, \frac{5\pi}{6}, \pi$

9. $\frac{\pi}{12}, \frac{\pi}{4}, \frac{5\pi}{12}, \frac{3\pi}{4}, \frac{13\pi}{12}, \frac{5\pi}{4}, \frac{17\pi}{12}, \frac{7\pi}{4}$ 11. $\frac{\pi}{2}, \frac{3\pi}{2}$ 13. $0, \frac{\pi}{3}, \pi, \frac{5\pi}{3}$

15. $\frac{\pi}{4}, \frac{3\pi}{4}, \frac{5\pi}{4}, \frac{7\pi}{4}$ 17. 3.57, 5.85 19. 0.262, 1.31, 3.40, 4.45 21. $0, \pi$

23. $\frac{3\pi}{8}, \frac{7\pi}{8}, \frac{11\pi}{8}, \frac{15\pi}{8}$ 25. 0.785, 1.25, 3.93, 4.39

Exercises 7-6, page 244

1. y is the angle whose tangent is x. 3. y is the angle whose cotangent is $3x$.

5. y is twice the angle whose sine is x. 7. $\frac{\pi}{3}$ 9. $\frac{3\pi}{4}$ 11. $\frac{3\pi}{4}$ 13. $\frac{\pi}{3}$

15. $\frac{5\pi}{6}$ 17. $x = \frac{1}{3}\arcsin y$ 19. $x = 4\tan y$ 21. $x = \frac{1}{3}\arcsec(y-1)$

Answers to Odd-Numbered Exercises

23. $x = 1 - \cos(1-y)$ 25. I, II   27. II, III

Exercises 7-7, page 250

1. $\dfrac{\pi}{3}$ 3. 0   5. $-\dfrac{\pi}{3}$ 7. $\dfrac{\pi}{3}$ 9. $\dfrac{\pi}{6}$ 11. $-\dfrac{\pi}{4}$ 13. $\dfrac{\pi}{4}$ 15. -1.309

17. $\dfrac{\sqrt{3}}{2}$ 19. $\dfrac{\sqrt{2}}{2}$ 21. 0.8542   23. -1.150 25. -1 27. $\sqrt{3}$

29. $\dfrac{x}{\sqrt{1-x^2}}$ 31. $\dfrac{1}{x}$ 33. $\dfrac{3x}{\sqrt{9x^2-1}}$ 35. $2x\sqrt{1-x^2}$ 37. $\dfrac{3}{\sqrt{10}}$

39. $\dfrac{12}{13}$ 41. $\dfrac{3\sqrt{5}+8}{15}$ 43. $-\dfrac{7}{5\sqrt{11}}$

Exercises for Chapter 7, page 252

1. $\sin(90°+30°) = \sin 90° \cos 30° + \cos 90° \sin 30° =$
   $(1)\left(\dfrac{\sqrt{3}}{2}\right) + (0)\left(\dfrac{1}{2}\right) = \dfrac{\sqrt{3}}{2} = 0.8660$

3. $\tan 105° = \tan(45° + 60°) = \dfrac{\tan 45° + \tan 60°}{1 - \tan 45° \tan 60°} = \dfrac{1+\sqrt{3}}{1-\sqrt{3}} = -3.732$

5. $\cos 2(90°) = \cos^2 90° - \sin^2 90° = 0 - 1 = -1$

7. $\sin 2(180°) = 2 \sin 180° \cos 180° = 2(0)(-1) = 0$

9. $\tan \tfrac{1}{2}(45°) = \sqrt{\dfrac{1-\cos 45°}{1+\cos 45°}} = \sqrt{\dfrac{0.2929}{1.7071}} = 0.4142$

11. $\cos \tfrac{1}{2}(90°) = \sqrt{\dfrac{1+\cos 90°}{2}} = \sqrt{\dfrac{1+0}{2}} = \dfrac{\sqrt{2}}{2} = 0.7071$ 13. $\sin 5x$

15. $4 \sin 12x$ 17. $2 \cos x$ 19. $2 \tan 2x$ 21. $-\dfrac{\pi}{2}$ 23. 0.2618

25. $-\dfrac{\sqrt{3}}{3}$ 27. 0   29. $\dfrac{1-\sin^2 \theta}{\sin \theta} = \dfrac{\cos^2 \theta}{\sin \theta}$

31. $\cos \theta \left(\dfrac{\cos \theta}{\sin \theta}\right) + \sin \theta = \dfrac{\cos^2 \theta + \sin^2 \theta}{\sin \theta}$

33. $\dfrac{(\sec^2 x - 1)(\sec^2 x + 1)}{\tan^2 x} = \sec^2 x + 1$

35. $\dfrac{1 + \dfrac{1-\cos x}{1+\cos x}}{1 - \dfrac{1-\cos x}{1+\cos x}} = \dfrac{1+\cos x + 1 - \cos x}{1+\cos x - 1 + \cos x} = \dfrac{2}{2\cos x}$

37. $2\left(\dfrac{1}{\sin 2x}\right)\left(\dfrac{\cos x}{\sin x}\right) = 2\left(\dfrac{1}{2\sin x \cos x}\right)\left(\dfrac{\cos x}{\sin x}\right) = \dfrac{1}{\sin^2 x}$

39. $\dfrac{1}{2}\left(2\sin\dfrac{\theta}{2}\cos\dfrac{\theta}{2}\right)$

41. $\dfrac{1}{\cos x} + \dfrac{\sin x}{\cos x} = \dfrac{(1+\sin x)(1-\sin x)}{\cos x(1-\sin x)} = \dfrac{1-\sin^2 x}{\cos x(1-\sin x)}$

43. $\cos[(x-y)-y]$ 45. $\sin 4x(\cos 4x)$

47. $\dfrac{\sin x}{\left(\dfrac{1}{\sin x}\right) - \left(\dfrac{\cos x}{\sin x}\right)} = \dfrac{\sin^2 x}{1-\cos x} = \dfrac{1-\cos^2 x}{1-\cos x}$ 49. $x = \dfrac{1}{2}\arccos\left(\dfrac{y}{2}\right)$

51. $x = \dfrac{1}{5}\sin\dfrac{1}{3}\left(\dfrac{\pi}{4} - y\right)$ 53. $\dfrac{\pi}{6}, \dfrac{5\pi}{6}, \dfrac{7\pi}{6}, \dfrac{11\pi}{6}$ 55. $0, \pi$ 57. 0

59. $\dfrac{\pi}{6}, \dfrac{5\pi}{6}, \dfrac{7\pi}{6}, \dfrac{11\pi}{6}$ 61. $2x\sqrt{1-x^2}$ 63. $\dfrac{2x\sqrt{1-x^2}}{1-2x^2}$ 65. $-\dfrac{3}{5}$

67. $\dfrac{9}{41}$

Exercises 8-1, page 262

1. $9i$ 3. $-2i$ 5. $2\sqrt{2}i$ 7. $\dfrac{1}{2}\sqrt{7}i$ 9. $-i$ 11. 1 13. 0
15. $-2i$ 17. $2+3i$ 19. $-2+3i$ 21. $3\sqrt{2}-2\sqrt{2}i$ 23. -1
25. $6+7i$ 27. $-2i$ 29. $x=2, y=-2$ 31. $x=10, y=-6$
33. $x=0, y=-1$ 35. $x=-2, y=3$ 37. It is a real number.

Exercises 8-2, page 266

1. $5-8i$ 3. $-9+6i$ 5. $7-5i$ 7. $-5i$ 9. -1 11. $7+49i$
13. $-8-20i$ 15. $22+3i$ 17. $-42-6i$ 19. $-18\sqrt{2}i$ 21. $3\sqrt{3}i$
23. $-28i$ 25. $-40-42i$ 27. $-2-2i$ 29. $\dfrac{-30+12i}{29}$ 31. $\dfrac{12+2i}{37}$
33. $-i$ 35. $\dfrac{-13+8\sqrt{2}i}{11}$ 37. $\dfrac{-1+3i}{5}$ 39. $\dfrac{-5+5i}{2}$
41. $(a+bi) + (a-bi) = 2a$ 43. $(a+bi) - (a-bi) = 2bi$

Answers to Odd-Numbered Exercises   365

Exercises 8-3, page 270

1. $8 + i$ 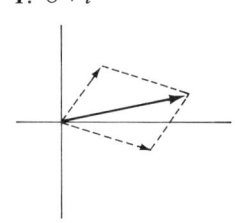
3. $-3i$ 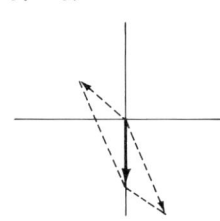
5. $-1 + 4i$

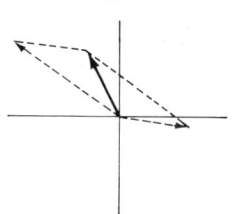

7. $-2 + 3i$ 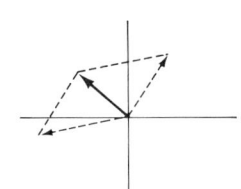
9. $7 + i$ 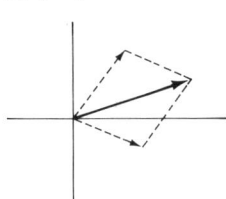
11. $-13 + i$

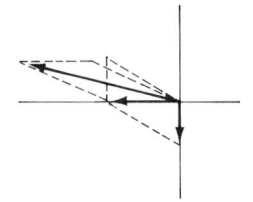

13.

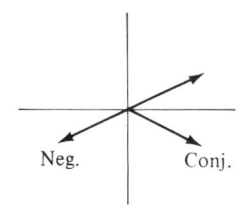

Neg. Conj.

15.

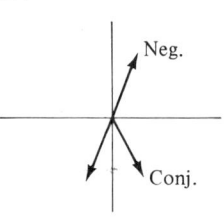

Neg. Conj.

Exercises 8-4, page 273

1. $10(\cos 36.9° + i \sin 36.9°)$ 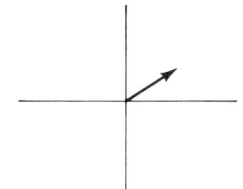
3. $5(\cos 306.9° + i \sin 306.9°)$

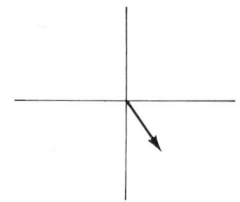

5. $3.61(\cos 123.7° + i \sin 123.7°)$ 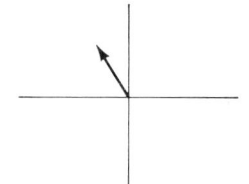
7. $5.39(\cos 201.8° + i \sin 201.8°)$

9. $2(\cos 60° + i \sin 60°)$

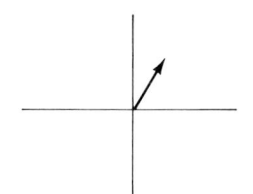

11. $3(\cos 180° + i \sin 180°)$

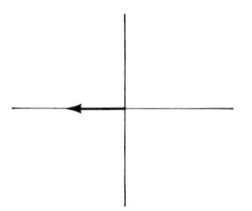

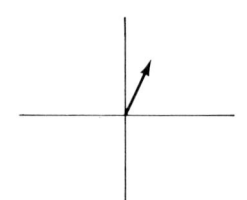

13. $2.94 + 4.05i$

15. $-1.39 + 0.80i$

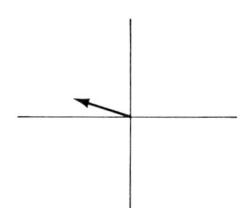

17. $9.66 - 2.59i$

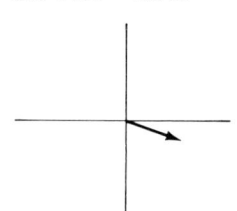

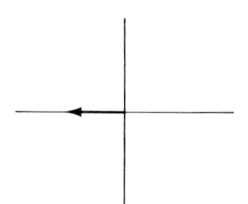

19. -6

21. $-3.76 - 1.37i$

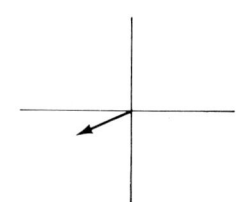

23. $-0.500 - 0.866i$

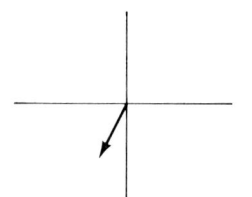

Exercises 8-5, page 276

1. $3e^{1.05i}$ 3. $4.5e^{4.92i}$ 5. $5.00e^{5.35i}$ 7. $3.61e^{2.55i}$ 9. $5.39e^{0.381i}$
11. $7.81e^{3.84i}$ 13. $3(\cos 28.6° + i \sin 28.6°); 2.63 + 1.44i$

Answers to Odd-Numbered Exercises

15. $4(\cos 106° + i \sin 106°); -1.10 + 3.85i$
17. $3.2(\cos 310° + i \sin 310°); 2.06 - 2.45i$
19. $0.1(\cos 137° + i \sin 137°); -0.073 + 0.068i$ 21. $0.55e^{0.415i}$

Exercises 8-6, page 283

1. $8(\cos 80° + i \sin 80°)$ 3. $3(\cos 250° + i \sin 250°)$
5. $2(\cos 35° + i \sin 35°)$ 7. $2.4(\cos 110° + i \sin 110°)$
9. $8(\cos 105° + i \sin 105°)$ 11. $256(\cos 0° + i \sin 0°)$
13. $65(\cos 345.7° + i \sin 345.7°) = 63 - 16i$
15. $0.385(\cos 120.5° + i \sin 120.5°) = \dfrac{-33 + 56i}{169}$
17. $625(\cos 212.4° + i \sin 212.4°) = -527 - 336i$
19. $2(\cos 30° + i \sin 30°), 2(\cos 210° + i \sin 210°)$
21. $-0.364 + 1.67i, -1.26 - 1.15i, 1.63 - 0.520i$ 23. $1, -1, i, -i$
25. $i, -\dfrac{\sqrt{3}+i}{2}, \dfrac{\sqrt{3}-i}{2}$

Exercises for Chapter 8, page 284

1. $10 - i$ 3. $6 + 2i$ 5. $9 + 2i$ 7. $-12 + 66i$ 9. $\dfrac{16 + 50i}{53}$
11. $x = -\dfrac{2}{3}, y = -2$ 13. $3 + 11i$ 15. $4 + 8i$

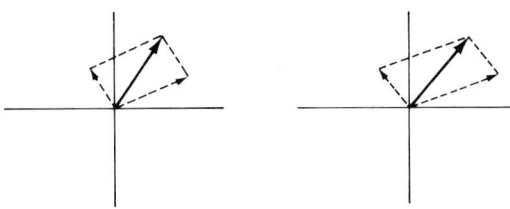

17. $1.41(\cos 315° + i \sin 315°) = 1.41e^{5.50i}$
19. $7.28(\cos 254.1° + i \sin 254.1°) = 7.28e^{4.43i}$ 21. $-1.41 - 1.41i$
23. $-2.72 + 4.19i$ 25. $1.94 + 0.495i$ 27. $-1.62 + 4.73i$
29. $15(\cos 84° + i \sin 84°)$ 31. $8(\cos 59° + i \sin 59°)$
33. $2^{10}(\cos 160° + i \sin 160°)$ 35. $81(\cos 216° + i \sin 216°)$
37. $32(\cos 270° + i \sin 270°) = -32i$ 39. $\dfrac{625}{2}(\cos 270° + i \sin 270°) = -\dfrac{625}{2}i$

41. $1.00 + 1.73i, -2, 1.00 - 1.73i$

43. $\cos 67.5° + i \sin 67.5°, \cos 157.5° + i \sin 157.5°, \cos 247.5° + i \sin 247.5°, \cos 337.5° + i \sin 337.5°$

Appendix A

Appendix A-1, page 291

1. x^7 **3.** $2b^6$ **5.** m^2 **7.** $\dfrac{1}{n^4}$ **9.** $8n^3$ **11.** a^8 **13.** $-t^{14}$

15. $\dfrac{8}{b^3}$ **17.** $64x^{12}$ **19.** $64g^2 s^6$ **21.** 1 **23.** $\dfrac{2}{a^2}$ **25.** $\dfrac{3}{c^2}$

27. $\dfrac{1}{a+b}$ **29.** 1 **31.** b^2 **33.** $\dfrac{5a}{n}$ **35.** $\dfrac{a^5}{y}$ **37.** $a^6 b^4$ **39.** $\dfrac{a}{x^2}$

41. $\dfrac{x^3}{64a^3}$ **43.** $\dfrac{x^2 y^6}{16}$ **45.** $\dfrac{1}{81x^8}$ **47.** $\dfrac{20}{t^{13}}$ **49.** $\dfrac{b^3}{3^3 \cdot 4^2} = \dfrac{b^3}{432}$

51. $\dfrac{x^8 - y^2}{x^4 y^2}$ **53.** $\dfrac{ab}{a+b}$ **55.** $-\dfrac{x}{y}$ **57.** $\dfrac{a^3 + x^3}{ax^2 + a^2 x} = \dfrac{a^2 - ax + x^2}{ax}$

59. $\dfrac{2x}{(x-1)(x+1)}$

Appendix A-2, page 295

1. $45{,}000$ **3.** 0.00201 **5.** 3.23 **7.** 18.6 **9.** 4×10^4

11. 8.7×10^{-3} **13.** 6.89×10^0 **15.** 6.3×10^{-2} **17.** 2.04×10^8

19. 4.85×10^9 **21.** 9.77×10^6 **23.** 3.66×10^5 **25.** 2.25×10^4 lb/sq in.

27. 0.00061 atm **29.** $5{,}000{,}000{,}000$ yr **31.** 1.0×10^{-4} in.

Appendix A-3, page 299

1. 5 **3.** -11 **5.** 5 **7.** -5 **9.** 2 **11.** 2 **13.** 5 **15.** 3

17. $2\sqrt{6}$ **19.** $3\sqrt{5}$ **21.** $4\sqrt{5}$ **23.** $8\sqrt{2}$ **25.** $2\sqrt[3]{2}$ **27.** $2\sqrt[5]{3}$

29. $\sqrt[4]{2}$ **31.** $\sqrt[15]{9}$ **33.** $\dfrac{\sqrt{6}}{2}$ **35.** $\dfrac{2\sqrt{6}}{3}$ **37.** $\dfrac{\sqrt{10}}{4}$ **39.** $\dfrac{2\sqrt{10}}{15}$

41. $\dfrac{\sqrt[3]{6}}{2}$ **43.** $\dfrac{\sqrt[4]{14}}{2}$

Appendix A-4, page 302

1. 5 **3.** 3 **5.** 16 **7.** 10^{25} **9.** $\dfrac{1}{2}$ **11.** $\dfrac{1}{16}$ **13.** 25 **15.** 4096

17. $-\dfrac{1}{2}$ **19.** $\dfrac{39}{1000}$ **21.** $a^{\frac{7}{6}}$ **23.** $\dfrac{1}{y^{\frac{9}{10}}}$ **25.** $s^{\frac{23}{12}}$ **27.** $2ab^2$

29. $\dfrac{27}{64t^3}$ 31. $\dfrac{b^{\frac{11}{10}}}{2a^{\frac{1}{12}}}$ 33. $\dfrac{2}{3}x^{\frac{1}{6}}y^{\frac{11}{12}}$ 35. $\dfrac{x}{(x+2)^{\frac{1}{2}}}$ 37. $\dfrac{a^2+1}{a^4}$

39. $\dfrac{a+1}{a^{\frac{1}{2}}}$

Appendix B, page 308

1. 24 is exact.   3. 3 is exact, 74.6 is approx.   5. 1063 is approx.
7. 100 and 200 are approx., 3200 is exact.   9. 3,4   11. 3,4   13. 3,3
15. 1,6   17. (a) 3.764, (b) 3.764   19. (a) 0.01, (b) 30.8
21. (a) same, (b) 78.0   23. (a) 0.004, (b) same   25. (a) 4.93, (b) 4.9
27. (a) 57,900, (b) 58,000   29. (a) 861, (b) 860   31. (a) 0.305, (b) 0.31
33. 51.2   35. 1.70   37. 431.4   39. 30.9   41. 62.1   43. 270
45. 160   47. 27,000   49. 5.7   51. 4.39

Index

Abscissa, 14
Absolute value of complex numbers, 271
Accuracy of number, 305
Addition: of complex numbers, 264; of ordinates, 203; of vectors, 105
Ambiguous case, 123
Amplitude, 182, 187
Angle, 35; of depression, 55; of elevation, 54; quadrantal, 36, 76; reference, 73; standard position, 36
Angular velocity, 87
Antilogarithm, 158
Approximate numbers, 303
Arc length, 85
Area: of circular sector, 86; of triangle, 132
Argument of a complex number, 271
Asymptote, 199
Axes, coordinate, 12

Base: of exponents, 287; of logarithms, 145

Characteristic of logarithm, 156
Circle, unit, 90
Circular functions, 90
Cofunctions, 50
Common logarithms, 156
Complementary angles, 49
Complex numbers, 260
Complex plane, 267
Components of vectors, 106
Computations using logarithms, 161
Conjugate of complex number, 262
Cosecant, 39; graph of, 199
Cosine, 39, 92; of double angle, 229; graph of, 180; of half angle, 234; of sum of two angles, 222
Cosines, law of, 125
Cotangent, 39; graph of, 198
Coterminal angles, 36
Cycle of sine curve, 195

Degree as measure of angle, 35
De Moivre's theorem, 279
Denominator: rationalizing of, 298
Dependent variable, 6
Displacement, 113; of sine curve, 191
Division: of complex numbers, 264, 278

Domain, 9
Double-angle formulas, 229

e (irrational number), 169
Equations: trigonometric, 238
Exponential complex number form, 274
Exponential functions, 145
Exponents, 287
Extraneous roots, 239

Fractions as exponents, 299
Functions, 6, 243; circular, 90; domain of, 9; exponential, 145; inverse, 242; logarithmic, 148; range of, 9; trigonometric, 39

Graph: of exponential function, 149; of function, 21; of inverse trigonometric functions, 248; of logarithmic function, 149; of trigonometric functions, 179
Graphical representation of complex numbers, 267

Half-angle formulas, 233

Identity, trigonometric, 213
Imaginary numbers, 259
Independent variable, 6
Initial point of vector, 105
Initial side of angle, 35
Interpolation, 44
Inverse functions, 242
Inverse trigonometric functions, 245

Law: of cosines, 125; of sines, 118
Linear interpolation, 44
Logarithmic function, 148
Logarithms, 145; to base 10, 155; to bases other than 10, 169; computations by, 161; natural, 167; properties of, 151; of trigonometric functions, 164

Mantissa of logarithm, 156
Metric system, 312
Minute (as measure of angle), 35
Modulus of complex number, 271
Multiplication: of complex numbers, 264, 277; of radicals, 297

371

Natural logarithm, 167
Negative angle, 35
Number: complex, 260; imaginary, 259

Oblique triangles, 177; ambiguous case of, 123
Order of radical, 298
Ordinate, 14
Ordinates, addition of, 203
Origin, 13

Period of sine curve, 186
Phase angle of sine curve, 190
Polar form of complex number, 271
Power, 287; of complex number, 279
Precision, 305
Principal root, 296
Product: of complex numbers, 264; by logarithms, 156
Pythagorean theorem, 16

Quadrant, 13
Quadrantal angle, 36, 76
Quotient: of complex numbers, 264; by logarithms, 151

Radian, 79
Radicals, 296
Radicand, 298
Radius vector, 39
Range, 9
Rationalizing denominator, 298
Rectangular coordinate system, 12
Rectangular form of complex number, 261
Reference angle, 73
Relation, 243
Resolution of vectors, 106
Resultant of vectors, 105
Right triangles, 48
Roots of complex numbers, 280
Rounding off, 17, 44, 306

Scalar, 103
Scientific notation, 293
Secant: of angle, 39; graph of, 199
Second (as measure of angle), 35
Significant digits, 17, 304
Signs of trigonometric functions, 67
Sine: 39, 92; of double angle, 229; graph of, 179; of half angle, 233; inverse, 242; of sum of two angles, 222
Sines, Law of, 118
Square root, 296, 313
Standard position of angle, 36
Subtraction: of complex numbers, 264; of vectors, 106

Tabular difference, 44
Tangent, 39, 92; graph of, 196; of double angle, 229; of half angle, 234; of sum of two angles, 223
Terminal point of vector, 105
Terminal side of angle, 35
Triangle: area of, 132; oblique, 117; right, 48
Trigonometric equations, 238
Trigonometric form of complex numbers, 271
Trigonometric functions, 39; of angles of right triangle, 38; of any angle, 71; graphs of, 179; inverse, 245; logarithms of, 164; of negative angles, 77; signs of, 67
Trigonometric identities, 213
Truncation, 307

Unit circle, 90

Variable: dependent, 6; independent, 6
Vectors, 103
Velocity, 87
Vertex of angle, 35

Zero as an exponent, 289

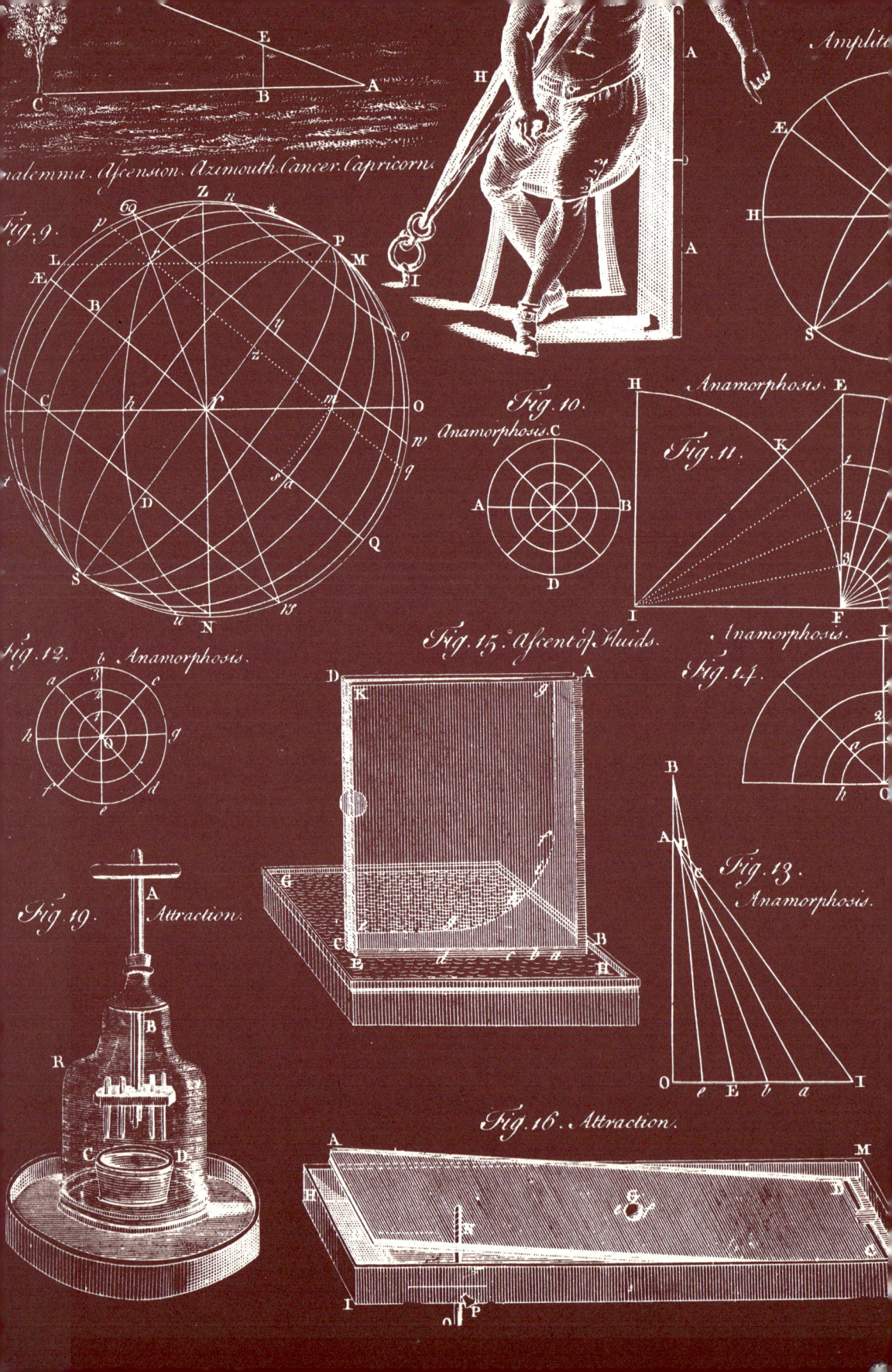